T0181183

Lecture Notes in Artificial Intelligence 638

Subseries of Lecture Notes in Computer Science
Edited by J. Siekmann

Lecture Notes in Computer Science

Edited by G. Goos and J. Hartmanis

A. F. Rocha

Neural Nets

A Theory for Brains and Machines

Springer-Verlag

Berlin Heidelberg New York
London Paris Tokyo
Hong Kong Barcelona
Budapest

Series Editor

Jörg Siekmann
University of Saarland
German Research Center for Artificial Intelligence (DFKI)
Stuhlsatzenhausweg 3, W-6600 Saarbrücken 11, FRG

Author

Armando Freitas da Rocha
Campinas State University
Dept. of Physiology and Biophysics, Inst. of Biology
P. O. Box 1170, 13100 Campinas, SP, Brazil

CR Subject Classification (1991): I.2, J.2, J.3, F.1.1

ISBN 3-540-55949-3 Springer-Verlag Berlin Heidelberg New York
ISBN 0-387-55949-3 Springer-Verlag New York Berlin Heidelberg

Typesetting: Camera ready by author/editor
45/3140-543210 - Printed on acid-free paper

Preface

The purpose of this book is to develop neural nets as a strong theory for both brains and machines. This means that the theory must be developed in close correlation with the biology of the neuron and the properties of human reasoning. This approach implies the following:

a) updating the biology of the artificial neuron: actually, the neural net is merely a vectorial calculus supported by what was known about the neuron in the 40's and about physiology of the synapsis in the 50's. But neuroscience has experienced a tremendous development during the last 50 years, and acquired an impressive body of knowledge about brain functioning. One of the main purposes of the present work is to incorporate this knowledge to both develop a stronger model of the artificial neuron (Chapters I, II, and III) and to introduce new concepts of learning (Chapter IV) in neural nets. The new formal model of the neuron introduced in this book renders neural nets as both numeric and symbolic computational structures, besides making them programmable and trainable (Chapters III and VI);

b) discussing the properties of expert reasoning: in order to have a better understanding of human reasoning, some of the properties of expert reasoning are investigated and discussed (Chapter V). For this purpose, a technique for knowledge acquisition based on fuzzy graphs is introduced, and experimental data on expert reasoning are presented and discussed. The results of this experimental research on expert reasoning show that this kind of thinking is approximate, partial and non-monotonic;

c) taking into consideration the new developments in the field of mathematics such as fuzzy set theory and fuzzy logic (Chapter X), which provide better tools for handling the approximate and partial reasoning of the expert; and

d) presenting artificial neural systems (Chapters VII, VIII and IX) supported by these new concepts on neural nets, which are able to mimic some characteristics of human symbolic reasoning. NEXTOOL (Chapter VII) is a neural environment using evolutive techniques to learn classification tasks. Evolutive learning uses both inductive and deductive tools to discover regularities in the world being modeled. JARGON (Chapter VIII) is a neural tool for acquiring knowledge from natural language data bases, which takes advantage of evolutive learning and of new features of the formal neural model introduced in this book, with the purpose of learning the specialized language used in the data base. SMART KARDS(c) (Chapter IX) is an object oriented environment where data base and neural net techniques are combined to create a very friendly intelligent system.

Because classic neural nets did not take into consideration the real physiology of the nerve cell, the artificial neuron remains a very simplified and restricted structure, and the performance of artificial neural nets on some very human capabilities like symbolic reasoning is very low in comparison with that of the human brain. On the contrary, the expanded model of neuron introduced in this book is proved to be an adequate tool for both numerical and non-numerical processing (Chapters I, II and III). As a matter of fact, the emphasis here will be in showing how to use the updated neuron for the development of intelligent systems mimicking human capabilities like expertise and text handling (Chapters VII, VIII and IX).

The updating of the biology of neural circuits makes fuzzy logic and neural nets two complementary theories to explain brain function and implement artificial intelligent systems (Chapter VI). In this line of reasoning, fuzzy logic (Chapter X) is considered the software or psychological implementation of human reasoning, and neural nets its hardware or physiological counterpart. Both theories provide a strong environment to treat the many types of uncertainty involved in approximate and partial reasoning, which are disclosed by the elicitation of expert knowledge.

In this context, the experimental and theoretical approach to human reasoning described in this book (Chapters IV, V and VIII) provides a true scientific theory to investigate the correlation between mind and brain. It provides the tools for investigating whether reality falsifies the theories it proposes. This approach defines artificial intelligence as an experimental science, where the purpose of knowledge elicitation is not only to provide knowledge bases for expert systems, but also to be a very important tool for scientifically investigating human reasoning.

The present book will be of interest not only to those like the biologist and psychologist who are interested in brain physiology and its emergent properties, or to those like the knowledge engineer and computer scientist who are interested in the development of artificial systems mimicking human capabilities, but also to those like the epistemologist and philosopher who are interested in understanding the tight correlations between brain and mind. Finally, the new models of the artificial neuron and synapsis opens research and development in artificial neural nets to many fields of mathematics such as formal languages, graph theory, etc., other than vectorial calculus.

The book was planned to have many different readings in order to address such a broad audience and to accomplish the different purposes of the many types of readers. If necessary, some chapters may be skipped by the reader not interested in details of fields far away from the domain of his expertise; otherwise, these same chapters may give him the opportunity to enlarge his knowledge for mastering the field of natural and artificial intelligent systems. For this purpose, the contents about biology decrease and the contents about mathematics and computation increase through cap. I to X. Also, each chapter is as self-contained as possible even if this implies some redundancy. In this way, basic information is repeated in different chapters to guarantee the freedom of the reading. Many cross references are provided in the text, so the reader may easily find complementary information. The index is another tool to help the many readings the book is supposed to have. All subjects are fully illustrated by clear diagrams whose purpose is to help the understanding of the topics by those readers unfamiliar with the particularities of the jargon of each field of specialization. Because of this, the book may be used as an advanced text in many types of courses in biology, artificial intelligence, etc. As a matter of fact, the general structure of the book was inspired by the lectures given by the author in the tutorials of the First International Conference on Fuzzy Logic and Neural Nets Iizuka, Japan, 1990 and in the postgraduate courses of physiology and engineering at the State University of Campinas, Brazil.

Chapter I is devoted to introducing the basic physiology of the nerve cell and to modeling it as a non-linear dynamic system. The characteristics of the phase space of the membrane ionic system is discussed and used to support the properties of the axonic encoding.

Chapters II and III deal with the electrical and chemical behavior of the synapse. Chapter II discusses the different types of numerical and non-numerical calculus supported by the electrical activity of the post-synaptic cell. Special attention is devoted to discussing the capability of the updated neuron in solving the Extended Modus Ponens. As a matter of fact, the close correlation between the physiology of the real neuron and the formal steps required for the solution of the Extended Modus Ponens are disclosed. Chapter III presents and discusses the most innovative aspects of our Multi Purpose Neural Net (MPNN) theory, because it introduces a formal language to support the complex chemical processing involved in different types of learning and reasoning. The genetics of the MPNN is the tool for programming different types of neurons and nets, and for specifying properties of learning. Basically, this is the interface linking the electrical processing with the DNA reading of the genetic specification of the properties of the neurons and synapses. Because of this, it is used not only to define and calculate learning, but also to characterize the whole process at the synapse.

Chapter IV introduces, discusses and formalizes the concept of evolutive learning, which allows the learning engine to continuously adapt the MPNN to a changing environment, and to support the creative power of both natural and artificial neural systems. This evolutive engine results from the combination of three different learning techniques: induction, deduction and inheritance. The properties of the inductive learning of the MPNN systems are discussed; and the emergent properties arising from the agglutination of the MPNN neurons promoted by this learning, are presented. Deductive learning is used to change the structure of the models the MPNN learned by observing the world, or inherited from the user in the case of artificial systems, or from the culture in the case of the human being. Learning by being told is introduced as one of the paradigms of knowledge inheritance whereas genetic programming is the other tool used for this purpose. Whenever some initial knowledge is acquired by a MPNN system, it can change the structure of this initial model by being instructed by other intelligent systems.

Chapter V describes a method of eliciting and encoding expert knowledge, which has been successfully used to investigate medical and engineering reasoning. In this approach, fuzzy graphs are used to represent the knowledge elicitated from the expert, and his knowledge is encoded as a net of knowledge graphs. The expert is shown as using two different kinds of knowledge: declarative and procedural knowledge, to handle at least three types of uncertainty in decision making: confidence (uncertainty of matching), relevance (uncertainty of frequency) and utility (uncertainty about cost/benefit). Linguistic quantifiers, fuzzy aggregation, and threshold reasoning are the tools for processing these kinds of uncertainty. Both declarative and procedural knowledge are read as a set of fuzzy productions and default knowledge required by non-monotonic reasoning is encoded by fuzzy linguistic variables. The basic properties of the expert knowledge nets are shown to be emergent properties from the MPNN nets.

Chapter VI introduces the concept of modular neural nets and discusses ways of programming the modules and programming the neural circuits as a net of modules. Such specification takes advantage of the chemical language introduced in Chapter 3. Also, the correlation between modular nets and both the physiology of the cortex and the structure of expert reasoning are analyzed.

Chapter VII introduces NEXTOOL, a MPNN system combining inductive and deductive learning to discover how to classify the events described in data bases. Inductive learning is used to modify the synaptic weight in the modules of neurons representing knowledge. Pruning is the tool for selecting the most successful of these modules in classifying the events. Deductive learning can change the structure of the remaining modules to increase the rate of learning and the quality of the acquired knowledge.

Chapter VIII presents JARGON, a hierarchical neural system of three modular nets, which can grasp the contents of a natural language data base. This system uses evolutive learning to discover the most frequent words and phrases in the data base, and questions the user about the meaning of these productions. In this way, it learns the semantics by "being told" by the user. This knowledge is used to organize the different possible summaries of the data base.

Chapter IX presents SMART KARDS(c), an object oriented language for programming modular neural nets, besides handling data base methods. It is an example of combined technology for complex intelligent systems. It uses JARGON to analyze the contents of its data base and evolutive learning to discover default rules.

Chapter X summarizes the basics of fuzzy sets theory and fuzzy logic necessary to understand of the contents of this book. Special attention will be paid to the discussion of the Extended Modus Ponens as the basic tool for implementing fuzzy reasoning.

To write a book is a good opportunity to discover how many good friends make one's life very pleasant. It was the moment I realized how important to the development of many of the ideas discussed in this book, was the assistance of my wife questioning my theories and introducing them to the real problems of daily life; of the many colleagues at my university and around the world who initiated me into the secrets of their fields of specialization and helped me to reduce my ignorance; of the many experts which shared their knowledge and scarce time with me in the attempt to grasp the dynamics of the human reasoning, and of the many students which during these last two decades inspired me with a lot of bright questions and comments about my reasons. The list of these friends is so long that I cannot name all of them here, but only express my thanks for their invaluable friendship, inspiration, help, support, etc. At the end of each chapter I mention some of these people whose contributions are tightly linked to the ideas discussed in that chapter.

I wish to thank Bárbara Theoto Lambert for her careful revision of the text. Finally, my gratitude to my father and mother for their efforts to make me grow up with an inquisitive mind, and to my wife and children for understanding my restless nature.

Brazil, March 1992   A.F. Rocha

Contents

Chapter VIII - JARGON: A Neural Environment for Language Processing

Chapter IX - SMART KARDS(c): Object Oriented MPNN Environment

CHAPTER I

THE NEURON

I.1 - Composition and properties of the cellular membrane

The cellular membrane is composed by proteins and lipids. Some of these proteins form a double layer in this membrane (Fig. I.1a) because of the chemical attraction among their hydrophobic radicals. The lipids bind to this proteic nucleus, supplying it with a hydrophilic cover. Thus the cellular membrane has a hydrophobic inner core enveloped by a hydrophilic surface. Because of this, the movement of many particles, e.g. ions, occurs across the cellular membrane at some special sites of this membrane, named pores (Fig. I.1).

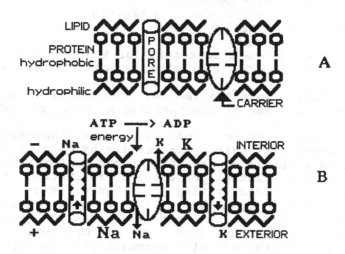

FIG. I.1 - THE CELLULAR MEMBRANE

The pores are composed of tubular proteins called channels, whose molecules cross the entire membrane, allowing specific particules to move from one side to the other of the membrane through their inner core. The channels exhibit different degrees of specialization according to the types of particles being allowed to move in their inner cores. The ionic channels specialize in different ions depending on the configuration of the inner core and the size and properties of the ion. As a consequence, the membrane of the nerve cell has different permeabilities to distinct ions such as sodium (Na), Potassium (K), Calcium (Ca), Chlorides (Cl), etc.

This selective permeability is responsible for the differential distribution of ions between the interior and exterior of the cell (Fig. I.1b), some of them (e.g. the potassium) being distributed predominantly in the intracellular space, and some others (e.g. the sodium) being maintained mainly outside the cell. This ionic separation accounts for the electrical gradient EM established across the cellular membrane, the intracellular space being negative with respect to the extracellular environment.

The dynamic movement of an ion i across the cellular membrane is dependent on both its permeability to this membrane and its electrochemical driven force. This force is calculated as the difference between EM and the ion's equilibrium potential E_i. E_i is the electrical force required to maintain a zero net flux of the ion i across the cellular membrane, and it is calculated as

$$E_i = (RT/FZ_i) \ln(I_o/I_i) \qquad (I.1a)$$

where   R is the gas constant
         T is the temperature in ºKelvin
         F is the faraday number
         Z_i is the ion's valence
         I_i is the ion's concentration inside the cell
         I_o is the ion's concentration outside the cell

The equilibrium potential E_{Na} for the sodium is around +40 mV, and in the case of the potassium E_K is around -95 mV.

Special portions of the ionic channels, called gates, govern the ionic permeability by opening or closing the pore to a specific ion. Hodgkin and Huxley, 1952, showed that the Na's channel possesses 3 activating (called m) gates and one inactivating or   gate (Fig. I.1b). The activating gates open while the h gate closes if EM decreases. The K's channel has 4 activating gates, named gates n. The conductance g_i of the ion i is the electrical measure of its permeability in the cellular membrane. This conductance is a function of the state of the gates of the ion's channel. Thus g_{Na} is a function of the state of the m and h gates, while g_K is dependent on the state of the n gates.

The movement of an ion i across the membrane generates an electric current calculated as (Hodgkin and Huxley, 1952):

$$i_i = g_i (E_i - EM) \qquad (I.1b)$$

i_{Na} is positive and i_K is negative, because EM is negative, E_{Na} is positive and E_K is negative.

These fluxes of Na toward the cell's interior and of K toward the extracellular space must be counteracted by transporting these ions in the opposite direction, in order to maintain their differential distribution between the two cellular spaces. This transport must be done against the ionic electrochemical gradients, thus it requires a supply of energy provided by the cellular metabolism. A special protein called carrier (Fig. I.1a) uses the metabolic energy provided by the ATP (adenosine triphosphate) to pump the Na and K back to their original sites. This process is called Na/K pumping. Metabolic energy is released when ATP is transformed into ADP (adenosine diphosphate) by the action of an enzyme called ATPase (Fig. I.1b). In the case of the Na/K pumping this enzyme is the Na-K-ATPase.

The amount of energy available at the membrane is crucial to determine the state of the ionic channels (Urry, 1971). Although there is some discussion about the structure of these channels (Armstrong, 1981), the model proposed by Urry, 1971, for gramicidin A (Fig. I.2) is useful for the understanding of the dependence of the cellular excitability on the energy available to the membrane.

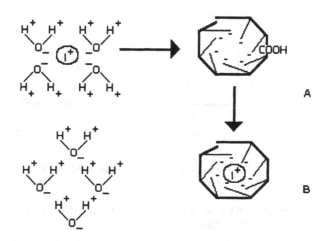

FIG. I.2 - A MODEL FOR THE CHANNEL'S GATE

It was proposed that the ionic channels are tubular proteic structures having the actual size of their inner core determined by the position of the acid (COOH) radicals (Fig. I.2a) of their proteins. The ions at both the intracellular and extracellular spaces have to exchange their hydratation molecules of water with the COOH radicals of the channel in order to move across its gates(Fig. I.2b). This exchange is dependent on electrostatic forces, so that the ease with which the ion does this exchange depends on the correlation between its diameter and the size of the inner core of the channel. In as much as these two diameters are similar the ion permeates the channel.

The diameter of the charged inner core of the channel is dependent on the amount of energy availabe to the protein (Urry, 1971), because the position of the COOH radicals is energy dependent (Fig. I.3). There is a diameter for the state of minimum energy E_M which characterizes the channel. The enhancement of the availabe energy moves the COOH radical predominately toward one of two possible directions, reducing (Fig. I.3a) or increasing (Fig. I.3b) the diameter of the inner core.

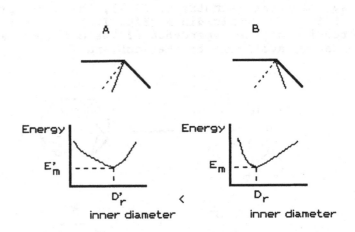

FIG. I.3 - THE ENERGETIC DEPENDENCE OF THE GATES

The available energy at the membrane is dependent on many factors. Among them:

1) the contents of cyclic AMP (cAMP) (McIlwain, 1977; Nathanson, 1977; Rasmussen and Barret, 1984): The enhancement of the intracellular contents of cAMP provoked by hormones, modulators, transmitters, etc. activates a protein kinase (PK) which phosphorylates the channel's proteins using metabolic energy derived from the conversion of ATP to ADP (e.g. Kandel and Schwartz, 1982; Nathanson, 1977). The main result of this chained enzymatic process is to change the membrane's permeability and threshold;

2) external sources of energy as in the case of the sensory

receptor membrane: Any sensory stimulus is an amount of
energy released from an external source and transferred to
the sensory membrane, and

3) the ionic flux across the membrane. The ions moving
according to their electrochemical gradient can transfer
part of their energy to the membrane. Because of this, the
conductance of the ions is EM and time dependent (Hodgkin
and Huxley, 1952). As a matter of fact, EM may be viewed as
a measure of the amount of energy at the cellular membrane.

Let A and B be the sets of measures in the
closed interval [0,1] of the opening of the activating and
inactivating gates of a channel, respectively. For a ∈ A
or b ∈ B equal 1 the gate is fully opened, and for a or
b equal to 0 the gate is closed. The conductance of the ion
i through its channel having x activating gates and y
inactivating gates, is proposed to be (Hodgkin and Huxley,
1952):

$$g_i = G_i \, a^x \, b^y \qquad (I.2a)$$

where G_i is the number of channels for i, and

f' : A x EM x T ---> [0,1] is a monotonically increasing
function concerning time T and EM

f'': B x EM x T ---> [0,1] is a monotonically decreasing
function concerning time T and EM

The notation used to define f' and f'' is the usual notation
in Fuzzy Sets theory, and it means that a degree of
possibility measured in the closed interval [0,1] is
associated to each point in the Cartesian space of these
functions. In the present case, a degree of possibility of
opening the gate is associated to each point in A x E x T
or B x E x T. In this way, a and b measures the
opening of the gates depending on the value of EM and t ∈
T;

a and b ---> 1 then the channel tends to be opened   (I.2b)

The opening of the activating gate increases as EM
decreases. The opposite is true for the inactivating gate.
In the case of Na channel there are 3 activating gates
called m gates and one inactivating gate named h. The 4
activating gates of the K channel are called gates n. The
dynamics of the dependence of m, h and n on time, classifies
(Plant, 1976) n and h as slow variables in respect of m (the
fast variable).

If the stimulus applied to the membrane increases
g_{Na} up to a threshold level (Fig. I.4a), fully opening the

m gate, the Na current reverses the value of EM almost
toward its equilibrium potential E_{Na}. This reduction of EM
closes the h gate and opens the K channel, resulting in a
net outward current of K (Fig. I.4b) which restores the
value of EM near to its resting value. This resting value is
achieved by a final adjustment of both the K and Na currents
to their resting level (Fig. I.4c). This sequence of events
describes the spike triggering at the axon induced by the
stimulation received at the synaptic sites or at the
receptor membrane. The possibility of triggering a new spike
changes during these different phases of the action
potential (Fig. I.4c) as a consequence from the
modifications on g_{Na} and g_K. After being activated, the
axon is said to be refractory, that is, it has a reduced
ability to generate a new spike.

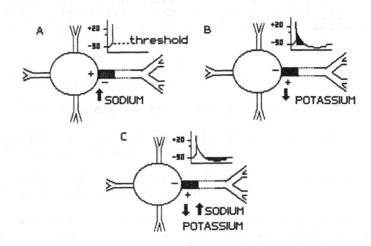

FIG. I.4 - THE SPIKE

I.2 - The Hodgkin-Huxley model

The conductances of Na (g_{Na}) and K (g_K) are
calculated as (Hodgkin and Huxley, 1952):

$$g_{Na} = G_{Na}\, m^3\, h \qquad (I.3a)$$

$$g_K = G_K\, n^4 \qquad (I.3b)$$

so that

$$dy/dt = \alpha_y(EM)\,(1-y) - \beta_y(EM)\,y \qquad (I.3c)$$

where y = m, n or h and the coeficients $\alpha_y(EM)$ and

β_Y(EM) are exponential functions of EM with relaxation times

$$\sigma_Y = 1 \ / \ (\alpha_Y + \beta_Y) \qquad \text{(I.3d)}$$

The dynamic system defined by eqs. I.3 is a complex system, difficult to analyze. The analysis of the geometry of the state phase space of a dynamic system provides very important information about its properties. The state phase space of the ionic system defined by eqs. I.3 is a 4 dimensional space: EM, m, h and n. The geometric analysis of a 4 dimensional space is complex. Plant, 1976, simplified this analysis in the case of the Hodgkin-Huxley model, studying the tridimensional space defined by EM, h and n (Figs. I.5,6,7). This was done by taking into consideration the space defined for the steady state values of m. This space is called the reduced space of the H-H model.

The results shown in Figs. I.5 to 11 were obtained by means of a digital simulation of the HH model (Rocha, 1981b) according to the approach proposed by Plant, 1976.

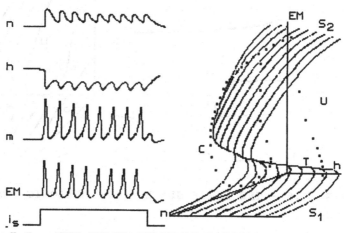

FIG. I.5 - THE PHASE SPACE OF A TONIC MEMBRANE

The reduced phase space of the HH model is composed of (Plant, 1976; Rocha and Bassani, 1982):

I.4a) two stable surfaces S_1 and S_2 (Figs. I.5,6 and 7) defined, respectively, around the E_K and E_{Na}. The values of g_K are higher than g_{Na} in S_1, and the opposite is true in S_2, and

I.4b) an unstable region U (Figs. I.5,6 and 7) in which g_{Na} and g_K are very similar.

The spike is generated as a limit cycle (c in Fig. I.5) in the HH phase space, because whenever the state point (dots in Figs. I.5,6 and 7) reaches the frontier T between S_1 and U, it jumps to the Na stable region S_2 as a consequence of the fact that it cannot remain in unstable region U, because of the huge opening of the m gates. The reduction of h during the stay of the state point in S_2 brings it back to the frontier T and to another jump to S_1, due to the increase of g_K. This cycle can be maintained stable by holding constant the stimulating (i_s) current (Fig. I.5). The intensity of i_s determines the length of c, thus the frequency of the spike firing. The frontier T between U and S_1 may be considered as the set of threshold states q_t of the HH system.

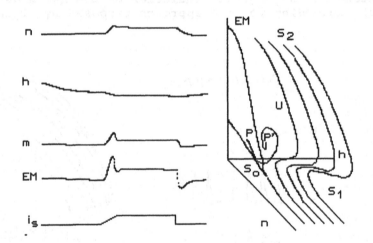

FIG. I.6 - THE PHASE SPACE OF AN ANALOGIC MEMBRANE

The frontier between S_1 and U is composed of those states for which the partial derivative $\delta(n_4)/\delta(EM)$ is equal to zero (Plant, 1976). This derivative is calculated as

$$\delta(n_4)/\delta(EM) = (g_{Na}/g_K)(\ \text{\$}(EM)\ h - \Omega\ ) \qquad (I.5a)$$

where

$$\Omega = (g_l\ (E_l - E_k) + i_s)/g_{Na} \qquad (I.5b)$$

$$\Phi(EM) = \Gamma m^3 / \Gamma EM.(E_{Na}-EM).(EM-E_k) - m^3 (E_{Na}-E_k)$$
$$\qquad\qquad\qquad \infty \qquad\qquad\qquad\qquad\qquad\qquad\qquad \infty$$

$$\tag{I.5c}$$

g_l is a leakage conductance; E_l is a leakage equilibrium potential; i_s is the stimulating current, and Γm_∞ is the partial derivative of m_∞ concerning EM.

The unstable region does not traverse the entire phase space, so that another stable region S_0 is defined for low values of h (Fig. 6). The behavior of the HH model at S_0 markedly differs from that on S_1/S_2, because S_0 is composed only of stable points. In this situation, any perturbation of the s ystem results in a trajectory from one (P) to another stable point (P') in S_0 (Fig. I.6). No limit cycle is established in S_0. Thus, the HH model may be viewed as a sustained oscillator (tonic system - Fig. I.5) for high and intermediate values of h (Rocha and Bassani, 1982; Plant, 1976), whereas it exhibits the properties of an analogic system for low values of h (Fig. I.6). The region of transition between S_0 and S_1, S_2 produces unsustained oscillations (phasic system - Fig. I.7), because after some spike triggering, the state points slides from a limit cycle (c in Fig. I.7) to a stable point P in S_0. The axonic membrane exhibits different filtering properties depending on the amount of available energy, since the state of the channel's gates is energy dependent.

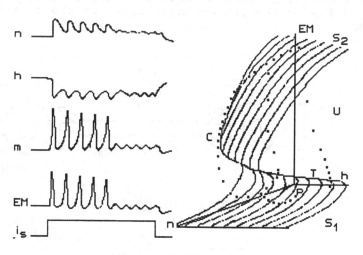

FIG. I.7 - THE PHASE SPACE OF A PHASIC MEMBRANE

It was hypothesized that the amount of Na and K moved by each spike is insufficient to promote noticeable modifications of either E_{Na} or E_K (Hodgkin and Huxley, 1952). However, not only the spike firing but also the resting currents can in the long run dissipate the differential distribution of these ions across the membrane, because they cannot passively move out of or in the cell against the electrical (EM) and their chemical (E_{Na}) gradients. The Na and K moved across the membrane according to their gradients have to be pumped back (Fig. I.1B) using the metabolic energy available from the conversion of ATP into ADP (see, Lauger, 1987). The active pumping serve as a recharger for the Na and K batteries providing the energy for maintaining the ionic currents across the membrane.

I.3 - The neural encoding

The sensory system is the most important source of information to the brain about the events on both the external world and the internal body environment. These two enviroments will be called here the sensory world S. There are two key elements in any sensory chain (Fig. I.8):

I.6a) the receptor cell: taking charge of sensing the energetic variations in S, and

I.6b) the sensory neuron: taking charge of distributing this information to the processing neurons in the central nervous system.

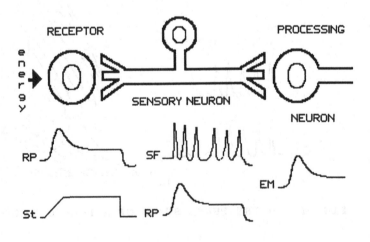

FIG. I.8 - THE SENSORY CHAIN

The response of the receptor cell to the energy of
the stimulus is an analogic variation of EM, because the
permeability of its membrane is very low when it concerns
the Na. The phase space of this type of membrane has a large
S_0 region, and the energy of the stimulus is used to move
the state point from one to another stable state in this
surface (Fig. I.6). This trajectory encodes both the amount
of transferred energy and its first derivative (Fig. I.9).
The EM response at the receptor membrane is called Receptor
Potential (RP).

In this way, it is said that the analogic receptor
membrane encodes both the amplitude and the velocity of the
stimulus, besides its duration. The different histologic
structure observed in nature for the distinct types of
receptors are merely adaptations to provide the most
efficient way to convey the energy of the stimulus to the
receptor membrane. This energy is used to modify the
conformational structure of the ionic channels, which
promotes EM changes correlated with the received amount of
energy (Figs. I.8 and 9).

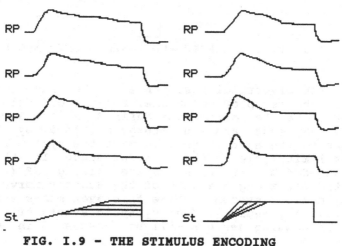

FIG. I.9 - THE STIMULUS ENCODING

The RP does not spread to the long distances
necessary to convey it to the brain where it must be
processed. This is because of the high energetic dissipation
provoked by the high electrical resistance of the analogic
membrane. The information recorded at the receptor sites
must be encoded into another form of EM variation in order
to travel to the central nervous system (Figs. I.8,10). This
encoding takes place at the axon of sensory neuron where the
analogic response is digitized into a sequence of spikes

(Fig. 8,10). This neuron is in general a bipolar cell, having one of the extremities of its axon innervating the receptor cell, and the other in touch with the processing neuron in the brain (Fig. I.8). In some cases, e.g. the pain receptor system, the receptor cell is absent, and the membrane of the peripheral endings of the sensory system plays the role of the receptor membrane.

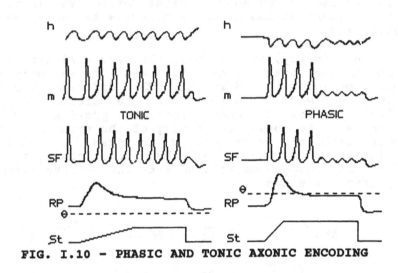

FIG. I.10 - PHASIC AND TONIC AXONIC ENCODING

The electrical resistance at the axon is lower than that at the receptor membrane. Because of this, the axon behaves like the oscillator defined in the S_1/S_2 surface of the HH model and the energy provided by the receptor membrane to the axon is used to move the system from one to another limit cycle in the HH phase space. In this way, the RP is encoded by the axonic spike firing SF (Figs. 8,10), which travels along the axon of the sensory nerve toward the central nervous system, where this axon makes synapsis with the processing neurons. The synaptic events triggered by SF at the processing neurons will be discussed in Chapters II and and III.

The axonic encoding depends on the amount of energy available to the axonic membrane. If the energy is high, then g_{Na} is also high, because the h gate is mostly maintained open. The encoding in this kind of membrane (Fig. I.10 - tonic) is supported by sustained oscillations obtained in those regions of the ionic phase space defined by high values of h and low cellular threshold Θ (Fig. I.5). On the contrary, if the energetic support for the axonic membrane is low, then the cellular threshold Θ is high, and the encoding (Fig. I.10 - phasic) is supported by unsustained oscillations obtained in the regions of the ionic phase space characterized by intermediate values of h (Fig. I.7). In the first case, the sensory chain is named tonic sensory system. In the second case, it is called phasic se nsory system. Intermediate types of encodings are defined as tonic/phasic or phasic/tonic systems, depending on the degree of the attenuation experienced by the spike firing during the trajectory of the state point from one stable limit cycle or point to the other. It must be remembered that these axonic filtering properties are energy dependent, thus the axonic encoding can be put under control of other neurons modulating the amount of energy available to the membrane (see Chapter II, section II.5).

Phasic and tonic systems are defined genetically according to the type of channels and metabolism of their nerve cells, but these properties can be either controlled by efferent nerves or modified by the use of the axon, conditions which may alter the amount of energy available to the axon. Thus, the neuron can be genetically programmed and functionally controlled for different readings of the input information. These readings are encoded in the spike frequency (SF in Fig. I.11) of the spike train travelling the sensory axon.

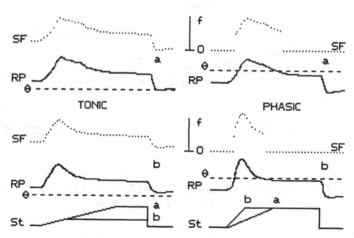

FIG. I.11 - THE FREQUENCY ENCODING

The physiological properties of the neuron discussed so far supports the following definitions and properties:

I.7a) The axonic code W is constructed upon an alphabet C of stable basins:

$$C = \{\ c_i \in C \mid c_i \text{ is a stable limit cycle or}$$
$$\text{stable point in the phase space }\}$$

I.7b) The set W of the axonic codewords generated by C is:

W = { w_i ∈ W | w_i is a trajectory in the basin c_i ∈ C }

I.7c) C is a totally ordered set because the position of the different basins in the phase space is energy dependent.

I.7d) W is also an ordered set.

The slow and rapid adapting stretch receptor organs of the Crayfish are examples of tonic and phasic systems, respectively (Kohn, Rocha and Segundo, 1981; Rocha and Buño, 1985). These organs are easily accessed for experimental research, and the stimulus necessary to activate them is easily controlled because it is the stretch of their muscle fibers acting as the receptor cell. The study of these sensory organs disclosed some interesting properties of the neural encoding (Figs. I.12 and 13).

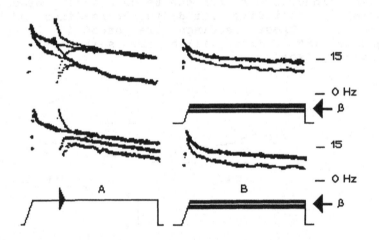

FIG. I.12 - THE ENCODING IN THE SLOW ADAPTING STRETCH ORGAN OF THE CRAYFISH

There is no correlated continuous augmentation of the spike firing at the sensory nerve if the amplitude of the stretch imposed upon the slow adapting receptor is continuously increased each time the stretch is applied (Fig. I.12b). As a matter of fact, this frequency remains the same while the stretch is smoothly increased up a threshold β (Fig. I.12b) is reached when a new spike frequency is triggered. This implies the corresponding limit cycles to define discrete stable basins in the HH phase space, and the necessity of some discrete amount of energy to move the encoding trajectory from one to the other of these stable basins.

These findings are corroborated by the fact that small perturbations (either small stretchings or releasings) added to the stimulating stretch (Fig. I.12a) can promote the same effect of changing the encoding trajectory from one stable limit cycle to another one (Rocha and Buño, 1985). A similar result is obtained in the case of the rapid stretching organ (fig. I.13), because small perturbations change the encoding trajectory toward different stable points in S_0.

These results imply that variables other than those considered by the HH model must promote secondary foldings in the phase space of the real neuron. These secondary foldings define a finite set of stable basins, each basin being characterized by a stable limit cycle in S_1/S_2 or by a stable point in S_0. Such stable limit cycles and points in the phase space define the possible symbols of C. The output of the real axon is a discrete rather than a continuous encoding of R as classically considered by artificial neural theory. W is not a binary set either. The axonic encoding W is supported by an alphabet C of a finite number of digits, each digit being characterized by a stable limiting cycle or point in its ionic phase space.

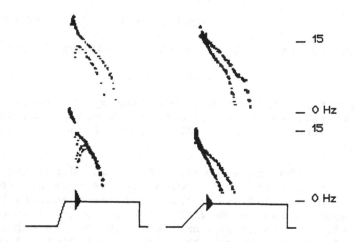

FIG. I.13 - THE ENCODING IN THE RAPID ADAPTING STRETCH ORGAN
OF THE CRAYFISH

C is a finite set of symbols because of the necessity of the axonic encoding to be resistant to the noise present in both the external and the internal environments. Error free encodings are supported by redundant codes (Shannon, 1974). The redundancy of the neural code is determined by the stability of its basins.

I.4 - Measuring the entropy of the neural code

        Shannon, 1974 proposed to measure the amount of information transmitted by a set S of messages by means of its entropy h(S) and proved that error-free encoding is possible if h(S) is maintained below the channel's capacity (h(C)). If S is

$$S = \{ s_1, \ldots , s_n \} \qquad (I.8a)$$

composed by a finite number n of discrete elements occurring with probability p(s$_i$) then:

$$h(S) = - \sum_{i=1}^{n} p(s_i) \log p(s_i) \qquad (I.8b)$$

this means that s$_i$ conveys no information if it is either frequent

$$p(s_i) \dashrightarrow 1 \qquad (I.8c)$$

or rare

$$p(s_i) \dashrightarrow 0 \qquad (I.8d)$$

On the contrary, s$_i$ conveys the maximum information if

$$p(s_i) = 1/n \qquad (I.8e)$$

The amount of information provided by a given message is directly related to the uncertainty about its occurrence. In general, the log in eq. I.8b is basis 2 and h(S) is measured in bits.

        In the same way, h(C) is dependent on the variability of its limit cycles. In the case of the neuron, h(C) is correlated with the axonic excitability because it must measure the ease with which the spike firing can be modified. The membrane excitability can be experimentally measured by means of the delay function (Fig. I.14), correlating the delay δ promoted on the spike firing with the phase ϕ of the firing cycle in which the stimulus was applied (Kohn et al, 1981; Segundo and Kohn, 1981). This delay is positive in case of inhibition and negative in case of excitation. The correlation between δ and ϕ is a linear function in the case of the inhibitory stimulus and a non-linear function in the case of excitatory stimulus (Kohn et al, 1981, Segundo and Kohn, 1981). The slope of the delay function is dependent on the magnitude of the stimulus.

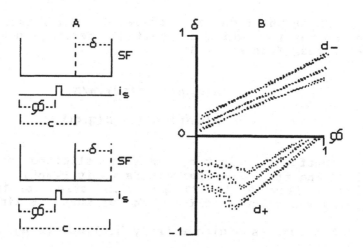

FIG. I.14 - SIMULATED DELAY FUNCTIONS

The area A between the curves c_{d+} and c_{d-} (Fig.
I14b), obtained when the stimulus intensity tends to values
D±d disrupting the spike firing, may be proposed to be
dependent on the entropy of the firing cycle:

$$h(C) = f(A) \qquad (I.10)$$

This hypothesis is confirmed by the following simulations
with the HH system.

The entropy of the limit cycle $c \in C$ is dependent on
the entropy of the state point q composing these cycles. The
entropy of the state point q is in turn dependent on (Rocha
and Bassani, 1982):

I.11a) the distance ($\mu(q,q_t)$) from the state point to the
jumping point q_t at the frontier T between the unstable
and stable regions, which measures the possibility of the
spike generation; and

I.11b) the distance from the state point to the nearest
(E_K or E_{Na}) equilibrium potential (q_0). This distance
measures the possibility of not spiking.

Thus, the entropy of the limit cycle c of period N
is obtained as

$$h(c) = \int_t^{t+N} -\mu(q,q_t).\log \mu(q,q_t) - \mu(q,q_0).\log\mu(q,q_0) \qquad (I.12a)$$

Since the frontier between S_1 and U is composed by those states for which the partial derivative $\delta(n_4)/\delta(EM)$ is zero, then from eq. I.5:

$$\mu(q,q_t) = \Phi(EM).h/\Omega \qquad (I.12b)$$

$$\mu(q,q_0) = 1 - \mu(q,q_t) \qquad (I.12c)$$

Different limit cycles were elicited from the HH model using the same parameters as in Plant, 1976 and by varying i_s from 10 to 70 μA with steps of 10 μA. The following was calculated for each of these 6 limit cycles:

I.13a) the corresponding entropy h(c) according to I.12 and

I.13b) the corresponding delay functions due to a random addition to i_s of brief pulses of either polarity and controlled amplitudes to obtain excitatory or inhibitory effects. The effect of the testing pulse was displayed in graphics (Fig. I.14b) correlating the phase ϕ of the stimulus application and the induced delay δ, which is negative in the case of excitation and positive in the case of inhibition. The stimulus phase tended to 0 or 1 if the testing pulse was applied just after or just before the spike firing, respectively.

The area between the maximal inhibitory and excitatory delay functions was calculated and correlated with h(c) by means of linear regression statistics. The maximal delay functions were obtained by increasing both the amplitude and duration of the testing pulse until the spiking firing was disrupted by overstimulation.

The delay functions simulated for the HH model (Fig. I.14b) exhibited a behavior similar to those obtained from the slow adapting stretch receptor organ of the crayfish (Kohn et al, 1981; Segundo and Kohn, 1981). They tended to be linear functions of the stimulus phase in the case of the inhibitory pulse and they showed a clear non-linear behavior as the intensity of the testing excitatory pulse decreased. The effect of increasing both the amplitude and duration of the testing pulse above a given limit was to silence the firing of the system, as a consequence of saturation of the constant h. Similar silencing of spiking was observed by Rocha and Buño, 1985 in the slow adapting stretch receptor organ of the crayfish by means of overstretching.

The area limited by the inhibitory and excitatory delay functions depended also on the spike firing. It decreased as the spike frequency increased. The entropy of the limit cycles decreased as the spiking frequency increased. This result is a consequence of the corresponding reduction of the length of the limit cycle and from the reduction of the distance between its state points and the threshold frontier T.

The correlation between the area delimited by the delay functions and the calculated entropy confirmed the hypothesis of a dependence of the delay area on the cellular entropy as calculated by I.12. The linear correlation between these two variables attained a correlation coefficient of .9 with a significance of $p < .01$. The calculated regression line was:

$$\text{Delay Area} = .98 + 3000 * h(c) \qquad (I.14)$$

The linear dependency between the entropy calculated according to the proposition of Rocha and Bassani, 1982, with the delay area delimited by the excitatory and inhibitory delay functions supports the conclusion that delay functions can be used to quantify the cellular excitability experimentally, and to provide a measure of the entropy $h(c)$ of the sensory codes.

The hypothesis that $h(c)$ calculated by I.12 is a measure of the redundancy of the axonic code C was further investigated by calculating the coefficient of variation (CV) of the spike firing of the HH model, triggered by distinct values of i_s, and different amounts of noise (n_1 to n_3 in Fig. I.15. n_0 means no noise) added to it.

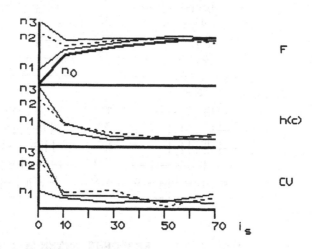

FIG. I.15 - CORRELATION BETWEEN REDUNDANCY AND ERROR ENCODING RESISTANCE

The main results are:

I.15a)   the augmentation of i_s is correlated with an increase of the spike firing frequency (F);

I.15b) h(c) decreases as F augments because of the reduction of the period N in eq. I.12. However,

I.15c) the resultant transmitted entropy $h(C) = h(c) * F$ increases as F augments. Besides,

I.15d) the addition of noise increases the encoding frequency, and

I.15e) low values of CV in presence of noise are observed for i_s greater than 15, independent of the amount of noise.

Error-free encoding is possible if the entropy of the code is greater than that of the source (Shannon, 1974). The enhancement of F induced by noise is accompanied by the augmentation of h(C) in order to allow the error-free encoding of RP in noise environments.

On the one hand, the reduction of h(c) reduces the possibility of the noise to change the length of N. On the other hand, the increase of F allows a better averaging of the length of c. This averaging guarantees the accuracy of the transmitted information.

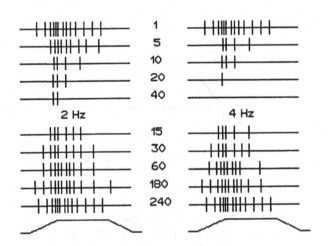

FIG. I.16 - RESPONSE ATTENUATION

These results are in agreement with the experimental results obtained by Buño et al, 1978, showing that noise

increases the spike firing of the stretch organs of the
Crayfish, while maintaining CV low. They also demonstrated
that the capacity of the axon in transmitting information
can be measured by h(C) calculated from eq. I.12.

I.5 - The plastic axonic encoding

Repeated stimulation can either increase or reduce
the spike firing in the sensory nerve of many sensory
systems. This phenomenon is called response sensitization in
the first case and response attenuation in the latter. Fig.
I.16 illustrates the response attenuation obtained by the
repeated stretching of the fast (phasic) adapting stretch
receptor organ of the crayfish. The receptor was stimulated
with ramp stretchings at rates of 2 and 4 Hz. The spike
firing induced by the stretching was drastically reduced by
the repeated stimulation, as illustrated in Fig. I.16, by
the spike trains associated with the 1st, 5th, 10th, 20th,
and 40th stretchings applied to the receptor's muscle.

The initial response to the strectch was recovered
after the stimulation was discontinued, as shown in Fig.
I.16, by the spike trains triggered by the stretchings
applied 15, 30, 60, 180 and 240 times after the end of the
repeated stimulation. Response attenuation may be the result
of a saturation of the Na/K pumping involved in restoring
the Na and K moved by the spikes, and it is a type of
elementary learning occurring on the axonic level.

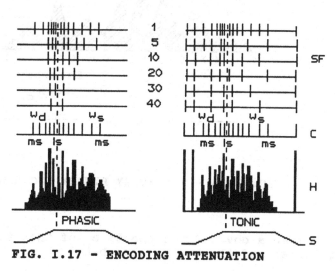

FIG. I.17 - ENCODING ATTENUATION

A very specific pattern of spike disapearance is

disclosed by a careful analysis of the different spike trains elicited during the series of stimulation provoking the response attenuation (Rocha,1981b). Briefly (Fig. I.17):

I.16a) the histogram (H in fig. I.17) of the spike trains shows the existence of distinct time epochs d of increased possibility of distribution of the spike over it;

I.16b) the spikes generated near the transition from the dynamic to the static phase of stimulation are less likely to disappear during the response attenuation than those triggered at the beginning or end of the stimulation, and

I.16c) a resistant spike may disappear occasionally during the attenuation sequence, to reappear later after a less resistant spike being eliminated from the spike train by the attenuation.

It may be concluded from these results that the spike distribution in the encoding trajectories (words) w_i ϵ W draining to the basin c_i is not a continuous distribution over the time continuum. Instead, it is a fuzzy distribution over defined epochs of this continuum

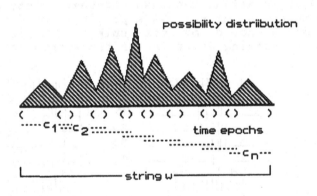

FIG. I.18 - THE FUZZY BINARY ENCODING

Given a cover C of epochs c of the time continuum T (Fig. I.18) and the binary alphabet

B = { 1 = spike occurrence, 0 = spike absence } (I.17a)

the axonic encoding W is a fuzzy point process Γ over C

(Rocha, 1981a) associating a possibility distribution $\pi_c(b)$ (see Chapter 10, section X.4) of $b \in B$ being distributed over the epochs $c \in C$:

$$\Gamma : C \times B \longrightarrow [0,1] \qquad (I.18b)$$

The possibility distribution $\pi_c(b)$ depends on the energetic time pattern $(\epsilon(t))$ of the stimulus to be encoded, that is,

$$\pi_c(b) = f(\epsilon(t)) \qquad (I.18c)$$

and it can be experimentally obtained from the histograms of the spike firing triggered by the stimuli. In this condition, $w \in W$ becomes a fuzzy distribution of B over the epochs $c \in C$:

$$w = \Gamma (\pi_c(b)) \qquad (I.18d)$$

or

$$w = \left[\begin{array}{l} w_i \text{ if } \epsilon(t) < \alpha_1 \\[2mm] w_u \text{ if } \epsilon(t) \geq \alpha_2 \qquad (I.18e) \\[2mm] g(\epsilon(t)) \text{ otherwise} \end{array} \right.$$

where α_i and α_2 are the thresholds defining the filtering properties of the sensory axon; w_i and w_u are the codewords associated, respectively, with these lower and upper thresholds, and g is the function relating the codeword $w \in W$ to $\epsilon(t)$ according to the numeric semantic in eqs. I.19b,c.

This process will be called the fuzzy binary encoding supported by the fuzzy point process Γ. W is a finite set when the time continuum is assumed to be finite. C is a totally ordered set. Thus, now it is possible to say that W is a finite and ordered set, too.

The spike trajectory during the stretch may be viewed as the composition of two codewords $w_d, w_s \in W$, one of them (w_d) encoding the dynamic phase and the other (w_s) the static phase of the stretch (Fig. II.17). The most significant (less frequent) digits (ms in Fig. II.17) of these codewords are, respectively, at the beginning and at the end of the stimulation, whereas the less significant (most frequent) digits (1s in Fig. II.17) are in the transition from the dynamic to the static phases of the stretch. In this way, the amount of encoded energy in both phases may be expressed as the numbers associated with these binary codewords: $w_d, w_s \in W$.

Each $w \in W$ is a string t of n epochs $c \in C$

$$t = \{ d_1, \ldots, d_n \} \qquad (I.19a)$$

ordered according to the significance of these epochs
(Rocha, 1981b). The significance of $c \in C$ is inversely
related to the spike possibility distribution $\mu_d(b)$
associated with it. In this way, the numeric decoding $\tau(w)$
of w can be obtained from binary to decimal numeric
conversion rules of the type (Rocha, 1981b):

$$\tau(w) = \sum_{i=1}^{n} 2^{i-1} b \qquad \text{(I.19b)}$$

where $b \in B$ is either 1 or 0 and i is the order of d in W.
In this context, W is a set of fuzzy numbers of the type
around $\tau(w)$ (e.g. around 6):

$$W = \{ \tau(w) , \pi_t(b) \mid d \text{ defines } w \} \qquad \text{(I.19c)}$$

where the semantics of around is given by $\pi_t(b)$.

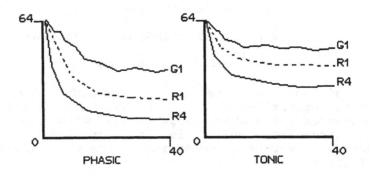

R1 – stretchs at a regular frequency of 1Hz
R4 – stretchs at a regular frequency of 4Hz
G1 – stretchs at a frequency of 1Hz and gamma 16
     distribution

FIG. I.19 - MEASURING ATTENUATION

Fig. I.19 shows the evolution of the response
attenuation of the dynamic words w_d of the stretching
organs of the Crayfish in the case of two different
frequencies and distributions of stretchings. The binary to
decimal numeric conversion was obtained according to I.19b
Using this approach, Rocha, 1981b, showed that the response
attenuation is dependent on (Fig. I.19):

I.20a) the number of the applied stretches;

I.20b) the frequency of stretching, and

I.20c) the distribution of these stretchings in the stimulation series. Also,

I.20d) the slope of these functions for the slow adapting organ is smaller than that for the fast adapting receptor.

The entropy $h(w)$ of $w \in W$ is obtained as

$$h(w) = \sum_{i=1}^{n} h(c_i) \qquad (I.21a)$$

with $h(c_i)$ calculated as in eq. I.12 to each of the n spike cycles composing w. The entropy $h(W)$ of W must be obtained as summation of $h(w_j)$, for all m codewords $w_j \in W$ taking into consideration the frequency w_j are used to encode some information in the sensory world S. If p_j is the probability of $s_j \in S$ encoded by w_j, then,

$$h(W) = \sum_{j=1}^{m} p_j \, h(w_j) \qquad (I.21b)$$

$$h(S) = \sum_{j=1}^{m} p_j \log p_j \qquad (I.21c)$$

Since error-free encoding in noise environment requires

$$h(W) > h(S) \qquad (I.22a)$$

then,

$$h(w_j) > - p_j \log p_j \qquad (I.22b)$$

In the case of response attenuation induced by the repeated stimulation with s_j:

$$p_j \longrightarrow 1 \text{ and } \log p_j \longrightarrow 0 \qquad (I.23a)$$

Consequently,

$$h(w_j) \longrightarrow 0 \qquad (I.23b)$$

This may explain the reduction of spiking in w_j encoding s_j as observed in Fig. I.17.

Similar results on response attenuation were observed by Rocha, 1980, for neurons in the cat's brain stem. Thus, the semantics I.19 may be associated with the the axonic encoding at neurons other than sensory nerve cells, and the relations I.21 to 22 hold for these other types of neurons, too.

I.6 - Controlling the axonic encoding

One of the main features of the sensory system is to provide a relative measure of the energy distribution in 8 according to some prototypical specification of this distribution associated with the physiology of the receptor cell. This is a consequence from both the structural and physiological organization of the receptor to better transmit the energy of s ∈ 8 to the receptor membrane. It imposes specific ranges of measurable energy. For example, light receptors are restrained to sense only part of the whole light spectrum; temperature receptors are activated by the small range of physiological temperatures, etc.

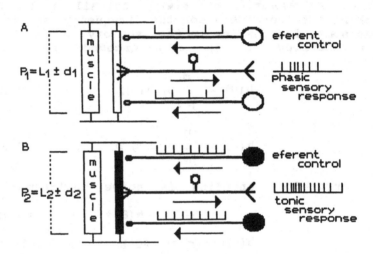

FIG. I.20 - Controlling the axonic encoding

It may be said that each receptor is specified to sense pre-defined patterns P of stimuli. Part of this specification is a phylogenetic characteristic of the animal, but part of it is controlled by the central nervous system. This is because the axonic encoding at the sensory nerve is mostly under control of the central nervous systems (Fig. I.20). For example, in the stretch receptor organ of the mammals the sensory encoding is controlled by 6 different types of efferent neurons, called efferent T1 to T6.

The efferent control of the peripheral sensory system can specify both:

I.24a) the range of measurable energy in 8: because the efferent control can transfer some amount of energy to the receptor membrane. For example, the stretch receptor organs

are composed of a receptor cell which is a muscle cell
adapted to this function. The central portion of this cell
loses the contractile apparatus while this machinery is
maintained at the end portions of this muscle cell. The
central portion of the receptor cell is in close relation
with the sensory axon, whereas the contraction state of the
end portions is controlled by the efferent τ axons. Because
of this, the tension in the central portion of the receptor
cell is dependent on both the length of the muscle and the
discharge of the τ control. In this way, the activity in the
τ axons determines the size $L \pm d$ of the measurable length
of the muscle displacement (Fig. 20), and

I.24b) the type of axonic encoding: because the efferent
control can modify the amount of energy availabe to the
sensory axon (Chapter II, section II.3). The different types
τ_1 to τ_6 are classified by their capacity to transform
the axonic encoding in the sensory nerve from tonic to
phasic, or vice versa. The type of axonic encoding is
partially dependent on genetic information, but it is also
partially specified by the amount of energy available to the
membrane. This amount of energy is influenced by the the
amount of cAMP in the cell, which is in turn controlled by
many types of transmitters released by the efferent control
neuron (Nathanson, 1977; Rasmussen and Barret, 1984).

The efferent control of the sensory encoding sets
the actual parameters of both f and Γ, defining the fuzzy
point process in I.18. In this way, the efferent control
adjusts the range of measurable energy in S and specifies
the type of axonic encoding. It must be remembered that the
axonic encoding is dependent on the axonic filtering
properties.

Given F as the set of availabe encoding functions f,
R as the set of available fuzzy point processes Γ, and W_E
as the set of outputs of the efferent control system, the
following control functions are defined:

$$g_f : W_E \times F \longrightarrow [0,1] \qquad (I.25a)$$

$$g_p : W_E \times R \longrightarrow [0,1] \qquad (I.25b)$$

so that

$$f = g_f \ (w \in W_E) \qquad (I.25c)$$

$$\Gamma = g_p \ (w \in W_E) \qquad (I.25d)$$

II.7 - The sensory world

Summarizing the previous sections, it may be said that the behavior of the axonic membrane depends on the position of the state point in the state phase space of its ionic system. This position is in part encoded by phylogenetic information, and in part modulated by the energy available to the membrane. This energy can, in turn, be dependent on the history of use of the axon, as well as on the control exercized by the central nervous system upon the sensory neuron. The amount of availabe energy specifies the filtering properties of the oscillator defined in the phase space of its ionic system and selects the range of measurable energy in both the internal and external environments of the sensory word S. In this way, both the range of measurable energies in S and the type of encodings to describe them, are phylogenetically inherited. The actual measurable range is, in turn, selected by the brain itself and modified by the use of the system, that is, by inductive learning.

The different ranges of measurable energy in S and the distinct axonic encodings of S, define patterns p of sensory information which may be expected or desired to be sensed in S. The set P of these patterns composes a prototypical knowledge inherited and/or learned about S.

In this context, the codewords $w \in W_i$ measure the degree of matching between the actual stimulus $s \in S$ and a prototypical pattern $p \in P$. Thus the semantic $\tau(w)$ is

$$\tau(w) = \mu_p(s), \quad p \in P \text{ and } s \in S \qquad (I.26a)$$

where $\mu_p(s)$ measures the similarity of p and s. Since W is a finite and totally ordered set:

$$\tau(w_i) \leq \tau(w_j) \qquad \text{if} \qquad w_i \leq w_j \qquad (I.26b)$$

The semantic of these labels can be obtained by means of the numeric decoding in eq. I.19. Thus

$$M_i \equiv W_i \qquad (I.26c)$$

where M_i is the set of finite fuzzy measures about W_i provided by the sensory neuron n_i.

The set of sensory neurons provides the brain with a set I of instruments to observe U. This set I supplies the brain with a set of fuzzy measures M about the matching between U and the prototypical patterns P of the expected behavior of S. P is genetically and functionally encoded in the definition of the neurons composing I.

Both C and W are finite and totally ordered sets of

symbols and strings, respectively. They are finite sets
because they must be redundant codes to cope with error-free
encoding in noisy environments. The capacity of this code
was estimated to be around 4 to 5 bits in the case of the
slow adapting stretching organ of the Crayfish (Rocha,
unpublished results). In this line of reasoning, it is
possible to assume that the axonic spike firing encodes
$\mu_p(s)$ in the close interval [0,1] with the precision
specified by the capacity h(C) of its code. This is the same
approach used in the computers, where the precision is
dependent on the size of the words manipulated by the
central processing unit.

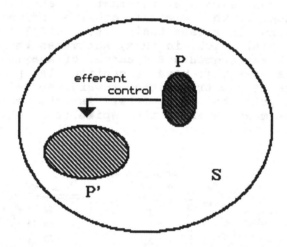

FIG. 1.21 - SPECIFYING THE RANGE OF MEASURABLE ENERGY

The efferent control can adjust the sensory encoding
in order to extend the capacity of the sensory systems to
adequately cover S with different patterns p of
prototypical knowledge, even if the sensory code has a low
capacity (Fig. I.21). This adjustment must be made in
accordance to I.21-22, in order to preserve the adequate
entropy relations between S and W. The combination of
defining patterns to match the sensory world with low
capacity controlled sensors results in powerful measuring
tools widely used by brains.

ACKNOWLEDGEMENT

I wish to express my gratitude to the following
former students which did most of the simulation studies
presented in this chapter: José Wilson Bassani, Ricardo
Mayer Aquino, Sotirus T. Pegos and Ernani A. L. Araujo.

CHAPTER II

THE SYNAPSIS
Electrical Properties

II.1 - Structure and physiology

Neurons exchange information at specialized sites called synapsis. The arrival of the spike at the pre-synaptic axons increases their intracellular concentration of Ca (Fig. 1a) which, in turn, activates the conversion of ATP into ADP and augments the amount of available metabolic energy. This energy is used by contractile proteins to move the vesicles of the transmitter toward the cellular membrane (Fig. II.1a,b). The transmitter is the molecule used to transmit the message about the spike from the pre- to the post-synaptic cell.

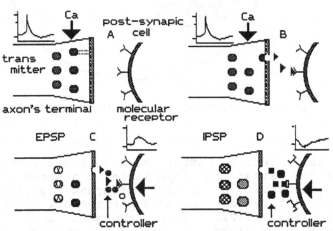

FIG. II.1 - THE SYNAPSIS

The transported vesicule fuses with the membrane and releases the transmitter t (Zimmermann, 1979) into the synaptic cleft (Fig. II.1b). This process is named exocytosis. The transmitter diffuses toward the post-synaptic cell and binds to its membrane at some specialized sites called receptors. The receptor r is a molecule having high chemical afinity with the transmitter t. The coupling of the transmitter to the receptor activates some post-synaptic molecules $\{a_i, ..., a_k\}$ here called actuators:

$$t \hat{} r \quad » \quad \{a_i, ..., a_k\} \qquad (II.1a)$$

where:

II.1b) ^ is the matching or binding operation between the transmitter and the receptor;

II.1c) » is the operation of the actuator activation, and

II.1d) the actuators exert an action over both the pre- and post-synaptic neurons, as well as over neighbouring cells.

The action exerted by the actuators depends on their structure and function. If $a_j \in \{a_i,...,a_k\}$ is some:

II.2a) ionic channel ($a_j = g$): the $t^\char94 r$ binding modifies the post-synaptic permeability and this promotes the modification of the membrane potential EM, either as a depolatizarion (EPSP in Fig. II.1c) or a hiperpolarization (IPSP in Fig. II.1d);

II.2b) enzyme controlling some metabolic chain ($a_j = e$): the $t^\char94 r$ binding changes the amount of energy available to the membrane. In general, this energy is used to modify membrane thresholds (Figs. II.4,5), and/or

II.2c) regulatory or control molecule ($a_j = c$): the $t^\char94 r$ binding triggers one or more of the following actions:

II.2c1) the modulation of the $t^\char94 r$ binding;

II.2c2) the specification of the DNA reading;

II.2c3) the activation of the DNA reading, and/or

II.2c4) the final specification of the proteic synthesis of defined molecules.

The chemical processing supported by II.2c will be discussed in Chapter III. Here, the attention will be focused upon the electrical processing supported by II.2a,b. Different types of neurons are specified by II.2a and b, respectively. The neuron of type II.2b will be called modulator neuron because its main role is to control the state of the ionic gates in the membrane. In general, it does not promote any noticiable EM variation. The neuron type II.2a corresponds to the classic neuron described in the physiology text books. No special designation will be used here to name it . Any reference to a neuron without further specification refers to a neuron type II.2a.

At least one controller of the type II.2c1 is activated by the $t^\char94 r$ binding in order to release the post-synaptic receptor for future bindings with the transmitter. As a matter of fact, the dynamics of this process mostly defines the temporal dynamics of the synaptic events. In

most cases, the controller II.2c1 is feedback released and
its action is to destroy the binded transmitter. However, it
can also be activated by the t^r binding in another
different synapsis in the neighborhood. In this case, its
action is to modify the transmission at one place depending
on the activity at another nearby site.

II.2 - Electrophysiology

If the post-synaptic actuator is some ionic channel
(action II.1b), the t^r binding changes the permeability of
the post-synaptic membrane and triggers a modification of
EM, by enhancing the conductance of either the Na (Fig.
II.1c) or K (Fig. II.1d). In the first case, an excitatory
post-synaptic polarization (EPSP) is generated, whereas in
the second case an inhibitory post-synaptic polarization
(IPSP) is the result. The actual response (EPSP or IPSP) is
dependent on the characteristics of both the transmitter and
the post-synaptic receptor (Fig. I.1).

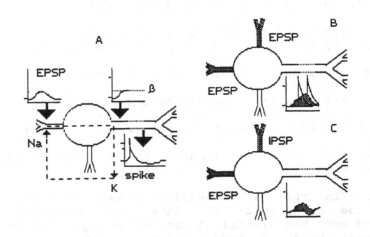

FIG. II.2 - ELECTRICAL PROCESSING AT THE SYNAPSIS

The EPSP or IPSP elicited at the dendrites induces
gating currents (Armstrong, 1981) at the initial portion of
the axon, called axon hill (Fig. II.2a). These gating
currents supply the energy to open ionic channels at the
initial portion of the axon, the axon hill. The EPSP
triggers an outward gating current to open the Na channel.
If the transferred energy is equal or above the axonic
threshold β, the Na channel is fully opened and spikes are

fired (Fig. II.2a,b). On the contrary, the IPSP enhances the axonic g_k provoking the hyperpolarization of the axon hill, which inhibits the spike firing (Fig. II.2c).

The different post-synaptic electrical activities triggered by the distinct pre-synaptic terminals are integrated at the axon hill, determining the degree of the axonic activation. This activation is encoded into spike trains. The spike interval in this codeword is the main source of information (Rocha, 1981a,b; Coon and Perera, 1989) about the axonic activation. This axonic encoding is similar to that defined by I.22 (Chapter I, section I.V) in the case of the sensory neuron.

II.3 - The early stages of the electrical processing

The amount q of transmitters stored in each pre-synaptic vesicle is more or less the same, such that the amount t of transmitters released at the synaptic cleft by each pre-synaptic spike is always an integer multiple m of this quantity q. The chemical release at the synapsis is a quantic process (Kuno, 1971):

$$t = m.q \qquad (II.3a)$$

Each quantum q of transmitters promotes a specified v_0 variation of the membrane potential EM in the post-synaptic cell. Thus, the total post-synaptic change v of EM provoked by the pre-synaptic spike is

$$v = m.v_0 \qquad (II.3b)$$

Let M(t) be the total amount of the transmitter t stored in the pre-synaptic neuron n_i, and M(r) be the total amount of the post-synaptic receptor r to bind t at the post-synaptic neuron n_j.

Each spike train in the pre-synaptic neuron n_i is a codeword w $\in$ W generated by the activation of its axon. The amount m of the transmitter released by the pre-synaptic neurons is dependent of both w and M(t):

$$m = w \cdot M(t) \qquad (II.4a)$$

The spike train w generates the calcium current required to release the contents of m vesicles of M(t) into the synaptic cleft. There are a minimum and a maximum Ca activating currents limiting the number of released vesicules. Let these currents be provided by w_l and w_u, respectively. The transmitter releases depend also on the distance between the actual position of the vesicles of M(t) and the cellular membrane. Vesicules near the membrane are easily released. The following are the minimum properties of operator •:

II.4b) no transmitter is released if the minimum calcium current is not provided

$$w_i \cdot M(t) = 0 \quad \text{if} \quad w_i \leq w_l$$

II.4c) there is a maximum amount $M_f(t)$ of transmitters which can be released by the pre-synaptic neuron

$$w_i \cdot M(t) = M_f(t) \quad \text{if} \quad w_i \geq w_u$$

II.4d) $M_f(t)$ is called the functional pool of the transmitter t. The difference

$$M_r(t) = M(t) - M_f(t)$$

is the reserve pool of the transmitter t.

FUNCTIONS

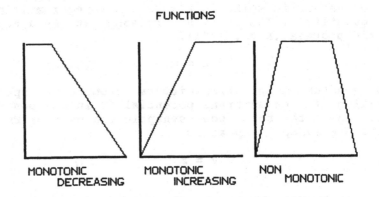

MONOTONIC DECREASING   MONOTONIC INCREASING   NON MONOTONIC

FIG. II.3 - MONOTONIC AND NON-MONOTONIC FUNCTIONS

Let $w_j \cdot M(t) = m_j$ and $w_k \cdot M(t) = m_k$ so that $w_j \leq w_k$. If (Fig. II.3):

II.4e) $m_j \leq m_k$ for all $w_j, w_k \in W$, then • is a monotonically decreasing function;

II.4f) $m_j \geq m_k$ for all $w_j, w_k \in W$, then • is a monotonically increasing function, or

II.4g) $m_j \leq m_k$ if $w_j \leq w_u$ and $m_j \geq m_k$ if $w_j > w_u$, • is a non-monotonic function.

If • obeys II.4b,f, then it may be considered a $\top$-norm. A triangular or $\top$-norm is (Dubois and Prade, 1982) a two place real-valued function whose domain is the square [0,1] x [0,1], and which satisfies the following conditions:

II.5a) $\top$ (0,0) = 0 ; $\top$ (a,1) = a   (boundary conditions)

II.5b) $\top$ (a,b) $\leq$ $\top$ (c,d)   if   a $\leq$ b; c $\leq$ d   (monotonicity)

II.5c) $\top$ (a,b) = $\top$ (b,a)   (symmetry)

II.5d) $\top$ (a,$\top$ (b,c)) = $\top$ ($\top$ (a,b),c)   (transitivity)

The minimum, product, bounded product operators are examples of $\top$-norms. Any continuous $\top$-norm satisfying the Archimedean property

$$\top (a,a) < a \qquad (II.5e)$$

can be rewritten as (Dubois and Prade, 1982):

$$\top (a,b) = f^{-1}(\ f(a)+f(b) \ ) \qquad (II.5f)$$

where f^{-1} is the pseudo-inverse of f, and f is the additive generator of $\top$. The product is an Archimedean $\top$-norm and it can be obtained as the anti-logarithm of the sum of the logarithm of each of the elements of the product.

A triangular or $\top$-conorm or S-norm is obtained if:

II.6a) $S(1,1) = 1$; $S(0,a) = S(a,0) = a$

II.6b) properties II.5b to d hold.

The t^r binding is not a crisp process. The degree $\mu(t,r)$ of binding depends on the chemical affinity between the two molecules. It can assume any value from null (0) to full (1) binding.

$$\mu: T \times R \ ---> \ [0,1] \qquad (II.7a)$$

Let T and R be the set of all transmitters at the pre-synaptic neuron and R the set of all post-synaptic receptors, and M(t) and M(r) be the pools of the transmitter t and the receptor r, respectively. Since, the post-synaptic electrical response depends both on the number of activated receptors and on the degree of this activation or binding:

$$v = m \ \hat{} \ M(r) * \mu(t,r) \ ' \ v_0 \qquad (II.7b)$$
so that
$$m \ \hat{} \ M(r) = M(r) \quad if \ \ m \geq M(r) \qquad (II.7c)$$

$$m \wedge M(r) = m \quad \text{if} \quad m \le M(r) \qquad \text{(II.7d)}$$

This means that the maximum number of activated post-synaptic receptor is $M(r)$ if m is very high; otherwise it is equal to m. The amount of activation of each of these binded post-synaptic receptors is governed by the actual value of $\mu_A(t,r)$. From eq. 4a:

$$v = w \cdot (M(t) \wedge M(r) * \mu(t,r) \;{}^{\shortmid} v_0 \qquad \text{(II.7e)}$$

where $\wedge$, $*$ and $\shortmid$ are, in general, $\top$ -norms.

The post-synaptic response v_0 is positive in the case of the EPSP or excitatory synapsis, and negative in the case of the IPSP or inhibitory synapsis. These electrical responses at the dendrites and the cell body are the source of energy for the gating axonic currents (Armstrong, 1981) controlling the axonic activation. There is always some energy dissipation in transmitting the dendritic and cell body currents to the axon hill. This dissipation is proportional to the total electric resistance between the activated site and the axon hill. Thus, the actual value of v_0 depends on:

II.8a) the type of channel activated by the $t \widehat{\;} r$ binding;

II.8b) the dynamics of this binding, and

II.8c) the relative position of $r \in R$ in respect to the axon hill A. This dependence is encoded in the behavior of the operator $\shortmid$.

As a consequence of the above R may be assumed to encode some pattern of axonic activation according to its spatial distribution over the post-synaptic cell. In this way, v may be viewed as the measure of the matching $\mu(w,R)$ between the actual activation w of the pre-synaptic neuron and the maximum possible post-synaptic activation represented by $M(t) \wedge M(r)$. Thus, eq. VII.7e may be rewriten as

$$v = w \cdot s \qquad \text{(II.9a)}$$

$$s = M(t) \wedge M(r) * \mu(t,r) \;{}^{\shortmid} v_0 \qquad \text{(II.9b)}$$

such that

$$\mu(w,R) = v \qquad \text{(II.9c)}$$

and v measures the similarity between w and the pattern encoded by the distribution of R over the post-synaptic neuron.

Eq. II.9a is very similar to the classic proposition in neural nets

$$v = w.s \qquad \text{(II.10)}$$

however, Eq.9b stresses the complexity of the real synaptic processing.

One of the main assumptions in artificial neural nets is that knowledge is stored in the weight or the strength s of the synapses of a given net, because the post-synaptic activation v is dependent on both s and w. Eq. II.9b stresses the fact that knowledge in natural neural nets may be stored in the synaptic weight in a very complex way, in contrast with the simple numeric encoding used in the classic artifical neural nets theory. Here, knowledge can be encoded in:

II.11a) the distribution of $M(t)$ inside the pre-synaptic terminal. This distribution defines the behavior of the operator $\circ$;

II.11b) the distribution of $M(r)$ over the post-synaptic cell. This distribution defines the behavior of the operator $\bullet$;

II.11c) the affinity $\mu(t,r)$ between the transmitter t and the post-synaptic receptor r, and

II.11d) the total amount of $M(t)$ and $M(r)$, besides the actual value of v_0.

This complexity of the knowledge encoding greatly increases the computational power of natural nervous systems compared with artificial neural nets. It also implies different types of learning (see Chapter IV) which increases the capacity of knowledge acquisition by natural neural nets.

II.4 - The axonic processing

Blomfield, 1974, has shown that the electrical activity triggered by different pre-synaptic cells at the post-synaptic neuron is usually summed at the axon hill. Only in special cases, when the inhibitory synapsis promotes a huge modification of the post-synaptic conductance, it can be used to implement the arithmetic division. The arithmetic product seems to be an unfeasible operation at the axon hill, because of the electrical properties of the nervous cells. Thus, it may be considered that, in general, the activity in all of the n pre-synaptic neurons is combined at the axon hill of the post-synaptic cell by means of a summation. Thus, if a_s measures the activity at the post-synaptic axon hill

$$a_s = \sum_{i=1}^{n} v_i \qquad (II.12a)$$

or from II.8b:

$$a_s = \sum_{i=1}^{n} w_i \cdot s_i \qquad (II.12b)$$

where $\cdot$ is a $\top$ -norm. In the case of artificial neural nets, from II.8a:

$$a_s = \sum_{i=1}^{n} w_i \, s_i \qquad (II.12c)$$

This activity is tranformed into the axonic spike firing by a fuzzy point process Γ similar to that discussed in Chapter I, section I.v. Thus, given a cover C of epochs of the time continuum T and the binary alphabet

B = { 1 = spike occurrence, 0 = spike absence }   (II.13a)

the axonic encoding W is a fuzzy point process Γ over C (Rocha, 1981a) associating a possibility distribution $\mu_c(s)$ of B being distributed over the epochs $c \in C$:

$$\Gamma : C \times B \dashrightarrow [0,1] \qquad (II.13b)$$

The possibility distribution $\mu_c(b)$ is dependent of the energetic time pattern of the axonic activation a_s to be encoded, that is

$$\mu_c(b) = f(a_s) \qquad (II.13c)$$

In this condition, $w \in W$ becomes a fuzzy distribution of B over the epochs $c \in C$:

$$w = \Gamma \, (\mu_c(b)) \qquad (II.13d)$$

Since w has a numerical semantic (see Chapter I, section I.V), eq. II.13d can be rewritten as:

$$w = \begin{cases} w_i \text{ if } a_s < \alpha_1 \\ w_u \text{ if } a_s \geq \alpha_2 \qquad (II.14a) \\ g(a_s) \text{ otherwise} \end{cases}$$

where α_i and α_2 are the thresholds defining the filtering properties of the sensory axon; w_i and w_u are the codewords associated, respectively, with these lower and upper thresholds, and g is the function relating the codeword $w \in W$ and a_s according to the numeric semantic in eqs. I.19b,c.

Another way of generically expressing the synaptic processing defined by eqs. II.12, 13 is

$$w = \overset{n}{\underset{i=1}{\Theta}} (w_i \bullet s_i) \qquad (II.14b)$$

where Θ and $\bullet$ are a $\overline{\top}$-conorm and a $\overline{\top}$-norm (Pedrycz and Rocha, 1992), respectively.

If $\alpha_1 = 0$ in Eq. II.14a, the neuron fires even if there is no pre-synaptic activation. This neuron is said to be an automatic neuron. The value of w_l defines this resting firing level. Synaptic inhibition (IPSP) reduces this activity, and synaptic excitation (EPSP) increases it. In the case of automatic neurons W is a subset of the fuzzy integers. In the other cases W is a subset of the fuzzy natural numbers. W is a finite set because it is supported by a finite alphabet and a finite grammar.

Given W_i as the set of codewords of a pre-synaptic inhibitory neuron and W_a as the set of codewords of a automatic post-synaptic neuron, negation (IPSP) is a

$$I: W_i \times W_a \longrightarrow [0,1] \qquad (II.15a)$$

so that

II.15b) $I(0) = w_l^a$;

II.15c) $I(w_u^i) = w_j^a = 0$;

II.15d) I is strictly decreasing, and

II.15e) I is continuous

Provided W^a is the inhibitory set to another post-synaptic code W^p and g^i as the inverse of g, then

II.15f) $I(I(a))=a$ if $w_l^p = w_u$ and $g_p = g_i$.

This condition is fullfilled if $W^p = W^i$ in the case of recurrent inhibition. All of this means that negation is a very complex operation in natural neural nets subjected to different types of definitions. Some strong properties, e.g. $a + \tilde{\ }a = 1$, of some types of negation require a very complex neuronal structure to guarantee the adequate relations between the different thresholds and encoding functions discussed above.

II.5 - Controlling the energy available to the membrane

If the post-synaptic actuator a_j in Eq. II.1 is an enzyme controlling a metabolic chain, then the effect of the coupling t^r is to control the amount of the energy available to the post-synaptic membrane (action II.2b). The most important example of this kind of mechanism is the control exercized by means of the cyclic AMP (cAMP) illustrated in fig. II.4,5.

The t^r binding activates the enzyme adenylate cyclase, whose action is to convert ATP into cAMP. This cAMP binds itself to another enzyme, named Protein Kinase. The role of this enzyme is to accelerate the conversion of ATP into ADP. This conversion results in the release of one energized atom of phosphor. This energy is transferred to the membrane, modifying the state of the ionic channels. The consequence is in many cases the change of the membrane excitability without any noticeable electrical activity at the post-synaptic cell. The importance of this type of synapsis, called modulator synapsis, is to control the various parameters of the function in eqs. II.4,7a and 14 as well as in eq. I.25 (Chapter I, section I.6) The type of action of this synapsis depends on its spatial position in respect to the pre- and post-synaptic cells.

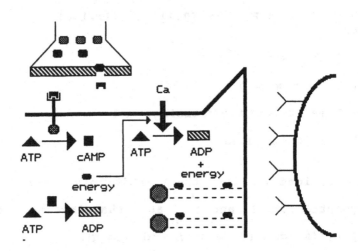

FIG. II.4 - CONTROLLING THE TRANSMITTER RELEASE

If the modulator synapsis is established over the pre-synaptic terminals of another synapsis (Fig. II.4), the augmented available energy is used to augment the currents of Calcium. The increase of this ion inside the cell augments the convertion of ATP into ADP and accelerates the movement of vesicles toward the cellular membrane. This results in the enhancement of the amount of transmitters released at the synaptic cleft. Thus the effect of the control exercized by the modulator synapsis is to change the decoding function in II.4.

If W_c is the modulator codeword, M_c is the total amount of modulator, and W_k is the codeword in the synapsis being controlled, then II.4 becomes:

$$m_k = w_k \cdot (w_c \cdot M_c) \qquad (II.16a)$$

where $\cdot$ and $\cdot$ are $\top$ -norms, and m_k is the amount of transmitters released at the pre-synaptic neuron n_k. This is because the function pool M_k of transmitter at n_k is under the control of the modulator neuron n_c.

The meaning of II.16a is that the decoding function at the pre-synaptic neuron is modified by the modulator neuron according to the synaptic strength and activity of this later cell. Kandel and Schwartz, 1982, showed that sensitization and associative learning in the Aplysia result from an increase of the concentration of the cAMP inside the terminal buttons promoted by the release of serotonin over the axonic terminals. The cAMP blocks the permeability of K and increases the intracellular concentration of Ca, which in turn augments the number of vesicules moved by each spike. Thus, II.16a can describe the medium term associative learning in the Aplysia.

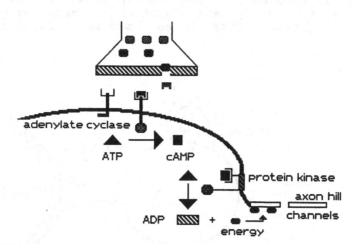

FIG. III.5 - CONTROLLING THE ENERGY AVAILABLE FOR THE MEMBRANE

If the modulator neuron makes contact with the post-synaptic cell near the axon hill (Fig. II.5) the effect is to control the encoding function g and the filtering thresholds α_1 and α_2 in eq. II.14a, because the released energy is used to change the state of the gates of the Na channel. Thus, if w_c is the modulator codeword, M_c is the total amount of modulators; w is the post-synaptic output word; w_i is the axonic pre-synaptic activation; and s_i is the weight of this synapsis:

$$w = (w_c \cdot M_c) * \overset{n}{\underset{i=1}{\Theta}} (a_i \cdot s_i) \qquad (II.16b)$$

where $\Theta,*$ and $\cdot,\cdot$ are $\top$ -norms and $\bot$ -conorms, respectively.

This mechanism can explain the role of the gating control exercized by some neurons in the brain making synaptic contacts near the post-synaptic axon hill. According to some authors (e.g., Allen and Tsukahara, 1974; Eccles, 1981), this type of neuron is used to control the output of the processing at the dendrites of the post-synaptic cell. The idea is that the function of the modulator neuron is to determine the moment the output of the post-synaptic processing is allowed to be transferred to other subsequent cells. This is done by controlling the axonic threshold. In a general view, the control expressed by II.16b specifies the filtering properties of the axonic membrane because the thresholds α_1 and α_2 and the encoding function g are under control of the modulator neuron. In conclusion, the role played by the modulator neurons is to specify and adjust the $\top$ -norms or $\bot$ -conorms (Dubois and Prade, 1982) to be used in II.4,14 and I.25.

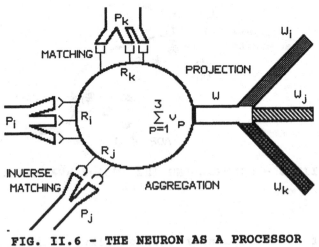

FIG. II.6 - THE NEURON AS A PROCESSOR

* * * *

II.6 - The neuron as a multipurpose processor

The events described so far about the electro-physiology of the neuron are related to the following actions:

II.17a) Matching: Let R be the total population of post-synaptic receptors of the neuron n_j. R is a family of subsets R_i of receptors specialized in binding different pre-synaptic transmitters t_i. Also, the electrical response v_o induced by the activation of $r \in R_i$ depends on their distribution in the post-synaptic membrane; on the type of receptor and on the dynamics of the binding $t \hat{\ } r$ (eqs. II.8). Thus, each subset R_i represents a possible pattern of activation of n_j (Fig. II.6), and its actual activation represents the degree of matching $\mu(w,R)$ between this pattern and the pre-synaptic activity w (eq. II.9c);

II.17b) Aggregation: The different activities at the distinct pre-synaptic terminals are aggregated at the axon hill as a consequence of the electrical properties of the neuron (Blomfield, 1974; Hodgkin and Huxley, 1952). The result a_j of this aggregation is generally obtained by the summation (eq. II.12a) of the different v_i triggered by the distincts pre-synaptic terminals, and

II.17c) Projection or encoding: the activity a_j aggregated at the axon hill is encoded into W. This encoding can be under control of modulator neurons, in the same way that the sensory encoding can be adjusted by the central nervous systems. This control specifies the filtering properties of the axonic membrane.

Different axonic branches can exhibit different filtering properties. This means that W can be partioned into different subsets W_i depending on the distinct filtering properties of the axonic branches (Fig. II.6). The result of this is a controlled encoding of the type

$$\text{if } \alpha_k < a_j < \alpha_{k+1} \text{ then } w \in W_k$$

where W_k is a subset of W. This kind of projection is quite different from the proposition in artificial neural nets that the axonic activation equally spreads over all terminals of each neuron in the net, and

II.17d) Inverse encoding (decoding): $w \in W_i$ is tranformed at any pre-synaptic terminal i into an amount m of transmitters released at the synaptic cleft. Any pre-synaptic terminal makes several contacts with the post-synaptic cell (Fig. II.6). This terminal branching pattern is one of the main factors determining the distribution of M inside the pre-synaptic neuron, because it determines the relative position of the vesicles of M in respect to the pre-synaptic membrane. The transmitter release is dependent of both this branching pattern and the dynamics of the activating ionic currents (see eq. II.4).

The above processing structure is called here MAPI paradigm. It supports the neuron as a general purpose processor, whose programming depends on the specification of:

II.18a) the decoding functions in eqs. II.4 and 7, and

II.18b) the encoding function g and the threshold encodings α in II.14a, that is, the axonic filtering properties. It must be stressed that different axonic branches can exhibit specific filtering properties.

The MAPI structure supports the neuron as a general processor for different types of calculations, in the same way that the central processor of a computer is able to handle different mathematical languages. In the following sub-sections some of these different processings are discussed without pretending to be an exhaustive analysis of the capacity of the neuron as a processsor.

II.6.a - The Neuron as a Numeric Processor

The result of the numeric processing at the axon hill is dependent of:

$$v_i = w_i \circ s_i \qquad (II.9a)$$

$$s = M(t) \,\hat{}\, M(r) * \mu(t,r) \,^{a} v_o \qquad (II.9b)$$

$$a_s = \sum_{i=1}^{n} v_i \qquad (II.12a)$$

$$w = \begin{cases} w_i \text{ if } a_s < \alpha_1 \\ w_u \text{ if } a_s \geq \alpha_2 \\ g(a_s) \text{ otherwise} \end{cases} \qquad (II.14a)$$

The numeric semantics of $w \in W$ can be provided by the:

II.19a) powered summation of the activity at the n pre-synaptic neurons n_p:

$$w = \sum_{p=1}^{n} w_p \circ s_p$$

if

$$V_p = W_p \circ S_p$$

$$\alpha_1 = V_0, \quad \alpha_2 = \infty$$

$$W = a_s$$

and ° is the product.

19b) **bounded powered sum** if: $\alpha_1 > V_0$, $\alpha_2 < \infty$ in II.19a

II.19c) **powered mean**: if II.19a holds and

$$\sum_{p=1}^{n} S_p = 1$$

Because the product is an Archimedian $\top$ -norm, the numeric semantic of $w \in W$ can also be that of the (eq.II.5):

II.19d) **powered multiplication of the pre-synaptic activity**: if ° in II.4 is the logarithmic function; g in II.14a is the power function, and $\alpha_1 = V_0$, $\alpha_2 = \infty$. In this case, W is considered to be the set of integers greater than 0. This is because the neuron can handle the product in a way to decode the pre-synaptic activity w_i into a logarithmic release of m_i, and then encode the summation of all pre-synaptic activities by means of a power function.

II.19e) **bounded powered multiplication**: if $\alpha_1 > V_0$, $\alpha_2 < \infty$ in II.19d;

II.19f) **a kind of geometric mean**: if condition II.19d holds, and

$$\sum_{p=1}^{n} S_p = 1$$

If the slopes of ° and g are equivalent, then w is the true geometric mean of the pre-synaptic activity.

Finally, other numeric semantics of $w \in W$ can be provided by:

II.19g) **other types of calculation**: if the adequate $\top$ -norms and $\top$ -conorms in II.4 and II.14 are chosen. Since the modulator neuron controls the behavior of these $\top$ -norms and $\top$ -conorms (II.16), these neurons can be used to program the required type of calculation.

The min and max operators do not have additive generators and require another approach to be implemented in the real neuron. This is the subject of the next section.

II.6.b - The neuron as a sequential processor

The neural nets are widely accepted as parallel distributed and processing systems. However, natural neural nets are sequential circuits, too. Hierarchy introduces sequentiality in neural nets. The brain is a hierarchy. One of the most important human cognitive function, language, is in practice a sequential task (see Chapter VIII).

Some kinds of neuronal processing require either a spatial or a temporal ordering as a fundamental pre-requisite. Max and min operations are the best examples of this kind of processing (Pedrycz and Rocha, 1992; Rocha and Yager, 1992). Different axonic filtering properties are used as the support for the spatial Max and Min processing in neural circuits (see Fig. VI.4 in Chapter VI). Here, the recurrent modulator synapsis is used to implement the temporal Max and Min processing (Fig. II.7).

The recurrent synapsis is established if the axon of the neuron n_j makes contacts with dendrites or with the cell body of n_j itself (Fig. II.7). If the recurrent modulator synapsis is located at the axon hill, then it can control the axonic thresholds as a function of the n_j activity itself.

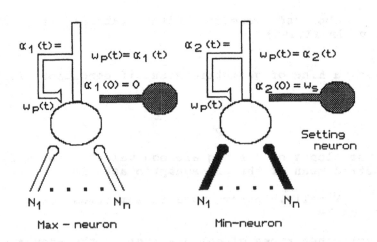

FIG. II.7 - THE NEURON AS A SEQUENTIAL PROCESSOR

Let the following types of neuron be defined:

II.20a) high threshold neuron (HTN): the bias of this type of neuron is set at a very high value, so that its output is maintained zero no matter what the value of its inputs. From time to time, this threshold is lowered under the control of another neuron, called setting neuron. At these specific moments, the actual value of the v in Eq. II.12a is encoded into w at the axon according to eq. II.14a, and transmitted to the HTN post-synaptic cells;

II.20b) low threshold neuron (LTN): the bias of this type of neuron is set at a low value, so that its output changes in time according to the temporal modification of its inputs. The neurons commonly used in artificial neural nets are LTN neurons.

Let the pre-synaptic activity over the post-synaptic LTN neuron n_p be timely ordered. This implies that the pre-synaptic neuron n_i always fires before another neuron n_j if $i < j$ (Fig. II.7). This temporal ordering is obtained by defining the pre-synaptic neurons as HTN neurons. Thus, the activity a_i of the pre-synaptic cell n_i is maintained equal to zero except at the moment $t=i$ of its firing. This firing moment is determined by the controlling setting neuron. On the contrary, the activity a_p of the post-synaptic cell n_p varies in time. This is denoted here by the notation $a_p(t)$. In this condition:

II.20c) the max-neuron (Fig. II.7) is defined if the axonic threshold $\alpha_1(t_0)$ at time $t_0=0$ in II.14a is set as equal to 0 by the setting neuron, and then at time t it is set equal to the firing level $w_p(t-1)$ at time $t-1$:

$$\alpha_1(t) = w_p(t-1)$$

Also, the output $w_p(t)$ is:

$$w_p(t) = \left[\begin{array}{l} \alpha_1(t) \text{ if } v(t) \le \alpha_1(t) \\ g(v(t)) \text{ otherwise} \end{array} \right.$$

where v(t) is the post-synaptic activation at time t:

$$v(t) = w_t \cdot s_t$$

In this condition, the output $w_p(t)$ of the neuron at time t encodes:

$$w_p(t) = \overset{t}{\underset{k=1}{V}} (w_k \cdot s_k)$$

If $s_k=1$ then

$$w_p(t) = \overset{t}{\underset{k=1}{V}} w_k$$

where V is the maximun operator.

II.20d) the min-neuron (Fig. II.7) is defined if the axonic threshold $\alpha_2(t_0)$ at time $t_0=0$ in II.14 is set as equal to w_s (the maximum axonic firing output) by the setting neuron, and then at time t it is set equal to the firing level $w_p(t-1)$ at time $t-1$:

$$\alpha_2(t) = w_p(t-1)$$

Also, the output $w_p(t)$ is:

$$w_p(t) = \left[\begin{array}{l} \alpha_2(t) \text{ if } v(t) \geq \alpha_2(t) \\ g(v(t)) \text{ otherwise} \end{array} \right.$$

where $v(t)$ is the post-synaptic activation at time t. In this condition, the output $w_p(t)$ of the neuron at time t encodes:

$$w_p(t) = \bigcap_{k=1}^{t} (a_k \cap s_k)$$

If $s_k=1$ then

$$w_p(t) = \bigcap_{k=1}^{t} a_k$$

where $\cap$ is the minimun operator.

But from II.4 and 14 15 and 16, it is possible to generalize the above temporal processing to any $\top$ -norm of $\top$ -conorm Θ (Pedrycz and Rocha, 1992):

$$w_p(t) = \mathop{\Theta}_{k=1}^{t} (a_k \circ s_k) \qquad (II.20c)$$

It is interesting to remark the necessity to reset the axonic thresholds to pre-defined levels at the beginning of the above calculations. This is the role of the Setting Neurons in Fig. II.7. These neurons play the role of synchronizers. Time synchronization is a very important task in sequential processing devices. This synchronization is also used to guarantee the temporal ordering of the pre-synaptic firing required by II.20. The same ordering can also be guaranteed by other processes mimicking time delay devices, like different axonic conducting velocities, different axonic lengths in a system of parallel fibres, different number of intermediate synapsis, etc.

II.6.c - The MAPI structure supports fuzzy logic

The solution of a fuzzy implication is proposed (Zadeh, 1983a) to be provided by the extended version of modus ponens (EMP):

if X_1 is A $^\wedge$ ... $^\wedge$ X_j is D then Y is E

$(X_1$ is A') $^\wedge$ ...$^\wedge$ $(X_J$ is D')

Y is E'   (II.21a)

which implies finding the fuzzy set E' given the fuzzy sets A'... D' and the implication function f relating the fuzzy sets A ... D and E (Godo et al, 1991; Katai et al, 1990a,b; Trillas and Valverde, 1987; Yager, 1990d; Zadeh, 1983a). $^\wedge$ is a $\top$ -norm.

This process is performed in 4 steps (Zadeh, 1983a):

II.22a) Matching: the compatibility σ between A and A' is assessed as a measure of the equality [A≡A'] between the fuzzy sets A and A' (Pedrycz; 1990a,b), so that (X is A') can be rewritten as (Godo et al, 1991)

$(X$ is A') $\equiv$ $(X$ is A) is σ

but   A $\equiv$ A' implies A = A'   and   A' = A

$\mu_A(x)$ $\leq$ $\mu_{A'}(x)$ and $\mu_{A'}(x)$ $\leq$ $\mu_A(x)$

The calculation of σ means to evaluate how equal are these two fuzzy sets taking into account all or some of their elements (Pedricz, 1990a; Rocha, 1991b). σ will here mean the confidence that (X is A') is (X is A). If A is considered to encode a prototypical knowledge, σ measures the confidence that A' is this knowledge.

II.22b) Aggregation: all compatibilities σ_i assigned to the arguments are aggregated into a unique value σ_a representing the total compatibility of the actual antecedents with the prototypical knowledge encoded by the antecedent part of the implication:

$$\sigma_a = \overset{n}{\underset{i=1}{\Theta}} (\sigma_i)$$

The aggregation function Θ can be the min function (Zadeh, 1983a) or any other $\top$ -norm (e.g., Castro and Trillas, 1990; Delgado et al, 1990a,b; Dubois and Prade, 1982; Mizumoto and Zimmermann, 1982; Mizumoto, 1989; Yager, 1984), or even a non-monotonic aggregation based on the geometric mean (Rocha et al, 1989, 1990a). The type of calculation represented by Θ depends on the degree of ANDness (ORness) of the ⊤ -norm ^ (see II.26e).

II.22c) Projection: the compatibility σ_c of the consequent is obtained as function of the aggregated value σ_a (Delgado et al, 1990b; Diamond et al, 1989; Godo et al 1991; Katai et al, 1990a,b)

$$\sigma_c = g(\sigma_a)$$

σ_c measures the compatibility of (Y is E') with (Y is E):

if X_1 is A ^ ... ^ X_j is D then Y is E

$(X_1$ is A) is σ_1 ^ ...^ $(X_j$ is D) is σ_1

(Y is E) is σ_c

II.22d) Inverse-Matching and Defuzzification: the problem now is first to find E' given E and σ_c, and then to obtain the singleton e ϵ E' or the subset D of the most representative elements of D' as the final output of the process. Many models of defuzzification are proposed in the literature (see Chapter 10).

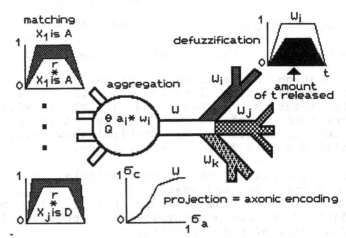

FIG. II.8 - THE NEURON AS A FUZZY DEVICE

The solution of a fuzzy implication is obtained with the very same MAPI structure used to support the neuron as a multiple purpose processor unit. In this way (Fig. II.8):

II.23a) the t^r coupling measures the matching between the incoming information and the prototypical knowledge encoded in R (see Fig. II.6 and eq. II.7e);

II.23b) the aggregation function θ in II.22b and the implication function in II.22c are correlated with the events at the axon hill, where the activity of all pre-synaptic terminals are combined to determine the axonic spike firing. Eq. II.22b is the same eq. II.12a, and eq. II.22c corresponds to the axonic encoding function II.14a. In this way, the neuron can perform a variety of types of aggregation (eq. II.14b) depending on the definition of the $\top$ -norms or $\bot$ -conorms used in the pre-synaptic decoding (eq. II.4) and post-synaptic encoding (eq.II.14a) functions. Thus, the neuron is able to perform any of the methods proposed in the literature to solve the EMP (see e.g., Greco and Rocha, 1987; Mizumoto, 1989; Rocha et al, 1989, Yager, 1988b; Zadeh, 1983a; Zimmermann and Zysno, 1980).

II.23d) the inverse matching and defuzzification is performed at the axonic terminals depending on the branching pattern of these terminals and on the dynamics of the transmitter release. The inverse matching can be described by any $\top$ -norm or $\bot$ -conorm assigned to II.4.

The relevance δ is the measure of the uncertainty about the frequency of occurrence of each argument of the antecedents of the fuzzy proposition. (Kacprzyk, 1988; Kacprzyk et al, 1990; Rocha et al, 1989; Sanchez, 1989). It expresses the importance of each argument in supporting the result of the implication. Confidence is defined as the matching between the actual value of the antecedant and the prototypical knowledge about it. Confidence is measured by σ. Relevance and confidence must be ANDed to express their influences upon the decision making. The implication

$$\text{if } [(X \text{ is } A) \cdot \delta] \text{ and } [(Y \text{ is } B) \cdot \delta'] \ldots. \text{ then } Z \text{ is } C \quad (II.24a)$$

will here be called powered implication. The operator $\cdot$ is a $\top$ -norm, in general the product. This kind of implication is used to describe knowledge of the type

$$\text{If } (X \text{ is } A) \text{ and relevant } \ldots. \text{ then } Z \text{ is } C \quad (II.24b)$$

The aggregation in the case of powered implication cannot be an ALL-NOTHING calculation like the min-function proposed by Zadeh, 1983a, because less relevant arguments may be either false or non-observed without damaging the results of the implication. The role played by the less significant argument must be to increase confidence in the conclusion if the most relevant arguments are not fully satisfied. In this line of reasoning, II.24a is rewritten as if $Q\{[(X$ is $A)\cdot\delta]$ and $[(Y$ is $B)\cdot\delta']$ ....$\}$ then Z is C

$$(II.24c)$$

where Q is a linguistic quantifier of the type MOST, AT LEAST N, etc. This kind of quantifier was defined by Zadeh, 1983b as a proportional quantifier. The quantified powered implication II.24b is used to describe knowledge of the type

If Most of the relevant (X is a) ... then Z is C   (II.24d)

Let the proposition be

$$QRXs \text{ are } A \qquad (II.25a)$$

where Q is a proportional quantifier, and R and A are fuzzy sets representing, respectively, relevance and prototypical knowledge (see chapter X, section X.6). The truth σ of this proposition is calculated in two steps (Zadeh, 1983b; Kacprzyk 1986a,b, 1988; Yager, 1990b)

II.25b) to obtain the relative sigma-counting (Σ-count) of the fuzzy set A given R as

$$s = \Sigma\text{-count}(A \text{ and } R)/\Sigma\text{-count}(R)$$

II.25c) to set the truth of the proposition as

$$\sigma = \mu_Q(s)$$

where $\mu_Q(s)$ measures the compatibility of s with the prototypical knowledge of the quantifier Q. In general, this membership function is of the type:

$$\mu_Q(s) = 0 \text{ for } s \leq \alpha_1$$

$$= g(s) \text{ for } \alpha_1 < s < \alpha_2$$

$$= 1 \text{ for } s \geq \alpha_2$$

If δ_i and σ_i represent, respectively, the relevance of and the confidence in the ith antecedent of proposition II.24c above, then the relative sigma-counting of the fuzzy set A given R is (Kacprzyck, 1988; Yager, 1990c):

$$\sigma_a = \sum_{i=1}^{n} \delta_i \cdot \sigma_i \Big/ \sum_{i=1}^{n} \delta_i \qquad (II.26a)$$

or

$$\sigma_a = \sum_{i=1}^{n} \Big(\delta_i \Big/ \sum_{i=1}^{n} \delta_i\Big) \cdot \sigma_i \qquad (II.26b)$$

where n is the number of antecedents in II.24c. The truth

σ_c of (Z is C) becomes a function of the compatibility of the antecedents σ_a with the prototypical knowledge of Q. Thus, σ_c is the compatibility function $\mu_Q(\sigma_a)$

$$\sigma_c = \mu_Q(\sigma_a) \qquad \text{(II.26c)}$$

$$\mu_Q(\sigma_a) = 0 \text{ for } \sigma_a \leq \alpha_1$$

$$= g(\sigma_a) \text{ for } \alpha_1 < \sigma_a < \alpha_2$$

$$= 1 \text{ for } \sigma_a \geq \alpha_2$$

The solution of II.26 requires the same type of calculation performed by the neuron, since II.26b is equivalent to II.22c, if the synapsis weights are normalized, and II.26c is the same eq. II.14a. Normalization is not required if Q in II.24c is not a proportional quantifier (Yager, 1990c).

In this way, W encodes Q, and Q measures the pre-synaptic activity (Rocha, 1991a; Yager, 1990c) as:

II.26d) if Q of the pre-synaptic neurons are connected and activated then the axon firing is w.

Now,

II.26e) if Q is assumed to be AT LEAST N, the post-synaptic neuron can play the role of an AND/OR device, whose degree of ANDness (ORness) is given by α_1 in both II.14a and II.26b and related to the semantics of N. If N tends to 1 the neuron tends to be an OR device and α_1 decreases. If N increases the neuron tends to be an AND device and α_1 augments. Once the filtering properties of the axon are under the control of modulator neurons, the actual semantics of Q can be learned and specified by the brain.

The variables in the fuzzy implication II.24a are in many instances better described as linguistic variables. A linguistic variable is characterized by the quintuplet (see chapter X, section X.5):

$$(X, T(X), U, G, M) \qquad \text{(II.27a)}$$

in which X is the name of the variable; U is the universe of discourse; T(X) is a set of terms of a natural or artificial language used to speak about X; G is the syntactic rule used to generate the terms of T(X); and M is the semantic rule defining the meaning of X. This semantic is defined according to the compatibility $\mu_{R(z)}(u)$ of u concerning the restriction R(z) defining the term $z \in T(X)$ (Zadeh, 1983a):

$$x = u : R(X), \quad x \in X, u \in U \qquad \text{(II.27b)}$$

so that

if $\alpha_z < \mu_{R(z)}(u) \leq \alpha_{z+1}$ then $x = \mu_{R(z)}/u$ (II.27c)

This partition is similar to that induced over W (eq. II.17c) by the different filtering properties of the axonic branches.

It is possible to conclude here that Fuzzy Logic and Neural Nets are equivalent languages, and the neuron may be viewed as a fuzzy logic unit (Rocha, 1991b) with the following equivalence between the steps of resolution of the fuzzy implication and the physiology of the nerve cell (Fig. II.7):

II.28a) the matching taking place at the post-synaptic receptors;

II.28b) the aggregation being performed at the cellular body;

II.28c) the projection being related to the axonic encoding;

II.28c) the inverse-matching and defuzzification being related to the transmitter release, and

II.28d) the thresholds of the axonic branches being related to the semantic of the linguistic variables.

II.6.c - The MAPI structure supports mathematical programming

The purpose of the Mathematical Programming (MP) is to maximize (or minimize) the objective function f used for decision making, given the set of constraints to be imposed upon this decision (e.g., Delgado et al, 1989; Verdegay, 1984):

$$\text{Max or Min}\quad y = f(x)\qquad \text{(II.30a)}$$

subject to

$$a_i x \leq (\text{or} \geq) b_i, \; i \in M = \{1,2,\ldots,m\}\quad \text{(II.30b)}$$

A Linear Mathematical Programming (LP) is obtained if f is a linear function and a_i, x and b_i are linear vectors. Fuzzy Mathematical Programming (FMP) implies (Verdegay, 1984):

II.30c: fuzzy objective function f'; and/or

II.30d: fuzzy constraints b'_i; and/or

II.30e: fuzzy coefficients a'_i.

MP and FMP are used for decision making in many situations where maximization of profits, minimization of losses, etc., are required. Let us see the following example (Verdegay, 1984):

" a decision maker who, for each of the products he sells, has a value interval to fix the public sale price, and, therefore, to assign the profit. Together with the price fixed for each item, he determines a function which represents the degree of accomplishment of his decision; this degree depends on factors as the season of the year, the quality of the competition in each product, etc. Due to these factors, the decision maker does not mark the items with the maximum profit, since in such a case the competition, for instance, could obtain a profit from his policy by reducing their prices for the same articles. In order to optimize his profit, the decision maker has an objective (in general a fuzzy objective) function, subject to (in general fuzzy) constraints."

As a matter of fact, deduction in fuzzy logic is based on the solution of a non-linear program (Zadeh, 1983a). If one is interested in determining the value of an unknown variable q which may be expressed as a function of a set of variables X which are constrained by a collection of propositions in the Knowledge Base, then one tries to maximize the confidence in q subject to the constraints in the data base.

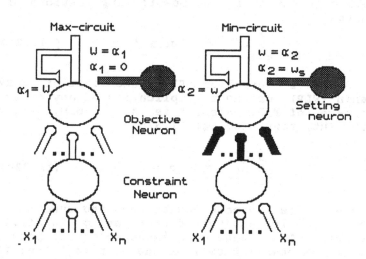

FIG. II.9 - NEURAL NETS AND MATHEMATICAL PROGRAMMING

A crisp rule like If X then Y implies that if X is true (1), Y is also true (1), but if X is false (0), Y can be either false (0) or true (1). This can be expressed by the linear constraint (Delgado et al, 1990a,b):

$$X - Y \leq 1 \qquad \text{(II.31a)}$$

Generalizing the above for the case in which

If X_1 and ..... and X_n then Y

II.31a becomes:

$$\Sigma \, X_i - Y \leq n \qquad \text{(II.31b)}$$

Thus, the calculation of:

if X_1 is A $\hat{}$ .... X_j is D then Y is C $\qquad$ (II.32a)

$$\frac{(X_1 \text{ is } A) \text{ is } \sigma_1 \hat{} \ .... \ (X_j \text{ is } D) \text{ is } \sigma_j}{(Y \text{ is } C) \text{ is } \sigma_c}$$

where $\hat{}$ is a $\top$ -norm, can be converted to the solution of the following constraint (Delgado et al, 1990a,b):

$$\overset{j}{\underset{k=1}{\Theta}} \ \sigma_k - \sigma_c \leq j \qquad \text{(II.332b)}$$

and taking in consideration powered implications of the type II.24b then

$$\overset{j}{\underset{k=1}{\Theta}} \ r_k \cdot c_k - \sigma_c \leq j \qquad \text{(II.32c)}$$

where Θ is a summation (Σ) in the case of Fuzzy Linear Programin (FLP), or the multiplication of any other function in the case of FMP. Also, $\cdot$ is a $\top$ -norm, in general the product. But, rewriting eq. II.31e:

$$\overset{j}{\underset{k=1}{\Theta}} \ r_k \cdot c_k - \alpha \leq \sigma_c \qquad \text{(II.32d)}$$

it shows to be the same neural encoding function 14a. The meaning of Θ is determined by $\cdot$ and g in II.4 and 14a, respectively. The value of α is dependent of both j and the degree of ANDness (ORness) of the $\top$ -norm $\cdot$ (see II.26e) in II.32d:

$$\alpha = \begin{bmatrix} j & \text{if } \cdot & \text{tends to be an AND operator} \\ \\ 0 & \text{if } \cdot & \text{tends to be an OR operator} \end{bmatrix} \qquad \text{(II.32g)}$$

Now, if neurons of the type II.20 are used to represent the object function, their value can be maximized or minimized according to the output of the constraint neurons defined by II.32d if these neurons are used as the pre-synaptic source of the objetive neuron (Fig. II.9).

II.7 - The formal neuron

Summarizing the previous sections, the Artificial Neuron N may be defined as the following structure

$$N = \{ \{ W_p \}, W_o, T, R, C, \Theta, \{ \alpha, g \} \}$$

$$(II.33a)$$

where:

II.33b) $\{ W_p \}$ is the family of pre-synaptic inputs conveyed over N by all its n pre-synaptic axons;

III.33c) W_o is the output code of N;

II.33d) T is the family of transmitters used by N to exchange messages with other neurons;

II.33e) R is the family of receptors to bind the transmitter released by the pre-synaptic neurons. The strength s_i of the synapsis with the i_{th} pre-synaptic neuron is

$$s_i = M(t) \,\hat{}\, M(r) * \mu(t,r) \,{}^{\shortmid}\, v_o$$

and the post-synaptic activity v_i is evaluated as

$$v_i = w_i \,{}^{\circ}\, s_i$$

where M(t) is the size of the functional pool of the transmitter t at the pre-synaptic cell n_i; M(r) is the amount of r available to bind t; $\mu(t,r)$ is the affinity of the $t\hat{}r$ binding, v_o is the standard EM variation triggered by this binding, and $\hat{}$, *, ${}^{\shortmid}$ and ${}^{\circ}$ are $\top$-norms or $\bot$-conorms.

II.33f) Θ is the function used to aggregate the actual pre-synaptic activity

$$a_n = \overset{n}{\underset{i=1}{\Theta}} (\, w_i \,{}^{\circ}\, s_i)$$

II.33g) $\{ \alpha, g \}$ is a family of thresholds and encoding functions

$$w = \begin{cases} w_i & \text{if } a_n < \alpha_1 \\ w_u & \text{if } a_n \geq \alpha_2 \\ g(a_n) & \text{otherwise} \end{cases}$$

II.33h) C is the set of controllers which can be activated by

$$t_i \,\hat{}\, r_i \,\gg\, c_i, \quad r_i \in R, \; c_i \in C, \; t_i \in T_p$$

where T_p is the set of pre-synaptic transmitters. Each c_i exercises one or more actions over N itself and over other neurons.

Condition II.33h supports a very important chemical processing involving the control of the axonic growth, of the physiology of synapsis, of the DNA reading, etc. This chemical processing is very important in learning and symbolic reasoning, and it is discussed in detail in the next chapter.

The formal neural introduced here exhibits the capabilities of a multi-purpose processing device, since it is able to handle different types of numerical calculations. This is in contrast with the simple processing capability of the classic neuron introduced by McCulloch and Pitts, 1943. To stress this difference, neural nets using the neurons defined in Eqs. II.33 will be called MultiPurpose Neural Nets (MPNN).

II.8 - Fuzzy logic control: An example

Let Fuzzy Logic Control (FLC) paradigm be used here to exemplify what has been discussed so far. The following is based in the ideas of Gomide and Rocha, 1991. A typical scheme of FLC is depicted in Fig. II.10.

The control to be exercised over the process is provided by the reasoning engine according to the fuzzy rules stored in the Rule Base (RB) of the Knowledge Base (KB) associated with the FLC, and the information provided by the Matching Interface about the output variables.

A typical RB in FLC is:

IF X_1 is C_1 and .... and X_m is D_1 .... THEN Z is F_1

.

.   (II.34a)

.

IF X_1 is C_n and .... and X_m is D_n .... THEN Z is F_n

That is, the antecedents and consequents of the different rules are composed by the same variables taking different linguistic values in T(X) or T(Z). For example:

IF Velocity is High and Acceleration is High THEN
Braking is High

.
.
.
(II.34b)

IF Velocity is Low and Acceleration is Low THEN
Braking is Low

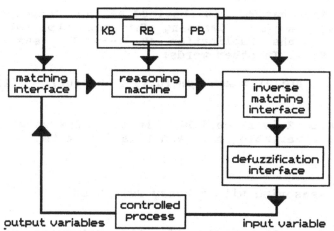

FIG. II.10 - THE STRUCTURE OF FLC
Modified from Gomide and Rocha, 1991

The output process variables are measured by sensory devices and matched to the prototypical patterns (set points) encoded in the corresponding linguistic variables (Matching Interface). This is the role played by the sensory neurons of the MPNN implementing the FLC. The prototypical knowledge encoded in the linguistic variables is part of the information stored in the Knowledge Base (KB) of the FLC. This prototypical knowledge refers to the compatibity function $\mu_{T(X)}$ defining the meaning of the terms of T(X) (see Chapter X, section X.V) The efferent control provided by MPNN over the sensory neurons (see Chapter I, section I.6) adjust the semantic of the sensory neuron according to the knowledge stored in the Pattern Base (PB) of the KB about the linguistic variables. This control is supported by the modulator neurons and is exercised according to eq. II.16b.

The reasoning machine uses the RB and the data provided by the matching interface for decision making about the control to be exercised over the process. The reasoning is supported by the following processing of the generalized modus ponens (GMP):

IF X_i is C_i .... then Z is F_i

X is C_i'...

_____

Z is F_i' (II.34c)

where

$F_i' = C_i' \cdot (C_i \dashrightarrow F_i)$ (II.34d)

which means that F_i' is obtained as the composition of the relations $R_{ci'}(X)$ defining C_i' and the relation R_{Fi} supporting the implication $C_i \dashrightarrow F_i$ (see Chapter X, section X.7). In other words:

$$\mu_{Fi'}(v) = \max_u \; (\mu_{RFi}(u,v) \; \Gamma \; \mu_{Rci'}(u))$$ (II.34e)

given that X is a variable in the universe of discourse U and Z is a variable in V, and Γ is a $\top$ -norm.

II.9a - Reasoning with a Fuzzy Rule Base

Let be considered the following fuzzy rule:

IF X is A then Y is B   (II.35a)

where A and B are fuzzy sets defined in the universes of discourse U and V, respectively. The following compatibility functions are associated with these fuzzy sets:

$\mu_A :$ U $\dashrightarrow$ [0,1]   (II.35b)

$\mu_B :$ V $\dashrightarrow$ [0,1]   (II.35c)

Let μ_A be described by a vector X of size n (Fig. II.11), so that

$x_i = \mu_{Ai}$ if $\alpha_i < u \leq \alpha_{i+1}$ (II.36a)

where i = 1, ..., n-1. Thus, the fuzzy set A is:

$$A = \begin{bmatrix} x_1 \\ \cdot \\ \cdot \\ \cdot \\ x_n \end{bmatrix}$$

(II.36b)

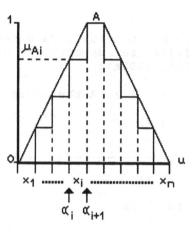

FIG. II.11 - THE MATCHING FUNCTION

Similarly, let μ_B be described by a vector Y of size m, so that

$$y_i = \mu_{Bi} \text{ if } \beta_i < u \leq \beta_{i+1}$$

(II.36c)

where $i = 1, \ldots, m$. Thus, the fuzzy set B is:

$$B = \begin{bmatrix} y_1 \\ \cdot \\ \cdot \\ \cdot \\ y_M \end{bmatrix}$$

(II.36d)

The implication A ---> B supporting the fuzzy proposition IF X is A THEN Y is B is defined by the following fuzzy relation (Zadeh, 1983a,b):

$$R: A \times B \longrightarrow [0,1]$$

(II.37a)

$$\mu_R(x,y) = \mu_A(u) \ \Gamma \ \mu_B(v)$$

(II.37b)

where Γ is a t-norm. From definitions II.36a to 36d, R becomes:

$$
\left[\begin{array}{c} x_1\ \Gamma\ y_1 \\ \bullet \\ \bullet \\ \bullet \\ x_n\ \Gamma\ y_1 \end{array}\right|\begin{array}{c} \cdots \\ \bullet \\ x_i\ \Gamma\ y_j \\ \bullet \\ \cdots \end{array}\left|\begin{array}{c} x_1\ \Gamma\ y_M \\ \bullet \\ \bullet \\ \bullet \\ x_n\ \Gamma\ y_M \end{array}\right] \quad \text{(II.37c)}
$$

The reasoning in a fuzzy data base composed by rules like that in II.35a is supported by the Generalized Modus Ponens (GMP) (Zadeh, 1983a):

IF X is A then Y is B

X is A'

_____

Y is B'   (II.38a)

where A' is defined by $\mu_{A'}$

$$\mu_{A'} : \quad U \longrightarrow [0,1] \quad\quad\quad \text{(II.38b)}$$

Hence, using the same vector notation as above to describe A':

$$x'_i = \mu_{A'i} \text{ if } \alpha_i < u \leq \alpha_{i+1} \quad \text{(II.38c)}$$

where i = 1, ..., n-1, the fuzzy set A' becomes:

$$
A' = \left[\begin{array}{c} x'_1 \\ \bullet \\ \bullet \\ \bullet \\ x'_n \end{array}\right]
$$
                                          (II.38d)

Similarly, B' becomes

$$
B' = \left[\begin{array}{c} y'_1 \\ \bullet \\ \bullet \\ \bullet \\ y'_M \end{array}\right]
$$
                                          (II.38e)

The most popular way to implement the GMP in II.38a uses the compositional rule of inference introduced by Zadeh, 1983a:

$$B' = A' \cdot R \tag{II.39a}$$

with R as in II.37. It follows from II.37c and II.38 that

$$B' = \begin{bmatrix} y'_1 \\ \cdot \\ \cdot \\ \cdot \\ y'_M \end{bmatrix} = [x'_1, \ldots, x'_n] \cdot \begin{bmatrix} x_1 \ \Gamma \ y_1 \\ \cdot \\ \cdot \\ \cdot \\ x_n \ \Gamma \ y_1 \end{bmatrix} \begin{array}{c} \cdots \\ \cdot \\ x_i \ \Gamma \ y_j \\ \cdot \\ \cdots \end{array} \begin{bmatrix} x_1 \ \Gamma \ y_M \\ \cdot \\ \cdot \\ \cdot \\ x_n \ \Gamma \ y_M \end{bmatrix}$$

$$\tag{II.39b}$$

where $\cdot$ is usually interpreted as the max-Γ operator.

III.9b) Mapping fuzzy rules into MPNN

The fuzzy neurons defined in II.14 and II.20 can be used to assemble MPNNs (Fig. II.12) to compute II.39b. Let it initially be discussed the simple case for which the dimensions n and m of the vectors X and Y are equal to 2 (Fig. II.12a). In this condition, let it be given:

$$A = \begin{bmatrix} x_1 \\ x_2 \end{bmatrix} \tag{II.40a}$$

$$B = \begin{bmatrix} y_1 \\ y_2 \end{bmatrix} \tag{II.40b}$$

In this condition:

$$\begin{bmatrix} x_1 \ \Gamma \ y_1 \\ x_2 \ \Gamma \ y_1 \end{bmatrix} \begin{bmatrix} x_1 \ \Gamma \ y_2 \\ x_2 \ \Gamma \ y_2 \end{bmatrix} \tag{II.40c}$$

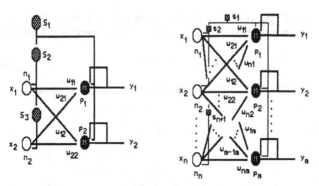

FIG II.12 - THE NEUROFUZZY CONTROLLER

Now let it be considered the MPNN displayed in Fig. II.12a, where:

II.41a) the input neurons n_1, n_2 are in charge of measuring $\mu_{A'}$, so that their activities x'_1 and x'_2 are calculated as:

$$x'_i = \mu_{A'i} \text{ if } \alpha_{i-1} < u \leq \alpha_i$$

In this condition, neurons n_1 and n_2 may be considered as sensory neurons in charge of matching the output variables with the desired set points (Fig. II.10);

II.41b) the output neurons p_1, p_2 are max neurons defined according to II.20, and

II.41c) the weight w_{ij} of the linkages of the input neurons n_i with the output cells p_j is set as:

$$w_{ij} = x_i \; \Gamma \; y_j$$

In this condition,

II.41d) the recoding v_{ij} at the synapsis between n_i and p_j becomes

$$v_{ij} = x'_i \; \Gamma \; w_{ij}$$

Now, if the firings of n_1 and n_2 are synchronized so that n_1 fires always before n_2, the output represented by the activity of the neurons p_1 and p_2 is

$$B' = A' \cdot R \qquad (II.41e)$$

because

$$a_{pj}(2) = v_{1j} \; V \; v_{2j} \qquad (II.41f)$$

$$a_{pj}(2) = (x'_1 \; \Gamma \; (\; x_1 \; \Gamma \; y_j)) \; V \; (x'_2 \; \Gamma \; (\; x_2 \; \Gamma \; y_j)) \quad (II.41g)$$

where V stands for the supremum..

The synchronization required by II.41f can be provided by the circuit of setting neurons s_i in Fig. II.12a which can control the sensory neurons besides the output neurons. The role played by these s_i cells is to control the initial value of the threshold of the post-synaptic max neurons p_i and the actual value of the threshold of the HTN neurons n_i. In this way, the first setting neuron s_1 can set the threshold of the max-neurons p_1 and p_2, besides activating the second setting neuron s_2. This second cell can trigger the reading of the output variable X by n_1, besides activating the third setting neuron s_3, whose role is to trigger the reading of X by n_2.

The same kind of structure can be used to construct the MPNN in Fig. II.12b, which generalizes the above processing when the size of the vectors X and Y are n and m, respectively. The synchronizer system is now composed by n+1 setting neurons disposed in a serial chain.

Now let it be considered the following RB of a FLC:

IF X is A_1 THEN Y is B_1 ELSE
.
.   (II.42a)
.
IF X is A_n THEN Y is B_n

where the connective ELSE is either a conjunction or a disjunction. In this condition, the entire RB can be considered a relation R obtained from the composition of the individual relations R_k associated with each rule k in RB:

$$R = \mathop{\tau}_{k=1}^{n} R_k \qquad \text{(II.42b)}$$

where τ is a $\top$-norm if RB is conjunctive or a $\bot$-conorm if RB is disjunctive. The most popular approach is to assume τ as the max-conorm. In this case:

$$R = \mathop{\tau}_{k=1}^{n} \begin{bmatrix} x_{k1} \; \Gamma \; y_{k1} \\ \cdot \\ \cdot \\ x_{kn} \; \Gamma \; y_{k1} \end{bmatrix} \begin{matrix} \cdots \\ \\ x_{ki} \; \Gamma \; y_{kj} \\ \\ \cdots \end{matrix} \begin{bmatrix} x_{k1} \; \Gamma \; y_{km} \\ \cdot \\ \cdot \\ x_{kn} \; \Gamma \; y_{km} \end{bmatrix} \qquad \text{(II.42c)}$$

This implies that the same MPNN in Fig. II.12b can compute

$$B' = A' \cdot R \qquad (II.42d)$$

if w_{ij} is assumed to be

$$w_{ij} = \mathop{\tau}_{k=1}^{n} (x_{ki} \; \Gamma \; y_{kj}) \qquad (II.42e)$$

The complexity of the MPNN implementing the RB in II.42a is mostly dependent on the degree of precision required by the input discretization. Let g_x be the size of the granule used to discretize the fuzzy set X, X = A, B, is equal to:

$$N(X) = U/g_x \qquad (II.43a)$$

where U is the range of the universe of discourse to be observed. In this way, the number of sensory neurons $n(S)$ in the matching layer is

$$N(S) = \sum_{i=1}^{a} N(X_i) \qquad (II.43b)$$

where a is the number of antecedents in the rules of RB. The number $N(P)$ of max-neurons in the output layer is

$$N(P) = N(X_p) \qquad (II.43c)$$

where X_p is the linguistic variable associated to the consequent of the rules in RB.

II.9c) Generalizing the Fuzzy-neuro controller

The most general rule structure in the RB of a FLC is of the type:

IF X is A and Y is B .... THEN Z is C   (II.44a)

where A, B, ... and C are fuzzy sets in the universes of discourse U, V, ... and W, so that given A' and B':

$$C' = (A' \; \hat{} \; B' \; \hat{} \; ... \;) \cdot R \qquad (II.44b)$$

$$R: A \times B \times ... \times C \longrightarrow [0,1] \qquad (II.44c)$$

$$\mu_R(x,y, ...,z) = (\mu_A(u) \; \hat{} \; \mu_B(v) \; \hat{} \; ... \;) \; \Gamma \; \mu_C(w) \quad (II.44d)$$

where $\hat{}$ and Γ are $\overline{\tau}$-norms.

$$R = \langle A' \wedge B' \rangle \qquad \circ \qquad \langle A \wedge B \longrightarrow C \rangle$$

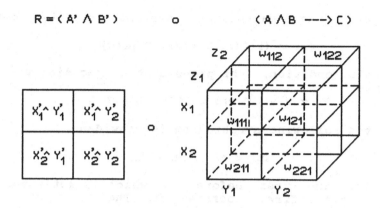

FIG II.13 - SOLVING THE FUZZY RULE

Fig. II.13 shows the processing of II.44a in the special case where the antecedents are restricted to X is A and Y is B:

$$\mu_R(x,y,z) = (\mu_A(u) \ \hat{} \ \mu_B(v)) \ \Gamma \ \mu_C(w) \quad (II.44e)$$

and

$$C' = (A' \ \hat{} \ B') \ \bullet \ R \qquad (II.44f)$$

The MPNN in Fig. II.14 implements the processing of II.44f if:

45a) the neurons n_i are used to convert the actual values of X and Y to the correspondent values of $\mu_{A'}(u)$ and $\mu_{B'}(v)$;

45b) the weight w_i of the synapsis between the sensory neuron n_i and the associative neuron h_{ij} is set as:

$$w_i = 1$$

45c) the neurons h_{ij} are used to compute:

$$a_{h_{ij}} = (\ x'_i \ \hat{} \ y'_j \ )$$

This is obtained by defining the ˆ as the encoding function of the neurons h_{ij}. If ˆ is defined as the min operator, then h_{ij} must be a min-neuron, and the sensory neurons must be defined as HTN neurons (Chapter II, section II.6) otherwise the h_{ij} will be declared HTN neurons;

45d) the weight w_{ijk} of the synapsis between the neurons

h_{ij} and the max-output neurons p_k is set as:

$$w_{ijk} = (\mu_A(u) \wedge \mu_B(v)) \; \Gamma \; \mu_C(w)$$

In this condition, the pre-synaptic recoding v_{ijk} of a_{ij} becomes

$$v = a_{ij} \; \Gamma \; w_{ijk} = (A' \wedge B') \; \Gamma \; R$$

45e) the following synchronism is provided:

$$h_{11} < h_{21} < h_{12} < h_{22}$$

that is, h_{11} fires before h_{21}, which is activated before h_{12}, which fires before h_{22}. The neurons n_i are supposed to read the input variables at the same time t, and to hold this information to the neurons h_{ij} for a period of time sufficient to assure the above synchronization. In this condition, the output of the max p_i neurons encodes

$$C' = (A' \wedge B') \cdot R$$

where $\cdot$ is the operator max-Γ. In this context, the axonic encoding at p_i plays the role of the inverse matching interface in fig. II.10.

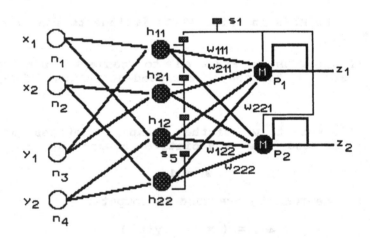

FIG. II.14 - COMPLEX NEUROFUZZY CONTROLLER

The MPNN in fig. II.14 can be generalized to support a RB composed by rules with any number of antecedents, since this requires only to adjust the number of neurons in the sensory and in the associative layers, and to define the adequate wiring between the neurons in these two layers, and between the h_{ij} and the max-neurons p_k. The number of neurons N(H) in the associative layer is:

$$N(H) = \prod_{i=1}^{a} N(X_i) \qquad (46f)$$

where $N(X_i)$ is the number of sensory neurons measuring the variable X_i, π is the product, and a is the number of antecedents in the rules on RB.

II.9d) Defuzzifying the FLC output

The activity of the output neurons p_i (Figs. II.12 and II.14) can now be defuzzified by a final neuron, called here controlling neuron (c in Fig. II.15).

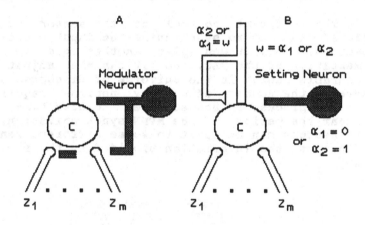

FIG. II.15 - DEFUZZIFYING THE CONTROL

The role played by the controlling neuron is to calculate the value of the input variable to control the process. This value depends on the actual values of the output vector Y and the chosen defuzzification method. These methods are part of the knowledge stored in PB (Fig. II.10). These methods are implemented in MPNN by means of the control of the recoding function (eq. II.4) of the p neurons and encoding mapping (eq. II.14) of the decoding neuron c. This control can be exercised by the modulatory neurons according to eq. II.16. The correspondent defuzzification function is obtained by adequately combining the recoding functions of the neurons p_i and the encoding function of the control neuron c. This is obtained by defining the neuron c as a numeric processor and programming the defuzzification function with procedures similar to those discussed in II.19 or II.20. Powered averaging functions (e.g. center of gravidity) are easily implemented since the neuron is basically an averaging device. Procedures based on max and min values can be implemented with neurons of type II.20a or II.20b, respectively.

II.9e) The complexity of the MPNN in FLC

The complexity of the MPNN used to implement the FLC is not dependent of the size of RB, but it is a function of the precision encoded in the vectors X, Y,... (eq. 43) and of the number of antecedents in the rules (eqs. 46f). The number of antecedents in the rules is dependent of the complexity of the process to be modeled by the FLC. The number of sensory neurons can be kept small by the efferent control of the sensory systems discussed in Chapter I.

The efferent control of the sensory neurons was introduced in FLC to synchronize the input required by the max-neurons. But, it can play another role in the MPNN implementation of the FLC, too. It can also adjust the range of measurable energy in the universe of discourse U in order to decrease the number of sensory neurons required by the FLC. If the process is not supposed to have time constants very near the period of the MPNN synchronization, then the sensory neurons can be reset to sense different ranges of U depending on the information provided by the same sensory neurons.

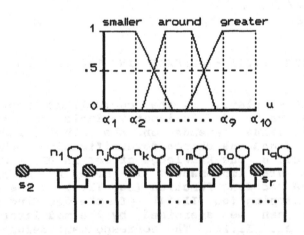

FIG. II.16 - CONTROLLING THE MATCHING INTERFACE

Let the matching functions of the sensory neurons be organized to provide the information depicted in Fig. II.16. In other words, the information provided by the neurons n_i can be combined to speak about the output variables being around, smaller or greater than the setting point required by the FLC. If the actual value of these variables is sensed as being around the set point, it can be supposed that the sensory neurons are covering the adequate range of U. On the contrary, if these values are matched as smaller or greater, the corresponding adjustment of the measurable range of U is obtained by a proportional modification of the sensory thresholds of the neurons n_1,

II.9f) Using the Neurofuzzy Controller as an exercize

The MPNN Fuzzy Controller described in sections II.9c and d was proposed as an implementation exercize the for the post-graduate students in the course of Neural Nets and Artificial Intelligence of the Faculty of Electrical Engineering, UNICAMP. Different groups of students use it implemented different types (PI, PID, etc) of control and obtained very good results in simulating the control of different standard problems (Figueiredo, Mazzeta, Rocha and Gomide in preparation).

ACKNOWLEDGEMENT

Most of the ideas discussed in this chapter concerning Fuzzy Control I have learned from F. Gomide. Some discussions I had with W. Pedrycz were important in developing the formal neuron presented here.

I am in debt with the students of the course of Neural Nets and Artificial Intelligence, Faculty of Electrical Engineering, UNICAMP, who worked many of the MPNNs proposed in this book as implementation exercizes.

CHAPTER III

THE SYNAPSIS
The chemical processing

III.1 - The production of proteins

The activation of the DNA reading at the cellular
nucleus produces molecules of RNA messengers, whose function
is to control the proteic synthesis in the cytoplasm
according to the specifications encoded in the gene (Fig.
III.1)

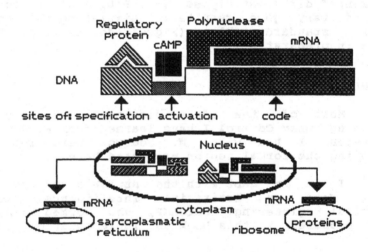

FIG. III.1 - THE GENETIC ENCODING OF PROTEINS

The DNA and RNA are strings of 4 different
nucleotides: Adenine, Thymine, Cytosine and Guanine or
Uracil. These nucleotides are the symbols of the genetic
alphabet. Two major properties of the genetic code are
explained by the chemical afinity between Adenine and Timine
and between Citosine and Guanine or Uracil:

a) the DNA double helix: two complementary strings are
combined into a single DNA structure, and

b) the DNA reading: the segments of DNA defining the genes
serve as template for the RNA synthesis (Fig. III.1). This
synthetized mRNA acts as a messenger of the DNA encoding to
the ribosomes and sarcoplasmic reticulum, the cytoplasmic
structures in charge of the production of the proteins.

Because of this, the synthesis of the mRNA is called here the DNA reading. The DNA reading starts with the opening of the double helix to expose the nucleotides to be copied into another sequence of mRNA nucleotides.

The proteins are molecules composed of amino acids. There are 20 amino acids. The proteins may be considered strings of these aminoacids, which in turn may be considered as symbols of a chemical alphabet. The sequence of the amino acids in the protein is encoded in the sequence of nucleotides in the DNA. Each 3 nucleotides specifies one amino acid. Thus, each codeword in the genetic code is composed by 3 symbols chosen among the 4 possible nucleotides. The genetic code has the capacity of 6 bits to encode 20 amino acids or around 4 bits. This redundancy is one of the biological mecanisms to guarantee the DNA reading as error-free as possible.

The mRNA synthesis is a controlled process. A very simple model of this control is considered in Fig. III.1 for the purpose of the present book. It is not intended as a complete description of the process.

The DNA reading depends on:

III.1a) the specification of the gene to be read: defined sites of the DNA specify or name the genes. These sites are activated by regulatory proteins (Fig. III.1) whose function is to specify which are the readable genes in the genetic code of a defined cell. This regulatory process is very important in specializing the cells to different activities in the multicellular animal. Only those genes coupled to a regulatory protein may be read in the DNA of a defined cell. This process is very dependent on the proteins produced by the same cell as well as by its neighboring cells. This causes the actual set of readable genes of each cell to be specified according to the necessities of the entire local population of cells.

III.1b) the activation of a specified gene: the activation of the reading of those readable genes is dependent on the coupling of the cAMP to some specific portions of the DNA, called here activation sites. This coupling activates the binding of the polynuclease to the inductive site of the gene and triggers the beginning of the RNA synthesis, which is guided by this enzyme (Fig. III.1). This synthesis copy the information stored in the DNA into the mRNA.

This synthetized mRNA migrates out of the cellular nucleus to control the proteic synthesis in the ribosomes and the sarcoplasmic reticulum (Fig. III.1). The proteins are produced as strings of amino acids according to their genetic specifications. These proteins will act either as:

III.2a) enzymes to control the production of other cellular products;

III.2b) structural elements of the cell; or

III.2c) messages to be exchanged among cells. As messengers these molecules may act as gene regulatory proteins to other cells. In this way, they define local environments for cellular specialization.

The proteins composing the ionic channels and post-synaptic receptors are examples of structural proteins in the nerve cell. The transmitters and controllers are examples of chemical messages and enzimes.

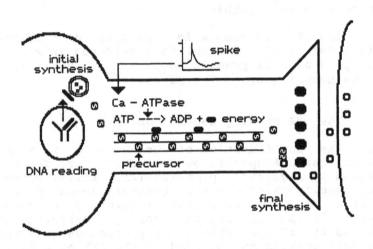

FIG. III.2 - THE ANTEROGRADE TRANSPORT

III.2 - Specifying transmitters and post-synaptic receptors

The production of the transmitter at the axonic terminals is dependent of both the pre- (Fig. III.2) and post-synaptic (Fig. III.3) cells (Black et al, 1974). The proteins produced by the pre-synaptic cell are called precursors (Fig. III.6), and those released by the post-synaptic neuron are called controllers. The genetic specification of the transmitter to be produced is dependent on signals provided by both the pre- and post-synaptic cells, and the activation of the reading of the specified genes is governed by the cAMP produced by the activation of the post-synaptic receptors (Fig. III.8).

The precursor produced at the pre-synaptic cell body must be transported (Fig. III.2) to the axonic terminals (Arch and Berry, 1989; Goodman et al, 1984; Grafstein and Forman, 1980; Holtzman, 1977; Landuron, 1987; Zimmerman, 1979). The same precursor may be used to produce different types of transmitters. This is because the final specification of the transmitter (Fig. III.3) is dependent on the controllers produced by the post-synaptic cell (Black et al, 1974).

The axoplasmic transport can be divided (Grafstein and Forman, 1980) into anterograde if it moves from the cell body to the terminals (Fig. III.2), or retrograde (Fig. III.5) if it moves in the opposite direction, and into fast and slow transport depending on its velocity.

The fast transportation is dependent of microtubular structures and microcontractile proteins to move particles in the cytoplasm (Grafstein and Forman, 1980). These particules are assumed to be vesicules of the sarcoplasmic reticulum (Alberts et al, 1983; DeDuve, 1984). The contractile proteins use metabolic energy derived from the ATP to move these vesicules along the axon (Fig. III.2). The conversion of ATP into ADP to produce a highly energized P (phosphor) is controlled by the Ca-ATPase, which is activated by the ion calcium (Grafstein and Forman, 1980).

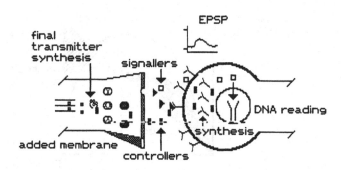

FIG. III-3 - THE POST-SYNAPTIC ACTIVATION

This is a quite interesting arrangement, because at the same time the spike runs toward the pre-synaptic sites where it will consume transmitter and vesicles, it promotes the entrance of Ca in the axoplasm in order to activate the transportation of the material necessary to replenish the transmitter and vesicules to be used.

The controllers required for the final synthesis of the transmitter are produced by post-synaptic cells and released into the synaptic cleft by the tˆr coupling (Fig. III.3). These substances are uptaken by the pre-synaptic terminal and used to guide the final steps of the synthesis of the transmitter (Black et al, 1974). Again, the use of the synapsis activates the mechanisms necessary to produce the required transmitter for the synaptic transmission.

The total contents $M(t)$ of the transmitter t in the post- synaptic terminals depend on the balance between its consumption and production. Because the synthesis of the transmitter depends on the axonic transport of the precursor, it is limited by the capacity of this transportation. The consequence is the existence of a limiting frequency of spiking at the pre-synaptic cell, below which the amount of stored transmitters may increase and above which the pool of this molecule will decline (Fig. III.4). In the first case, the synaptic strength may increase since it depends on the amount of stored transmitters (see eq. II.7 in Chapter II, section II.3). In the second case, the efficiency of the synaptic transmission will be reduced, and the phenomenon is called synaptic fatigue.

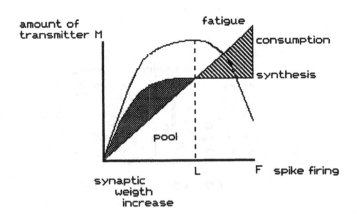

FIG. III.4 - THE DYNAMICS OF THE POOL OF TRANSMITTERS

The aggregation of the vesicules to the cellular membrane to release the transmitter (or exocytosis) is one of the mechanisms related to the growth of the pre-synaptic terminal associated with the increase of the sinaptic strength (Fig. III.3). This process is balanced by the removal of pieces of the cellular membrane in the process of endocytosis (Fig. III.5a), by means of which, pieces of the pre-synaptic membrane are invaginated and cut to form other vesicules (Arch and Berry, 1989; Zimmerman, 1979). The growth of the axonic membrane is dependent of the net balance between the exo and endocytosis. If the first predominates the membrane growths, otherwise it is reduced.

The endocytosis plays another role in the long term modifications of the synaptic strength. The endocytotic vesicles incorporate the post-synaptic controller bound at the pre-synaptic membrane and they are retrograde transported to (Fig. III.5b) the cell body (Arch and Berry, 1989) where the controller may be used to modify the DNA reading. By this mechanism, chemical information exchanged at the pre-synaptic terminals may be transmitted to the nucleus to regulate the DNA reading (Arch and Berry, 1989). These proteins specify the readable genes at the pre-synaptic cell. This control of the pre-synaptic genetic information will result in the production of new proteic material used for the synthesis of transmitter as well as of vesicules and channels. Thus, the pre-synaptic synthesis is under control of the post-synaptic activity.

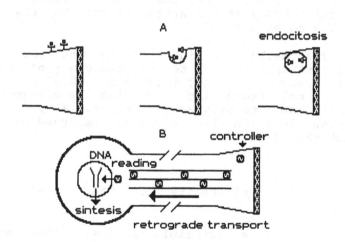

FIG. III.5 - THE RETROGRADE TRANSPORT

The pre-synaptic vesicule stores molecules other than the transmitter. These other molecules are released into the synaptic cleft together with the transmitter and they are called here signallers (Fig. III.3). These signallers are uptaken by the post-synaptic cell and can be used to control its DNA reading. In this way, the proteic synthesis at the post-synaptic neuron is influenced by the pre-synaptic activity. Thus, the production of post-synaptic receptors, controllers and channels may be put under control of the pre-synaptic neurons, too.

Chemical information can also be exchanged between different synapsis. The activity in one of them may release controllers to bind to the receptors of many neighboring cells. This mechanism accounts for both the heterosynaptic cooperation and competition (Cowan et al, 1984; Goodman et al, 1984; Hyvarinen, 1982; McConell, 1988; Sidman and Ravick, 1973).

In the first case, the development of one synapsis facilitates the growth of the contact with another source of information, providing the basis for associative learning or the guided embryogenic wiring of the neural circuits (e.g. Byrne, 1987).

In the second case, the development of one synapsis inhibits the growth of another pathway. This establishes a mechanism of competition between different paths for the control of the post-synaptic cell (Cowan et al, 1984), providing the basis for reducing the connectivity inside the brain. Regressive processes are also involved in learning (Cowan et al, 1984; Rocha, 1982 a,b), in order to adjust the structural entropy of the neural circuit to the entropy of the system to be modeled.

The mechanisms involved in the control of the proteic synthesis related to the physiology of the synapsis support a very complex and integrated chemical processing. This chemical processing specifies the characteristics of the pre- and post-synaptic cells, according to the environment in which these cells are immersed and the manner in which they are used.

III.3 - The plasticity of the chemical encoding

The Dale's principle proposed that each neuron synthesizes only one type of transmitter. However, it was discovered that many neurons are able to produce and to use more than one type of transmitter (Bursntock, 1976). The precursor produced at the cell body may specify different transmitters at different axonic terminals (Fig. III.6) depending on the enzymatic action of the controllers released by the post-synaptic cells (Black et al, 1974). It is now well established that the post-synaptic environment is very important for the final differentiation of the neurons during the embryogenic growth of the brain (McConnell, 1988).

Another important finding about the physiology of the synapsis was the discovery that the use of the different transmitters produced by the same neuron is dependent of the level of activity at the axon (LaGamma et al, 1984; Lam et al, 1986; Starke et al, 1989). In this condition, low spiking activity may be associated with the use of one of the transmitters, whereas high spike firing may release the other one (Fig. III.6). It is interesting to remember that different pre-synaptic terminals may have different filtering properties. This implies distinctive spreading of the spike train to different terminals, a fact to explain the differential effect of the axonic activity on the transmitter release (Fig. III.6). In this way, different post-synaptic cells are activated if the pre-synaptic spiking changes. This control of the spreading of the information in the real brain is quite different from the assumed homogenous spread of the pre-synaptic activation in the case of the artificial neural nets.

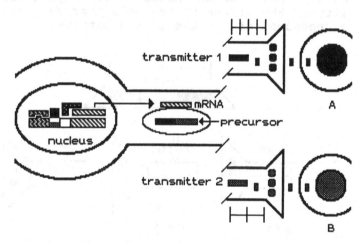

FIG. III.6 - PLASTICITY OF THE TRANSMITTER

This mechanism can be used as the simplest way to implement control knowledge in neural nets, because distinct activations of the pre-synaptic cell may travel different axonic branches and may activate distinctive pathways involved in controlled inferences (see Chapter V and VI).

It is now accepted that hormones like ACTH, vasopressin, etc., can play two different roles in the central nervous system, besides their traditional actions upon the somatic cells (Wied and Jolles, 1982). Many hormones in the blood could reach the nervous system and exert a broad action on those cells having receptors for them. Many of these hormones are also produced at some sites in the brain, and transported to specific areas by means of modulator neurons, to act upon a group of other neurons. This action is called modulatory control of the synapsis. As a matter of fact, not only hormones but many other neuropeptides (Fig. III.7) are known to exert modulatory effects upon the synaptic activity (Arch and Berry, 1989; Kandel and Schwartz, 1982; Thoenen, 1980, Starke et al, 1989).

Modulators are very plastic messengers, because their actions are exercised by part of their molecules called active sites, and each molecule possesses different active sites (Arch and Berry, 1989). More interesting is the fact that different sites of the same neuropeptide may exert antagonic effects upon the synapsis (Arch and Berry, 1989). What site of the neuropeptide will influence the synapsis (Fig. III.7) depends either on the receptors available for binding, or on the chemical environment which could select the different active portions of the peptide. Since this environment depends on the axonic activity, there is the possibility of the effect of the modulator upon the synapsis becoming dependent of the use of the pathway.

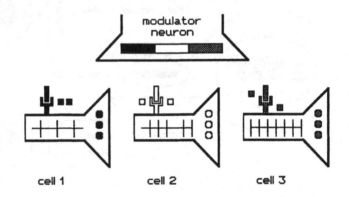

FIG. III.7 - THE PLASTICITY OF THE MODULATORS

This mechanism opens interesting possibilities to the control of learning by means of motivation, besides punishment and reward. Brain structures (e.g. the Limbic System) involved in processing motivation, punishment, reward, etc., are also producers of neuropeptides. Distinct activities in these centers can release different modulators upon the circuits being learned. Also, the action exercised by the neuropeptide could be dependent of the activity in these circuits because its action is also dependent of the

local environment. In this way, motivation, arousal, etc.,
could influence long term memory and learning.

        The major differences between transmitters and
modulators are:

III.3a) transmitters act locally upon the post-synaptic
cell, promoting modifications in the ionic conductances as
the main post-synaptic event, and

III.3b) modulators act upon a neighborhood of cells,
promoting modifications of the available energy of the DNA
reading in the post-synaptic cell.

III.4 - Modulator learning control

        The DNA reading is activated when the cAMP bindings
to the genes activating site (Fig. III.1). The contents of
cAMP in the neuron is under control of some special
transmitters called modulators. The modulator binds to the
post-synaptic receptors and activates the adenylate cyclase
to convert ATP into cAMP (Fig. III.8). The cAMP in turn is
used to activate the coupling of the polynuclease to the
inductor sites of the readable genes. This binding of the
polynuclease begins the synthesis of the corresponding mRNA.
These mRNAs guide the proteic synthesis in the cytoplasm
which results into the production of channels, transmitters,
controllers, etc. involved with the maintenance of the
physiology of the neuron and of the synapsis.

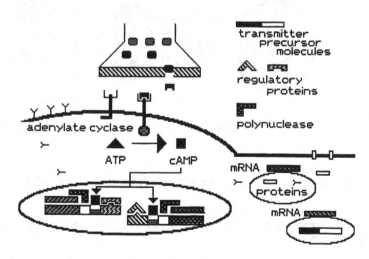

FIG. III-8 - CONTROLLING THE DNA READING

Modulators are produced by many neurons in the brain. Attention will be focused here upon those neural centers involved in the control of motivation, arousal, punishment and reward, etc. These neurons produce neuropeptides or modulators to be released upon neurons in other cerebral areas where learning may be in course. In this way, learning may be controlled according to the necessities of the animal. This basic process of learning control MLC is called here Modulator Paradigm (Fig. III.9).

The basic steps in MLC are:

III.4a) the synaptic activation results in chemical signals which serve as markers of the neural pathways activated during the information processing;

III.4b) the results of these processing steps are evaluated according to some goals at some specified areas of the brain, called here learning control areas. Depending on this evaluation:

III.4c) the activated synapses may be rewarded or punished by means of the release of the adequate modulators from the control learning areas. These areas are those related to arousal, motivation, punishment and reward, etc., in the brain of animals.

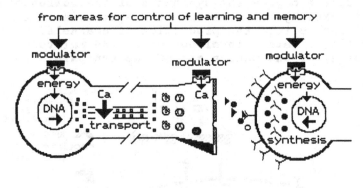

PRE SYNAPTIC CELL   POST SYNAPTIC CELL

FIG. III-9 - THE CONTROL OF LEARNING BY MODULATORS

Learning in the real brain is a complex process depending on:

III.4d) the specification of the pools of transmitters and signallers at the pre-synaptic sites;

III.4e) the specification of the post-synaptic receptors and controllers;

III.4f) the regulation of the amount of the specified transmitters, signallers, receptors and controllers;

III.4g) the adjustment of the axonic threshold and encoding functions, and

III.4h) the adjustment of the pre-synaptic decoding function relating the axonic spiking and the transmitter release.

The amount of transmitters and receptors determines the strength of the synapsis. Condition III.4g is acomplished either by changing the amount of energy available to the membrane, or by altering the channel composition and distribution in this membrane. Condition III.4h is the result of the regulation of the concentration of ion Calcium inside the pre-synaptic terminal.

All these activities are supported by a complex form of chemical processing specifying and activating the DNA reading, and/or controlling the amount of available energy. This chemical processing, in turn, is under the control of specified neural circuits (learning control areas) producing modulators according to the attained goal satisfaction. The importance of MLC concerning the MPNN learning strategies is discussed in chapter IV, section IV.10.

III.5 - A formal genetic code

Let the alphabet D be a set of symbols

$$D = \{ d_1, \ldots , d_k \} \qquad (III.5a)$$

An alphabet is said to be simple if none of its symbols d_i is allowed to be repeated in D, otherwise D is redundant. The number of instances the same symbol d_i is repeated in D (or the frequency of d_i in D) is said to be the amount $a(d_i)$ of $d_i \in D$. In the case of a simple alphabet D:

$$a(d_i) = 1 \text{ to all } d_i \in D \qquad (III.5b)$$

Let the fuzzy grammar Φ (Negoita and Ralescu, 1975) be given:
$$\Phi = \{ D, G, \phi, D_0 \} \qquad (III.6a)$$
where (Fig. III.10)

III.6b) D is a simple input alphabet

$$D = \{d_1, \ldots, d_j\};$$

III.6c) G is the output vocabulary;

III.6d) ϕ is the syntax of $\maltese$:

$$\phi : D^n \times G \longrightarrow [0,1]$$

generating strings g_k of the type

$$g_k(e_k) = d_0, \ldots, d_j$$

of length n or less, as the productions of G. Also, e_k measures the possibility of g_k to be read as a production of $\maltese$ or the membership of g_k to G;

III.6e) $D_0 \subset D^n$ is the set of start symbols for generating g_k:

$$\phi : (D^s)^n \times G \longrightarrow [0,1]$$

This means that all productions of G must be composed by n s-types of $d_i \in D$.

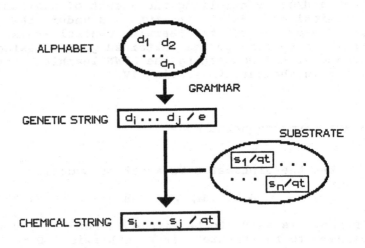

ALPHABET

$d_1 \quad d_2$
$\ldots$
d_n

GRAMMAR

GENETIC STRING $\quad d_i \ldots d_j \,/\, e$

SUBSTRATE

$s_1/qt \ldots$
$\ldots \quad s_n/qt$

CHEMICAL STRING $\quad s_i \ldots s_j \,/\, qt$

FIG. III-10 - THE GENETIC ENCODING

In the case of the genetic code:

III.7a) $\qquad D= \{d_1; d_2; d_3, d_4\}$

where d_1 = adenine; d_2 = thymine; d_3 = cytosine, and

d_4 = guanine or uracil;

III.7b) $D_0 = D^3$

since 3 bases are needed to encode the 20 amino acids used to synthesize the proteins;

III.7c) G is the set of genes generated by the genetic code ⚡;

III.7d) e_k is the degree of expression of the gene g_k, whose actual value is dependent both of the available regulatory proteins to specify g_k as a readable gene, and of the available amount of cAMP to activate this reading, and

III.7e) n is related to the size and complexity of the proteins encoded by g_k.

Let P be another set of strings (Fig. III.10)

$$p_i = (s_j, \ldots, s_k) \qquad \text{(III.8a)}$$

over a redundant alphabet

$$S = \{s_1 \ldots s_1, \ldots, s_M \ldots s_M\} \qquad \text{(III.8b)}$$

The amount $a(p_i)$ of the string $p_i \in P$ is the number of instances of p_i in P, and it is dependent of the minimum (Ω) available amount of its symbols or substrates in S:

$$a(p_i) = \overset{k}{\underset{i=j}{\Omega}} s_i \quad , \; s_i \in p_i \qquad \text{(III.8c)}$$

P is a redundant set of strings. In the case of the cell, P is the set of proteins used as enzymes, transmitters, controllers, channels, etc., and S is the set of amino acids used in the synthesis of these proteins.

G is said to encode P (Fig. III.10) if there is a grammar ⚡(G)

$$⚡(G) = \{ G, S, P, \delta, \beta \} \qquad \text{(III.9a)}$$

where G is as in III.6 and S and P are as in III.8, and

$$\delta : G \times S \times P \longrightarrow [0,1] \qquad \text{(III.9b)}$$

converts strings $g_i \in G$ into strings $p_k \in P$

$$\delta : g_k \longrightarrow p_k \qquad \text{(III.9c)}$$

or

$$\delta : (d_i, \ldots, d_k / e_k) \longrightarrow (s_i, \ldots, s_k / a(p_k))$$

because there is a map β from D_0 to S

$$\beta : D_0 \times S \text{---> } [0,1] \qquad (III.9d)$$

recoding $(d_i) \in D_0$ in g_k into s_i in p_k.   P is said to be the set of products of S under the description of G. Equally, G is said to be the genetics of P.

The amount $a(p_k)$ of p_k in P being dependent of both the amount $a(s_i)$ of its substrates s_i in S and of the degree e_k of expression of $g_k \in G$:

$$a(p_k) = e_k \cdot \overset{k}{\underset{i=j}{\Omega}} s_i \qquad (III.9e)$$

where $\cdot$ is a $\top$ -norm, in general the product, and Ω stands for the minimum.

The synthesis of p_k from g_k and S

$$\delta : (g_k,..., g_k / e_k) \text{---> } (s_i,..., s_k / a(p_k)) \quad (III.9f)$$

updates the amount of substrate $a_t(s_i)$ in S at the instant t in regard to that $s_{t-1}(s_i)$ at the instant $t-1$ by

$$a_t(s_i) = a_{t-1}(s_i) - a(p_k) \qquad (III.9g)$$

for all $s_i \in p_k$

The process described by III.9 corresponds to the genetic control of the proteic synthesis. The sequence of amino acids (substrate S) composing the proteins $p \in P$ is encoded in the sequence of nucleotides (D) in the genes $g \in G$. The map β corresponds to the genetic encoding of each of the amino acids in S, and the map δ is the encoding of the proteins $p \in P$ produced by the cell.

III.6 - An example of formal genetic encoding

Let the following be an example of a formal genetic encoding (FGC) to be used in a formal modeling of the neuron. The substrate dictionary S will be enhanced in order to take advantage of some facilities provided, in general, by the actual computer programming languages, e.g. the matching of strings. Let S be composed of all letters, capital letters, numbers and the characters # and ?:

$$S = \{ a, b, ..., z, A, B, ..., Z, 0, 1, ..., 9, \#, ? \}$$
$$(III.10a)$$

The number of elements of S is 64. The dictionary S may also be considered the union

$$S = L_1 \cup L_2 \cup L_3 \cup L_4 \qquad \text{(III.10b)}$$

of the following subsets

$$L_1 = \{ a, b, \ldots, l, \#, ? \} \qquad \text{(III.10c)}$$

$$L_2 = \{ m, n, \ldots, z, \#, ? \} \qquad \text{(III.10d)}$$

$$L_3 = \{ A, B, \ldots, Z, \#, ? \} \qquad \text{(III.10e)}$$

$$L_4 = \{ 0, 1, \ldots, 9, \#, ? \} \qquad \text{(III.10f)}$$

Let the genetic alphabet D

$$D = \{ d_1, d_2, d_3, d_4 \}$$

used to encode S be composed of 4 basic digits as in the case of the natural genetic code (NGC). In this condition, the dimension of the set of initial symbols D_S must be 4, in order to preserve the same redundancy of NGC. The cardinality of D_4 is 128, which provides 1 bit of redundancy if D_4 is used to encode S. In this condition, the 4th digit of the encoding substring may be used as a control digit. All elements of S may be specified by the 3 first digits of each encoding substring. In this way, a structure similar to that of the NGC may be used to define the FGC.

Sequences of tuples of D_0 is the hardware used by nature to encode and duplicate the genetic syntax ϕ. A similar implementation of ϕ may be computationally expensive in the actual computers. Tools like graphs and formal grammars were developed with the same purpose of representing a syntax in the actual computational languages. The use of these tools to describe the genetic syntax ϕ in the simulation of FGC in the computer is more efficient than trying to mimic nature in every detail.

In this context, ϕ will be described by means of fuzzy graphs or fuzzy grammar. This syntax will be used to specify the set of transmitters, receptors and controllers of a given MPNN, according to the problem to be solved by this net. For example, the syntax of natural languages is used in Chapter VII as the genetic syntax of the FGC supporting MPNNs devoted to natural language processing.

TRANSMITTERS

$$\phi : T \times L_1^n \longrightarrow [0,1] \qquad L_1 = \{a, b, ..., l\}$$

$$t_k = \boxed{ABC ....}$$

RECEPTORS

$$\phi : R \times L_3^o \longrightarrow [0,1] \qquad L_3 = \{A, B, ..., Z\}$$

$$r_k = \boxed{ABC .... MOS}$$

CONTROLLERS

$$\phi : C \times L_2^p \times L_5^s \longrightarrow [0,1] \qquad L_2 = \{m, n, ..., z\}$$

$$L_5 \text{ is } L_1 \text{ and } L_3 \text{ and/or } L_4 = \{0, 1, ..., 9\}$$

$$c_i = \boxed{m\ o\ s\ ....\ 1abc\ /\ qt} \qquad c_j = \boxed{ptw\ ....\ AJL\ /\ qt}$$

FIG. III.11 - A FORMAL GENETIC CODE

The set P of strings encoded by FGC is (Fig. III.11)

$$P = T \cup R \cup C \qquad \text{(III.11a)}$$

where:

III.11b) the subset T of transmitters is composed of those strings t formed only with digits provided by L_1:

$$\phi : T \times L_1^n \longrightarrow [0,1]$$

III.11c) the subset R of receptors is composed of those strings r formed only with digits provided by L_3:

$$\phi : R \times L_3^o \longrightarrow [0,1]$$

III.11d) the subset C of controllers is composed of those strings c formed by two substrings c_l and c_r. The first one is composed only with digits provided by L_2, and the second substring is formed only with digits provided by $L_5 = L_1 \cup L_3 \cup L_4$ or $L_5 = L_1 \cup L_3$:

$$\phi : C \times L_2^p \times L_5^s \longrightarrow [0,1]$$

Thus, $c = c_l \cup c_r$, where $c_l \in L_{p2}$ and $c_r \in L_{s5}$. Here, c_l is called the matching substring and c_r the target substring. The target substring specifies the address (letter substring) of the target of the controller c_i and the type of action (numeric string) it must exercise over the target (Fig. III.11);

III.11e) the length of the strings t, r and c obey the following order

$$n < o < p + s$$

which means that the transmitter strings are smaller than the receptor strings which are smaller than the controller strings, and

III.11f) the maximal length p+s of the strings of P encoded by FGC is defined by the complexity of the problem to be solved.

This FGC will be used as the genetic G of a chemical processing language in this book. The meaning of the wild characters # and ? will be discussed below.

III.7 - A formal chemical language

Let the elements s_i of some subsets L_i of S have the property of binding to elements s_j of the other subsets L_j of S. Let this property be called binding affinity and be measured by $\mu(s_i, s_j)$:

$$\mu: L_i \times L_j \longrightarrow [0,1] \qquad \text{(III.12a)}$$

Two symbols $s_i \in L_i$ and $s_j \in L_j$ are said complementary symbols if

$$\mu(s_i, s_j) > .5 \qquad \text{(III.12b)}$$

Complementary symbols are used to concatenate strings of P. Condition III.12b is guaranteed in the case of the present FGC if $s_j = l_j \in L_3$ is the corresponding capital letter of $s_i = l_i \in L_1 \cup L_2$. As a matter of fact, in this condition:

$$\mu_s(l_i, l_j) = 1 \qquad \text{(III.12c)}$$

Also

$$\mu_s(\text{\# or ?}, l_j) = 1 \qquad \text{(III.12d)}$$

$$\mu_s(l_i, \text{\# or ?}) = 1 \qquad \text{(III.12e)}$$

for any l_i or l_j.

A concatenation substring b is a set of complementary symbols

$$b = s_n, \ldots, s_r \qquad \text{(III.13a)}$$

used to bind a given string p_i to other strings of P. In other words, $b \in S_k$ is a concatenation substring if for each $s_i \in b$ there exists a complementary symbol $s_j \in S$. The length of b is k. In the case of the present FGC:

III.13b) $k \leq n$ in the case of the transmitters, and

III.13c) $k \leq p$ in the case of the controllers.

The binding between two concatenation substrings $b_j \in R$ and $b_i \in T \cup C$ is determined by the matching between their symbols (Fig. III. 12). If b_i and b_j share no complementary symbols they do not bind. If all symbols of b_i and b_j are complementary then the matching is said to be complete, otherwise the matching is said to be a partial matching. The strength $\mu(b_i,b_j)$ of the binding between b_i and b_j is defined by the degree of matching of their symbols s_i, s_j:

$$\mu(b_i,b_j) = \sum_{t=1}^{k} \mu_s(s_{i+t}, s_{j+t}) \ /n \quad (III.13d)$$

If $s_i = l_i \in L_1 \cup L_2$ and $s_j = l_j \in L_3$ then

$$\mu(b_i,b_j) = m/n \quad (III.13e)$$

where m is the number of symbols $l_i \in b_i$ matching the symbols $l_j \in b_j$ This is a consequence from eq. III.12c.

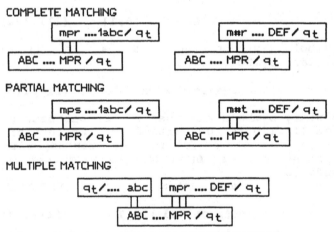

COMPLETE MATCHING

PARTIAL MATCHING

MULTIPLE MATCHING

FIG. III-12 - STRING CONCATENATION

The same string p_i may have a set of concatenation substrings $b_k, \ldots, b_p$, each one to be used to bind p_i with different strings of P. The concatenation substrings correspond to the binding sites of transmitters; receptors, controllers, etc. Each binding site is composed by different chemical radicals. The strength of the binding between two molecules (e.g., the t^r binding) is dependent of the degree of maching between their binding sites.

In this context, the following is defined (III.13):

III.14a) the operation ^ of concatenation of strings in T and R

$$\hat{} : T \times R \longrightarrow [0,1]$$

so that given $t_i \in T$ and $r_j \in R$ then

$$t_i \hat{} r_j = p_{t/r} \in P$$

The possibility $\pi(p_{t/r})$ of $p_{t/r}$ being a string of P is set equal to the matching compatibility $\mu(t_i, r_j)$ between t_i and r_j

$$\pi_P(p_{t/r}) = \mu(t_i, r_j)$$

The concatenation operation corresponds to the $t\hat{}r$ binding discussed before.

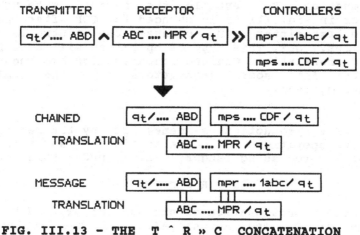

FIG. III.13 - THE T ^ R » C CONCATENATION

III.14b) the triggering or translation operation » as a special case of concatenation between $p_{t/r}$ and $c \in C$, $c = c_l + c_r$. The result of this translation operation depends on the type of the substring c_r:

III.14c) if $c_r \in L_{s5} = L_1 \cup L_3 \cup L_4$

$$p_{t/r} \hat{} c = c_l + c_r$$

In this case, c_r is released from c as a message to activate some further computational process. In this condition, c_r is used to specify the output of the $t\hat{}r$ binding. For example, c_r is used to control the proteic synthesis of the substrate dictionary S; or to specify the readable genes of G; or to control the amount of available energy to different cellular processes, etc. This type of operation will here be called message translation. The message is composed by the target address (letter substring) and the type of action (numeric substring) the controller must exercise over the target.

III.14d) if $c_r \in L_{s5}$, $L_5 = L_1 \cup L_3$

$$p_{t/r} \, \hat{} \, c = p_j$$

where p_j is a string of T, R or C. In this condition, c_r is used to condition the t^r binding itself (Fig. III.12). The same protein or production $p_j \in P$ can exhibit different active sites or concatenation substrings b_k, the actual b_k being specified by the local conditions or context. This is implemented here with the chained concatenation. For example, the receptor site MPR of the receptor in Fig. III.13 is changed to CDF after the binding of the transmitter ABD, because the controller c is the string mps...CDF. This type of operation will here be called chained translation. Chained concatenation has been recently described for some transmitters in the real brain (Teichberg, 1991).

The productions P generated by the genetics G and the above operations of concatenation ^ and translation » provide a processing language L(G) under this genetics G

$$T \, \hat{} \, R \, » \, C \qquad\qquad (III.15a)$$

where the concatenation of strings of T and R activates strings of C. The amount $a(c_i)$ of the controller c_i triggered by the t^r binding is calculated as:

$$a(c_i) = (\, a(t) \, \hat{} \, a(r) \,) \, \cdot \, \mu(t,r) \qquad\qquad (III.15b)$$

where $a(t)$ is the amount of transmitters t; $a(r)$ is the amount of the receptors r; $\mu(t,r)$ is the degree of matching in the t^r binding, and ^ and · are, in general, $\mp$ -norms or $\mp$ -conorms. The activated strings $c_i \in C$ exert one or more defined actions $a \in A$ over strings $p_i \in P$ or $g_i \in G$, or over the substrate dictionary S. A is said to be the semantic of L(G) and it is defined by a set A_R of minimum functions called the restricted semantic of L(G) which is related with the maintenance of L(G), and a set A_E of auxiliary functions or the expanded semantic of L(g), which is specified according to the problem to be solved.

The following actions define the restricted semantic A_R of any L(G) (Fig. III.14):

III.15c) affinity control (function 1):

$$\mu(r_i, t_i) = f_a(a(c_i)), \; t_i \in T, \; r_i \in R$$

The degree of matching $\mu(t_i, r_i)$ between the transmitter t_i and its receptor r_i becomes a function of the amount of the controller c_i, either because c_i changes the affinity between the complementary symbols of t_i and r_i or because it changes the number of the complementary symbols.

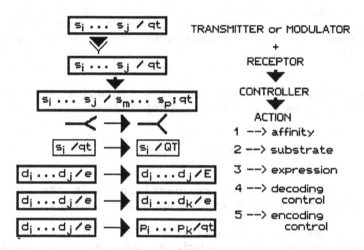

FIG. III-14 - THE RESTRICTED SEMANTICS OF THE CHEMICAL LANGUAGE

III.15d) substrate control (function 2):

$$a(s_i) = f_s(a(c_i)), \; s_i \in S$$

The amount $a(s_i)$ of the available substrate $s_i \in S$ becomes a function of the amount of the controller c_i. In the case of the real neurons, c_i may be an enzyme controlling the final synthesis of any protein.

III.15e) expression control (function 3):

$$\text{given } g_i \in G \text{ then } e_i = f_E(a(c_i))$$

The degree of expression of the gene g_i becomes controlled by c_i. In the case of the nervous cell, c_i may control the amount of availabe cAMP, which is one of the parameters governing gene expression or may be the regulatory protein which specifies the readable genes. The gene activation is a fuzzy process, whereas the gene specification is a crisp operation.

III.15f) control of the decoding function (function 4): c_i specifies the decoding properties of eq. II.4, because the size $M(t)$ of the functional pool of transmitters t becomes a function of $a(c_i)$:

$$m = a(c_i) \cdot (w \cdot M(t))$$

where $\cdot$ is, in general, a $\top$ -norm. In other words, the controller may participate in the specification and regulation of the transmitter release at pre-synaptic terminals.

III.15g) control of the encoding function (function 5): c_i specifies the encoding properties of eq. II.14:

$$w = a(c_i) \cdot \overset{n}{\underset{i=1}{\Theta}} (a_i \cdot s_i)$$

that is, the controller may participate in the specification and regulation of the axonic encoding function.

III.15h) deconcatenation (function 6): c_i is used to separate concatenated strings $p_i p_j$ and to destroy the target string p_i or p_j:

$$p_i p_j + c_i = p_i \text{ or } p_j$$

The amount OF deconcatenation is the minimum Ω

$$a(p_i) \text{ or } a(p_j) = a(c_i) \; \Omega \; a(p_i p_j)$$

The special application of this function is to release and destroy the transmitter from its binding to its receptor.

All these actions are supported by the translation operation defined by III.14c, whereas the operation defined by III.14d supports the

III.15i) context specification: the chained concatenation is used to condition the $t^\char`^r$ binding according to the local environment or the actual context.

Any program $P(G)$ in $L(G)$ is a set of statements of the type

$$t \char`^ r » c \longrightarrow a \in A \qquad (III.16)$$

The set A of actions of $L(G)$ is composed by the set A_R defining the restricted semantic of $L(G)$ and by the set A_E

adapting A to the task to be processed. The set A_E is composed by any computational procedure necessary to solve the problem under consideration.

III.8 - Updating the formal neuron

The spike train w_i arriving at the pre-synaptic cell n_i is recoded into pulses m_i of transmitters $t_i \in T$ to act upon the receptors $r_i \in R$ at the post-synaptic neuron n_j. This recoding is dependent of the size M_i of the functional pool of t_i. The $t^\wedge r$ coupling triggers the modification of the membrane potential of the post-synaptic cell, which is the m_i multiple of the unitary value v_0. Thus, the action v_i the pre-synaptic may exert upon the post-synaptic cell depends:

III.17a) on the amount m_i of transmitters t_i released by w_i;

III.17b) on the amount $M(r_j)$ of available receptors r_j to bind t_i;

III.17c) on the affinity $\mu(t_i, r_j)$ between t_i and r_j, and

III.17d) on the spatial position of the synapsis, which is represented by v_0. Thus:

$$v_i = w_i \circ M(t_i) \wedge M(r_i) * \mu(t_i, r_j) \mathbin{\text{\textbardbl}} v_0 \quad \text{(III.17e)}$$

where $\circ$, $*$ and $\mathbin{\text{\textbardbl}}$ are $\overline{\top}$ -norms and/or $\overline{\top}$ -conorms.

The weight or strength s_i of the synaptic contact between n_i and n_j is defined as:

$$s_i = M(t_i) \wedge M(r_i) * \mu(t_i, r_j) \mathbin{\text{\textbardbl}} v_0 \quad \text{(III.18a)}$$

and the post-synaptic activity v_i triggered by n_i is considered as

$$v_i = w_i \circ s_i \quad \text{(III.18b)}$$

where M_i is the available amount of t_i at the pre-synaptic cell; $M(r_i)$ is the amount of available post-synaptic receptor; $\mu(t_i, r_j)$ is the $t^\wedge r$ binding affinity, and v_0 is the electrical activity triggered at the post-synaptic cell by one quantum of transmitter, and $\circ$, $\wedge$, $*$ and $\mathbin{\text{\textbardbl}}$ are $\overline{\top}$ -norms and/or $\overline{\top}$ -conorms.

The different electrical activities elicited by the distinct n pre-synaptic cells n_i upon the post-synaptic n_j are aggregated into a total activity v_j

$$v_j = \sum_{i=1}^{n} v_i \qquad \text{(III.18c)}$$

which is encoded in $w_j \in W$ as:

$$w_j = \begin{cases} w_i & \text{if } v_j < \alpha_1 \\ w_u & \text{if } v_j \geq \alpha_2 \\ g(v_j) & \text{otherwise} \end{cases} \qquad \text{(III.18d)}$$

where w_i and w_u are specific parameters of the nerve.

Eqs. III.18c,d may be combined in:

$$w_j = \Theta \ (\ a_i \ {}^{\bullet} \ s_i) \qquad \text{(III.18e)}$$
$$ i=1$$

with the semantic of Θ depending on the type of the functions g and ${}^{\bullet}$ (See Chapter II, section II.6).

The spike train w_j travelling the axon may not necessarily spread upon all the terminal branches of n_j, since the axonal membrane is not a homogeneous structure, allowing the distinct terminal branches to exhibit different filtering properties. In other words, the different branches of the same axon may have different encoding functions:

$$\text{if} \quad \alpha_k < v_j < \alpha_{k+1} \quad \text{then } w \in W_k \qquad \text{(III.18f)}$$

The activation of the post-synaptic cell n_j due to the transmitter released by the pre-synatpic cell n_i, activates some control molecules c_j

$$t_i \ \hat{} \ r_j \ {}_{\text{»}} \ c_j \qquad \text{(III.19a)}$$

The same neuron synthesizes precursors for different transmitters. The synthesis of a specific transmitter at a defined axonal branch depends on the post-synaptic activity signalled by c_j. The consequence is that different transmitters can be alocated to distinct terminal branches of the same pre-synaptic neuron contacting different post-synaptic cells as the result of the controller action III.15i. The same post-synaptic cell will produce different receptors r_j to combine with different pre-synaptic transmitters t_i. Each specific coupling between a pre-synaptic transmitter and a post-synaptic receptor, in turn, activates different types of controllers c_j. These controllers exercise different types of action over the pre- and post-synaptic cells, which define the semantic of the chemical language L(G) supported by T, R and C. A statement in this language is:

$$t_i \hat{} r_j » c_j \text{ ---> action} \quad\quad (III.19b)$$

A program P(G) in this language is a set of these statements.

Thus, the neuron n is the complex processor defined in Chapter II, section II.7 as having its computational power increased by the L(G) described in this chapter. Thus:

$$N = \{ \{ W_p \}, W_0, T, R, C, \Theta, \{ \alpha, g \}, L(G) \}$$
$$(III.20a)$$

combines two different types of processing:

III.20b) electrical processing: supported by the MAPI structure discussed in Chapter II, and

III.20c) chemical processing: supported by the language L(G) defined by the genetics G, the supporting dictionary S and the grammar ꝑ. The semantic of L(G) is defined by the set A of actions supported by its controllers. The semantic A_R defined in III.15 is the minimum semantic of L(G). It may be expanded into other functions necessary to solve a specific problem.

Both kinds of processing are coupled by the dynamics of the t^r binding which triggers both electrical modifications in the membrane of and the release of control molecules $c \in C$ by the post-synaptic neuron.

Programs P(L) supported by L(G) may be used both to specify the topology and program a MultiPurpose Neural Net (MPNN), and to enhance the processing capacity of the MultiPurpose Neuron (mpN). In this way, L(G) expands the programming capacity of MPNN providing it with algorithm processing capability. The next sections explore some of the properties of L(G).

III.9 - Growing a neural net

Neurons are created in the brain by means of cellular division. This process imposes both temporal and/or spatial orderings in the development of the neural circuits, and it is called embryogenesis. Embryogenesis describes the temporal and/or spatial order neurons, layers and connections between neurons which are constructed. Any MPNN may be fully specified given its genetic and embryogenic process.

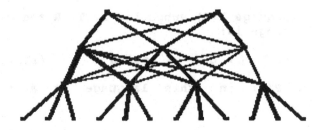

TYPE OF GROWTH: CONTINUOUS

LAYER 1
Number of neurons: 12
Transmitter: (abc/1)
Receptor: none
Controller: none
Substrate: (a/1)(b/1)(c/1)

LAYER 2
Number of neurons: 4
Transmitter: (def/1)
Receptor: (ABC/1)
Controller: none
Substrate: (a/3)(b/3)(c/3)
(d/3)(e/3)(f3)

LAYER 3
Number of neurons: 3
Transmitter: (abc/1)
Receptor: (DEFMNO/1)
Controller: (def2MNO/2)
Substrate: (a/3)(b/3)(c/3)
(d/4)(e/4)(f/4)

LAYER 4
Number of neurons: 2
Transmitter: (abc/1)
Receptor: (ABC/1)
Controller: none
Substrate: (a/3)(b/3)(c/3)

FIG. III.15 - CONTINUOUS EMBRYOGENESIS

The genetics of a MPNN is specified by:

III.21a) the processing language L(G) under a genetic G;

III.21b) the number of its layers, and for each layer:

III.21c) the number of neurons to be created;

III.21d) the amount of the substrates in the dictionary S of the genetic G, and

III.21e) the readable genes, or the strings $g_i \epsilon$ L(G) for which $e_i > 0$.

The genetic G defined in the case of Figs. III.15 to 17 is based on the FGG defined in III.10 and 11. The values of the code lengths are n = 3; o = 6, p = 3 and s = 4. The concatenation and translation operations are those defined in III.14. The semantic A of L(G) is set equal to the A_R in III.15. The type of transmitters, receptors and controllers to be produced by the activation of the corresponding genes are shown in the figures together with the degree of expression of these genes. E.g.: (abc/1) in Fig. III.15 means that the receptor abc must be produced by the activation of its gene, which has the degree of expression equal to 1. The amount of available substrate at each layer is specified in the figures. The amount assigned to each symbol corresponds to its amount in both L3 and L1 or L2. E.g.: layer 1 in Fig.III.15 has the substrates (a/1), (b/1) and (c/1) available in unitary quantities for both a,b,c and A,B,C. The different topologies in Figs. III.15 and 16 are partialy dependent of the different amount of the available substrates at layers 2 and 4, besides the different controllers and receptors assigned to layers 2 and 3.

The embryogenesis describes the order in which each layer is constructed and how the connections between neurons of the different layers are established. The embryogenesis may be continuous or recursive. In the first case, for each layer i of the net, connections grow up from layer i to all other layers j>i. In the second case, connections grow up from layer i to all other layers j>i, and then from all other layers k<i to all layers of the neural net. The nets in figs. III-15 and 16 were built by a continuous embryogenesis and that of Fig. III-17 was obtained by recursive embryogenesis. The MPNN embryogenic process is the algorithm to implement the processing of L(G) according to defined temporal and spatial orderings, to generate one specified neural net. It looks like:

```
BEGIN
    .
    .                       CREATING THE NEURONS
    .
    FOR LAYER 1 TO N
        .
        .
        .
        Provides each neuron with its appropriate genes
        .
    NEXT
    .
    .                       GROWING CONNECTIONS
    .
    FOR I=1 TO N
        .
        .                   DEFINING THE GROWTH
        .
        IF GROWTH IS CONTINUOUS THEN K=I, OTHERWISE K=1
        .
        .                   TESTING CONNECTIONS
        .
        FOR J=K TO N
            .
            .
            .
            CALL CONNECTIONS
        NEXT J
    NEXT I
END
```

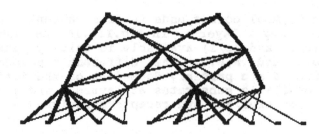

TYPE OF GROWTH: CONTINUOUS

LAYER 1
Number of neurons: 12
Transmitter: (abc/.9)
Receptor: none
Controller: none
Substrate: (a/1)(b/1)(c/1)

LAYER 2
Number of neurons: 4
Transmitter: (def/1)
Receptor: (ABCPQR/.9)
Controller: none
Substrate: (a/5)(b/5)(c/5)
(d/4)(e/4)(f/4)
(p/5)(q/5)(o/5)

LAYER 3
Number of neurons: 3
Transmitter: (abc/.9)
Receptor: (DEFMNO/1)
Controller: (mno2MNO/6)
Substrate: (a/3)(b/3)(c/3)
(m/3)(n/3)(o/3)

LAYER 4
Number of neurons: 2
Transmitter: (abc/.9)
Receptor: (ABC/.9)
Controller: none
Substrate: (a/8)(b/8)(c/8)

FIG. III.16 - THE INFLUENCE OF THE SUBSTRATE AND CONTROLLERS

Connections are established if:

III.22a) the cell m under consideration in the layer j can produce the receptor r_j required by at least one of the transmitters t_i produced by the neuron p being considered in the layer i. Thus

if $\mu(t_i, r_j) > 0$ for at least one t_i, r_j
and

III.22b) both neurons in layer i and j have enough substrate to sinthesize the correspondent t_i and r_j, respectively. Thus

if $a(t_i) > 0$ and $a(r_j) > 0$

If the connection is established then:

III.23a) its power w_i is calculated as

$w_i = \mu(t_i, r_j) (a(t_i) \, \Omega \, a(r_j))$

where Ω is the minimum;

III.23b) the substrate dictionary S at each layer is updated by

$$a_t(s_i) = a_{t-1}(s_i) - a(t_i), \quad s_i \in S_i$$

and

$$a_t(s_j) = a_{t-1}(s_j) - a(r_j), \quad s_j \in S_j$$

III.23c) the controllers triggered by the $t_i \, \hat{} \, r_j$ are produced if their substrates are available, and then they are allowed to exert their actions. Thus

$$t_i \, \hat{} \, r_j \, » \, c_k \, ---> \, a \in A$$

The action of the control molecules may modify the genes of other layers; the amount of substrate in the actual and other layers; as well as the affinity between transmitters and receptors at already established synapses.

Thus:

```
BEGIN
    .
    .                        CONNECTIONS
    .
    FOR NEURON 1 TO M
        .
        .                CREATING THE CONNECTION
        .
        TEST COMPATIBILITY μ(Ti,Ri)
        .
        .
        .
        IF μ(ti,rj) > 0

            TO ANY ti ∈ Ti AND rj ∈ rj THEN
                .
                .
                .
                Calculate the amount of ti and rj
                .
                .
                Update the dictionaries Si and Sj
                .
                .
                Calculate the correspondent wi
                .
                .
                Produce the controller ck if possible
                .
                .
                Update the correspondent dictionary S
                .
                .
                Process the action triggered by ck
    NEXT
END
```

Temporal order is important in recursive embryogenesis. E.g.: a synapsis not previously established between the antecedants and the actual layer (Fig. III-17a) may be allowed to grow up if the embryogenesis is recursive and the action of the controller turns, in a specific instant of time t, the conditions favorable to the synthesis of the required transmitter and receptor (Fig. III.17b). By a similar mechanism, a synapsis previously defined may be inhibited, giving rise to structures restricted to some period of the embryogenic process.

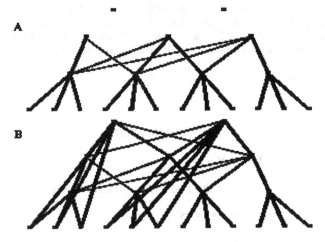

A

B

TYPE OF GROWTH: RECURSIVE

LAYER 1
Number of neurons: 12
Transmitter: (abc1)(ghi/0)
Receptor: none
Controller: none
Substrate: (a/1)(b/1)(c/1)
(g/1)(h/1)(i1)

LAYER 2
Number of neurons: 4
Transmitter: (def/1)
Receptor: (ABCMNO/1)
Controller: none
Substrate: (a/3)(b/3)(c/3)
(d/4)(e/4)(f/4)
(m/3)(n/3)(o/3)

LAYER 3
Number of neurons: 3
Transmitter: (abc/1)
Receptor: (DEF/1)
Controller: (pqr3DEF/6)
Substrate: (a/3)(b/3)(c/3)
(p/4)(q/4)(r/4)

LAYER 4
Number of neurons: 2
Transmitter: (abc/1)
Receptor: (ABC/1)(GHI/0)
Controller: (tuv4ghi/1)(tuv4GHI/1)
Substrate: (a/3)(b/3)(c/3)
(g/3)(h/3)(i/3)

FIG. III.17 - RECURSIVE EMBRYOGENESIS

Spatial order may influence positively or negatively the affinity $\mu(T,R)$ between T and R and/or the substrate production induced by the control molecule c_j. This may be used, if necessary, to turn the weight w of the synapsis depending on its spatial position in the net.

Finally, the actual value of the axonic threshold α and the actual axonic encoding function g may be defined as dependent of the actual value of w, besides being eventually specified by means of specific controllers c_s. In this way, different filtering properties may be assigned to distinct branches of the same axon depending on both spatial order and learning.

The complexity of L(G) may be enhanced by augmenting the complexity of both G and A to account for the complex embryogenic growth of the real brain. For example, Eldeman, 1988 shows that cell (CAM) and substrate (SAM) adhesion molecules play a central role in mediating the developmental mechanisms producing variation in the neuronal connectivity during the embryogenesis. CAM and SAM operate in a fashion similar to that of transmitters and receptors. SAMs are marks provided by neighboring cells used by the growing axon or the moving cell in order to find its pathway. The moving element matches these SAMs to its CAMs and uses the results of these matchings to orient its movement. In this condition, CAMs and SAMs may be genetically encoded by G and their actions inserted in A, so that statements of the type

$$CAM \ \hat{} \ SAM \ » \ C \ ---> \ movements \ \epsilon \ A \qquad (III.23d)$$

may be used to describe the role played by these molecules in the wiring of the real and artificial MPNNs.

Embryogenesis is used as the basic process of programming MPNN circuits, either as only one big net or as a family of processing modules (see Chapter VI). The choice depends on the complexity of the problem to be processed. The modular architeture of MPNN (see Chapters V and VI) is a strong paradigm allowing the implementation of parallel multitask processing or the creation of different solutions for the same task (see Chapter V). Controlled or random alterations of the composition of the available substrate dictionary S and/or of the structures of the genes of G may produce slightly different modules in the MPNN circuit, which may then be selected according to their performance in solving the task. This constitutes the base of the evolutive learning discussed in Chapter V and it is similar to the propositon of Eldeman, 1988, to explain the biological variability of the nervous system of different individuals of the same species.

III.10 - The algorithmic chemical processing

The chemical processing supported by L(G) greatly

enhances the computational power of the MultiPurpose Neural Net (MPNN) because it simplifies the implementation of symbolic reasoning. Also, L(G) allows the implementation of both crispy and fuzzy logics, because the assertions of the type:

$$t \char`^ r » c \longrightarrow a \in A \qquad (III.24a)$$

support either:

III.24b) a crisp semantic of the type

$$t \char`^ r » c \longrightarrow \{yes,no\}\ a$$

if the triggering function f associating a with a(c) in III.15 is a crisp mapping:

$$f : C \times A \longrightarrow (0,1)$$

$$d = \begin{cases} 1 \text{ if } a(c) \geq \alpha \\ 0 \text{ otherwise} \end{cases}$$

$$t \char`^ r » c \longrightarrow d.a,\ a \in A$$

In this case, A is a α-cut of C (Negoita and Ralescu, 1975); or

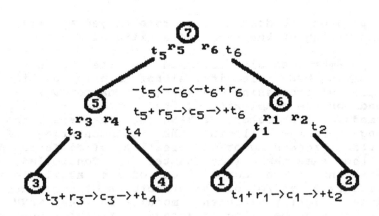

FIG. III.18 - SYMBOLIC PROCESSING

III.24c) a fuzzy semantic of the type

$$t \char`^ r » c \longrightarrow q.a$$

if the triggering function f is a fuzzy mapping:

$$f : C \times A \dashrightarrow [0,1]$$

$$q_i = f(c) \text{ if } \alpha_i \leq a(c) \leq \alpha_{i+1}$$

$$t \,\hat{}\, r \text{ » } c \dashrightarrow q.a, \ a \in A$$

In this case, A is a α-level set of C (Negoita and Ralescu, 1975).

Fig. III.18 shows an example of symbolic processing in a MPNN. This net runs the following algorithm:

IF N_1 IS ACTIVATED (or $t_1 + r_1 \dashrightarrow c_1$) THEN
.
.
   Sets the transmitter of N_2 as t_2
   because $t_2 = t_1 + r_1$
   thus N_2 may now couple to N_6
   .
   IF N_6 IS ACTIVATED (or $t_6 + r_6 \dashrightarrow c_6$) THEN
     .
     .
     Sets the transmitter of N_5 as t_5
     because $t_5 = r_6 + c_6$
     thus N_5 may now couple to N_7
     because N_7 has r_5
     .
     IF N_3 IS ACTIVATED (or $t_3 + r_3 \dashrightarrow c_3$) THEN
     .
     Sets the transmitter of N_4 as t_4
     because $t_4 = t_3 + t_3$
     thus N_4 may now couple to N_5
     because N_5 has r_4
     .
     .
     IF N_5 IS ACTIVATED (or $t_5 + r_5 \dashrightarrow c_5$) THEN
       .
       .
       Sets the transmitter of N_6 as t_0,
       because $t_0 = t_5 + r_5$
       thus N_6 may now not couple to N_7
       because N_6 has not r_0
       .
      END IF
     END IF
   END IF
END IF

This may be a crisp algorithm if the amount of transmitter is always set to 1 or 0, otherwise it is a fuzzy procedure.

III.11 - Combining numeric and symbolic processing in a MPNN

The MPNN nets may be hierarchicaly organized in MPNN circuits to process complex tasks requiring both numerical and symbolic processing. Let the automatic process control be the example to be discussed.

Automatic control systems challenge the designers to integrate symbolic and numeric computation in robust models (Handelman et al, 1990) in order to account for both a low level adaptative control and a high level classification and planning of the control.

Let the low level adaptative control be supported by a modular MPNN, e.g. like those MPNNs used to implement Fuzzy Logic Control (FLC) in Chapter II. Each module provides a building block of a set B of possible actions. These building blocks are used by the high level to plan the desired control according to the actual classification of the desired task. This classification is the result of the matching between the actual state of the system with some desired goals. The plan to attain the desired goals is a composition of actions processed by blocks of B. The programming of the MPNN circuit to execute the plan is to combine and to activate each MPNN module supporting the planned actions.

Let the discussion of the motor control of an animal or robot illustrate the problem. In this case, low level circuits (e.g. the circuit in Fig. III.19) are used to process muscle or motor actions such as holding a position p; changing position from p to p'; oscillating with period T (e.g. walking), etc. These actions provide the building blocks used to plan complex motor controls involved in walking, attaining and maintaining a specific posture, etc.

The structure of the low level MPNN module in Fig. III.19 is the following:

III.25a) two sets of sensors s_i and l_i measure the actual value of the two output variables: the muscle length (or the motor rotation angle) and tension, respectively. In the case of natural system, s_i is the stretch receptor discussed in Chapter I, and l_i is the Golgi receptor located at the muscle tendon;

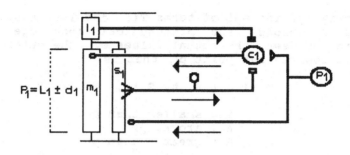

FIG. III.19 - THE LOW LEVEL MPNN MOTOR CONTROL

III.25b) the information provided by these sensory neurons is used by the circuit C_1 to adjust the muscle length (or motor angle) to perform a desired action specified by the planning module P_1;

III.25c) P_1 programs the desired action in this low level MPNN by specifyiing the set point to C_1 and by adjusting the sensors (e.g., s_1) to function either as a proportional (tonic receptor) in the case of posture control or as a differential (phasic) sensor in the case of movement control (see Chapter I). Movements to adjust posture require PD (tonic-phasic) sensors;

III.25d) the adjustment of the muscle length (or motor angle) is then locally calculated by a classic feedback control implemented in C_1. E.g. if the muscle length increases compared to the set point, the sensor measures this increases and activates C_1, which increases the muscle contraction (or motor power) to bring the system back to the desired position p. This muscle contraction is modulated by the information about the tension in the system, provided by l_1. If the actual value of the tension may damage the system, the ouput of C_1 is reduced even if this implies losing the position p. This describes the basic spinal cord processing of the muscle system in many animals. This low level control can be implemented as a FLC (see Chapter II, section 8).

Let X and Y be linguistic variables associated with the measures provided by the sensors s_i and l_i, respectively. The set of terms $T(X)$ and $T(Y)$ associated with X and Y encodes the prototypical knowledge about the matching between the actual values of the output variables and the desired set point p_i, that is:

$$T(W) = \{ S, A, G \} , \quad W = X, Y$$

$$S = \text{smaller than } p_i$$
$$A = \text{around } p_i$$
$$G = \text{greater than } p_i \qquad \text{(III.25e)}$$

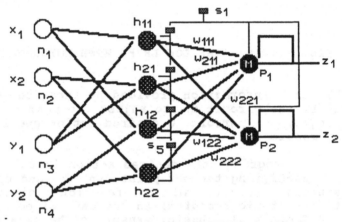

FIG. III.20 - A MPNN FOR IMPLEMENTING C_1 OR C_2

Now, let Z be the linguistic variable associated with the input variable controlling the state of muscle contraction or motor power. If the input variable is the increment of the muscle contraction (motor power) required to maintain the system around the set point, $T(Z)$ may be defined as:

$$T(Z) = \{ NB, NM, NS, Z, PS, PM, PB \}$$

where:

$$NB = \text{negative big}$$
$$NM = \text{negative medium}$$
$$NS = \text{negative small}$$
$$NU = \text{null}$$
$$PS = \text{positive small}$$
$$PM = \text{positive medium}$$
$$PB = \text{positive big} \qquad \text{(III.25f)}$$

The rule base RB of FLC may be composed by rules of the type:

If X is S and Y is A then Z is PS or else
.
.
If X is A and Y is A then Z is NU or else
.
.
If X is G and Y is A then Z is NZ or else
.
.
If X is G and Y is G then Z is NU
.
.

                                                    (III.25g)

This FLC may be implemented by a MPNN with a structure
similar to that in Fig. III.20 (see Chapter II, section 8).

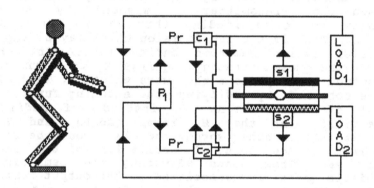

FIG. III.21 - HYBRID CONTROL OF THE MOTOR SYSTEM

        Antagonistic muscles or motors may be used to
control the position of each one of the articulations of a
given animal or robot (Fig. III.21). Let these antagonist
systems be called flexor F and extensor E. Different low
level MPNN modules C_1 (F) and C_2 (E) are used to
control each of these antagonistic motors (muscles) handling
the articulations of the robot (animal). Let the control of
one of these articulations to hold a set of positions P
under different load conditions (see Fig. III.21) be
discussed:

III.26a) both loads L_1 and L_2 are maintained equal and
constant to w: C_1 and C_2 must learn to hold the desired
position $p \in P$. This implies to learn the adequate weights
between the associative layer (h_i neurons) and the output
neurons p_i of the MPNN nets implementing C_1 and C_2
(Fig. III.21). This corresponds to the learning of the

implication function of the rules in III.25g. Many different algorithms may be used for such a purpose (e.g. the methods discussed in Chapter V). The alternative to this learning is to use the knowledge provided by an expert to program the wiring of C_i.

But now let the load condition change:

III.26b) the load L_1 is held constant and equal to w_1 while the actual value w_2 of L_2 is variable: P_1 may know about this condition by analyzing the information provided by the load sensor l_1. One possible strategy to cope with this new situation may be to maintain the output of C_1 constant because w_1 is constant, and to let C_2 in charge of the adjustments required to compensate the modifications of w_2 in order to maintain the desired position p. The first step in this process is to disconnect s_1 from C_1. P_1 may use a modulator m_1 whose action is drastically reducing the affinity (action III.15c) between the transmitter t_1 released by n_i and the receptor r_1 in C_1 binding this t_1 at the associative layer H. This is a crisp decision which may be supported by the L(G) logical semantic defined in III.24b. The second step in the chosen strategy requires C_2 to learn to adjust its output to compensate the extra weight w. This adjustment may be obtained by modifying the encoding function g (eq. II.14) of the neurons p_i instead of modifying the connectivity at the H layer (Rocha and Yager, in preparation). In other words, the task could be to learn to calculate a new amount of activation of the same muscles being used. This means adjusting the semantic of the linguistic quantifier associated to the output neurons. This learning may be accomplished if P_1 uses a modulator m_2 to control the axonic encoding function (action III.15g) of the output neurons of C_2. This kind of decision is fuzzy and may be supported by the L(G) logical semantic defined in III.24c. In this way, the amount of m_2 becomes a function of w_2:

$$m_2 = f(w_2)$$

which is implemented by the connectivity between the sensory neurons l_i and the MPNN implementing P_i (Fig. III.21). Thus, this kind of learning may be supported by the same algorithms used in III.26a;

III.26c) a sudden modification in the environment now turns L_2 constant and equal to w_2 while w_1 becomes variable: P_1 can now easily revert the plan in III.26b, blocking the coupling between n_i and h_i in C_2 and modulating the output of C_1;

III.26d) finally, another modification in the external environment turns both L_1 and L_2 variable loads: in this condition, P_1 can assign different relevances to the processings in C_1 and C_2 according to the magnitudes of the mean variations of L_1 and L_2. These distinct relevances result into different degrees of affinity between t_1 and r_1, and between t_2 and r_2; as well as on distinct semantics associated with the output of the linguistic quantifiers of C_1 and C_2. This control is obtained if the amount of modulators m_1, ..., m_n released upon C_1 and C_2 are governed by the matching between the actual values of L_1 and L_2 and the knowledge acquired under the conditions of III.26b,c.

This kind of hybrid systems combining numeric and symbolic calculations has been used in the literature, but the expert system technology was used to implement the symbolic reasoning and the neural nets technology to implement the numerical control. Here, MPNN takes charge of both tasks in a single type of computational structure. The structure discussed above may also be applied to areas other than robotics. For example, if the variable L_1 is associated to Product Supply, L_2 to Product Consumption or Production, and p_r is the Desired Stock, the above discussion holds for the control of Production Processes.

III.12 - Mail and broadcasting

The chemical language L(G) supports also the physiology of the hormones. Hormones are chemical products released in the blood stream by neurons at the limbic systems or by glands located in different parts of the organism. These chemicals in the blood stream may reach any site in the whole organism. The role played by the hormones is to activate some special functions in the organism or to control the activity of glandular cells. The biggest difference between hormones and transmitters is the broad distribution of the hormones in the body due to its transportation in the blood, in contrast to the localized distribution of the transmitters due to its release by the axonic terminal branches. Modulators or neuropeptides are considered to have a distribution over the brain which is intermediate to that of hormones and transmitters.

From the computational point of view, transmitters are mailed to specific receivers, while modulators and hormones are broadcasted by means of general buses (Fig. 3-22). The target of the modulators is in general a MPNN, while the target of hormones is in general the non-neural set of actuators under control of a MPNN circuit. The hormone broadcasting is wider than the distribution of the modulator. Because of this, the hormonal brodcasting will be called here general broadcasting, while in the case of modulators it will be referred to as restricted broadcasting.

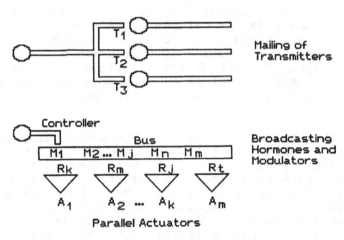

Mailing of
Transmitters

Broadcasting
Hormones and
Modulators

Parallel Actuators

FIG. 3-22 - MAILING AND BROADCASTING

In this line of reasoning, hormones may be assumed to be general messages released by some control MPNN and broadcasted by means of a bus connecting many parallel independent processes in order to coordinate their actions. The role played by these control MPNNs is to coordinate the activities of such multi-task systems (MTS). L(G) may support many types of numerical and non-numerical control related to the planning and coordination of the activity in these MTS. It provides also a strong mechanism for message recognition in the bus, which is the h^r binding. Here, h stands for hormone.

In the same way, modulators or neuropeptides may be used to coordinate the activities of neural systems composed by many MPNNs opperating in a parallel fashion and/or according to some hierarchy. Modulators will also be very useful to control learning in MPNNs, either by reinforcing (reward) or inhibiting (punishing) the synaptic growth of those activated circuits related to the task being learned, according to the success in reaching the specified goals. All these tasks are easily supported by a restricted broadcasting system provided by the m^r binding. Here, m stands for modulator.

Broadcasting may be the best alternative if the purpose in the examples III.25-26 turns to be to control not only one single articulation of the robot, but its global posture. In this case, the decisions provided by the symbolic reasoning at P1 may be broadcasted to the control building blocks at all articulations.

III.13 - Consequences and future research

The language L(G) introduced here may be used to implement both symbolic and algorithmic processings in neural circuits, as a consequence of the fact that concatenation ^ is defined as specific operation depending on the existence of affinity among the elements of T, R and C; and from the fact that

$$T \; \hat{} \; R \; » \; C \longrightarrow actions \; \epsilon \; A \qquad (III.27a)$$

may be used to control the processing flow inside the neural as follows

if t_i ^ r_j then c_k and w_i --> 1 (or 0)

or else if t_p ^ r_l then c_s and w_p ---> 1 (or 0)

$$\begin{array}{c} \cdot \\ \cdot \\ \cdot \end{array} \qquad (III.27b)$$

otherwise w_i --> 0 (or 1)

The modifications introduced here in the physiology of the artificial neuron drastically change the field of Neural Nets because it turns the neuron into a strong computational element, able to handle both numerical and symbolic calculations. Numerical calculus is electrically processed, while symbolic transactions are chemically encoded. More interesting is the fact that both processings are associated because modifications of the post-synaptic permeability are among the possible actions in A. Also, the chemical language introduced makes MPNN programmable besides treinable. Thus, MPNN has the most pleasant properties for a tool in the Artificial Intelligence field. As a matter of fact, L(G) is a processing language founded on a message passing and processing paradigm very similar to the Object Oriented Paradigm used nowadays in AI. The modularity of the MPNN circuits is another point of similarity with the object paradigm, since each module may be viewed as an object inheriting its own variables and methods. SMART KARDS(c) described in Chapter IX is an object oriented environment specially developed for the construction and processing of MPNN circuits. JARGON in Chapter VIII users the symbolic capabilities of MPNN to encode the syntax of natural languages.

The operations of concatenation (III.14a) and deconcatenation (III.15h) are key tools for implementing temporal dependencies in MPNN. This is because the time relations between the coupling and the corresponding decoupling of strings in P are established when the timing between these 2 operations is specified. While the string coupling is maintained, some information is retained in L(G). Thus, the timing between the coupling and decoupling operations is associated with some types of memory (e.g. medium term memory). The amount of time the coupling is allowed to be maintained conditions the MPNN processing. Time relations are key issues in the case of procedural reasoning (see Chapter V).

The definition of the grammar $ is crucial to minimize the dictionary S, and to maintain the cardinality of T, R and C as low as possible, in order to reduce the computational cost of the model. By the same reason, the set of actions defined for C is also required to have low cardinality. However, the computational power of the MPNN so defined is directly related to the semantic power of the language defined by $. Thus, $ must be optmized in order to guarantee a strong semantic from dictionaries of low cardinality and from a small set of semantic primitives (actions). This is what nature has obtained so far, and what will constitute an interesting research problem in the MPNN research.

The chemical language discussed here is also important if learning is considered. It supplies a strong tool for calculating the changes of the synaptic weights triggered by inductive learning, no matter if hebbian or associative procedures are taken into account. This is discussed in the next chapter. But it is also a language to formalize evolutive learning, which is able to really modify the structure of the net, by creating new connections among neurons either by means of association of circuits (crossover) or by inclusion or deletion (mutation) of cells. In this regard, it provides a language to formalize the heuristic used to guide the evolutive learning, a characteristic that is missing in the Genetic Algorithm approach (Booker et al, 1989) The possibility of developing a language to represent heuristic is another interesting research problem to embrace in the future in MPNN research.

IV.1 - Modeling

To learn is to model the observable world in order to understand it. A model is a set of relations between data or evidence obtained with a set of intruments, and actions performed by a set of effectors. Understanding requires the model to fulfill some defined purpose which may be simple survival or a complex subject like pleasure, religion, science, etc. From a general point of view, to understand is to provide a set of adequate responses to adapt the system to the surrounding world or, in other words, to maintain its identity in a changing environment.

The first step in modeling is to collect a set E_0 of evidence or data from the surrounding universe U with a set I_0 of available instruments. This set I_0 may be the set of sensory or input neurons of a MPNN. To observe is:

$$O: U \times I_0 \times E_0 \dashrightarrow [0,1] \quad (IV.1a)$$

that is, to obtain a set E_0 of measures about U with the set of instruments I_0. Generally E_0 is a set of redundant measures about U.

The second step of the process is to discover the set M_0 of the relations between the collected pieces of evidence:

$$M_0: E_0 \times E_0 \times \ldots \dashrightarrow [0,1] \quad (IV.1b)$$

These relations are used to construct a family C_0 of coverings or classifications of U:

$$M_1: E_0 \times E_0 \times \ldots \times C_0 \dashrightarrow [0,1] \quad (IV.1c)$$

This is the task of the associative neurons of a MPNN.

In the third step, the discriminant categories of the topology C_0 are associated to the set A_0 of actions the system may exert upon U to attain a set G of defined goals:

$$M_2: C_0 \times A_0 \times G \dashrightarrow [0,1] \quad (IV.1d)$$

The actions A_0 are the outputs of the effector neurons of MPNN or other non-neural elements called here actuators. In the case of the natural systems, these actuators are muscle cells, glands, the phonetic system, etc. In the case of artificial MPNNs, the actuators may be any other computational structure. This opens the possibility of combining the neural net technology with other computational tools for building hybrid systems (e.g. see Chapters VII, VIII and IX).

E_0 is in general a set of redundant measures, this means that M may be refined by eliminating this redundancy. The refinement $\Gamma(M)$ of M implies, therefore, to find the most significant pieces of evidence $e \in E$ supporting M. If j > i, Γ must:

IV.2a) reduce the cardinality of E_i:

$\Gamma(M_j)$: E_j x E_i ---> [0,1] , $E_i \subset E_j$

IV.2b) decrease the complexity of C_i

$\Gamma(M_j)$: C_j x C_i ---> [0,1] , $C_i \subset C_j$

IV.2c) optimize the performance of A_i

$\Gamma(M_j)$: A_j x A_i ---> [0,1] , $A_i \subset A_j$

where $\subset$ denotes contained by. These steps correspond to the following phases of learning:

IV.3a) discovery of patterns in U;

IV.3b) classification of these patterns;

IV.3c) association of the classifications to procedures, and

IV.3d) consolidation of the learned model M;

which are events taking place almost at the same time in the real brain.

If $P \subset U$ is the result of the action of A over U according to M, the efficiency $\delta(M)$ of M is determined by the matching $\mu(P,G)$ between P and the set of goals G:

$$\delta(M) = \mu(P,G) \quad (IV.4)$$

In the case of well learned models M fulfilling conditions IV.2a,c:

$$\delta(M) ---> 1 \quad (IV.5)$$

If A is allowed to modify U

$$A: U \times U' ---> [0,1] \quad (IV.6)$$

learning becomes a partially closed process. This type of learning will be called here σ-learning. Learned models provided by σ-learning are named σ-models. Both the instruments I; the actions A, and the goals G in the case of σ-learning are learnable σ-models, too.

σ-Models are recursive

$$M_n = \Gamma(M_{n-1}) \qquad (IV.7a)$$

because

$$A_{n-1}: U_{n-1} \times U_n \longrightarrow [0,1] \qquad (IV.7b)$$

so that

$$M_n: U_n \times I_n \times E_n \times \longrightarrow [0,1] \qquad (IV.7c)$$

and

$$M_n: C_n \times A_n \times G_n \longrightarrow [0,1] \qquad (IV.7d)$$

The limiting value of n is given by

$$\delta(M_n) - \delta(M_{n-1}) \longrightarrow 0 \qquad (IV.7e)$$

In the case of Well Learned σ-Models, condition IV.7e must imply

$$\delta(M_n) \longrightarrow 1 \qquad (IV.7f)$$

Recursiveness is the consequence of the fact that the σ-MPNN may act upon U to modify it, and to select the evidence it wants to collect about U.

If A is allowed to provide alternative modifications U_a of U

$$A: \quad U \times U_1 \times U_2 \times \ldots \longrightarrow [0,1] \qquad (IV.8a)$$

different σ-models M_{ta} can be created at time t. If only those M_{tk} for which

$$\delta(M_{tk}) \geq \alpha \qquad (IV.8b)$$

are allowed to survive or to produce new models at the moment t+1, these M_{tk} are said to be evolutive σ-models. Eqs. IV.8 implements natural selection of learned σ-models. The process of obtaining evolutive σ-models is called here Evolutive Learning. The main goal of evolutive learning is to obtain well learned σ-models.

IV.2 - The evolutive reasoning machine

Let, the learning process defined in IV.1 be called inductive learning and that defined in IV.7 be named deductive learning.

Inductive learning means that the regularities of U strengthen relations between some elements of the structure S modeling U and reduce the connectivity among some others, in such a way that the relations M to be modeled are represented by the resultant relations among the elements of S. For instance, the regularities of the surrounding environment strengthen the connectivity between some neurons and reduce it between some other cells in both real and artificial neural nets (Rocha, 1990b,c, 1991b). The modeling structure S is called here a connectionist machine. Neural Nets are connectionist machines.

Simply because man observes the regularities of the surrounding world, he accepts them as true and his mind incorporates them as heuristic knowledge. Conditioning is the most popular paradigm of inductive learning, but many other physiological mechanisms like habituation, response attenuation, sensitization, etc. allow man to learn from the repeated observation of the same facts. Inductive learning is the most important and often the exclusive type of learning of the artificial neural nets. It is one of the learning paradigms of MPNN.

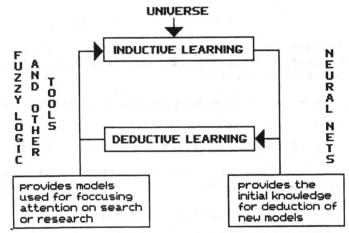

FIG. IV-1 - THE EVOLUTIVE REASONING MACHINE

Deductive learning means that the structure of already existing models M are changed either to improve their performance or to create new ideas to guide new observations of U (Fig. IV.1). Search and research are processes of observation guided by hypotheses deduced from previous knowledge, whose purpose is to confirm or reject these very same hypotheses. Deductive learning is used here not in the usual sense of a logical operation, but meaning any modification of a previous refined knowledge, no matter the formal tool used to calculate this modification. To deduce is, therefore, to derive the truth or a conclusion from something known (Webster Dictionary). Deduction may be supported either by a logical formal system like Fuzzy Logic, or by a random process such as the Genetic Algorithm (Booker et al.,1989), or any other formal tool. It may also be implemented in Neural Nets by means of Associative Learning and Modulatory Control of Learning (see Chapter II, section II.5 and Chapter III, section III.IV). Deductive learning is the other important learning paradigm of MPNN.

Inductive and Deductive learning support σ-models because they constitute the complementary parts of the σ-reasoning machine (Fig. IV.1). This machine uses some initial knowledge M_0 provided by inductive learning from the observation of U to create new models M_n, which are, in turn, used to guide a new exploration of this very same universe. Such reasoning machine is partially closed learning paradigm because inductive learning opens its recursive behavior to new unknown regularities of U. This kind of device is called here Evolutive Reasoning Machine (ERM). ERM is a connectionist machine. MPNN is an Evolutive Reasoning Machine since the synaptic physiology supports inductive learning and the modulatory control of the synapsis provides the tool for implementing deductive learning. It must be remembered that the multipurpose neuron is both a fuzzy logical and numerical device.

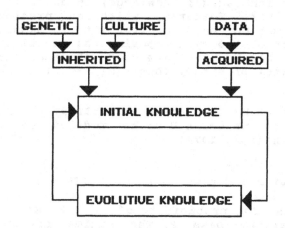

FIG. IV-2 - KNOWLEDGE INHERITANCE

To obtain all the initial knowledge (K_0) required by ERM from inductive observation of U may be a very time consuming task. In the case of man, this means that each one of us would have to repeat the history of mankind. An impossibility imposed by the mortality of the human being. Inheritance of part of this K_0 of ERM is the solution to this problem (Fig. IV.2).

Inheritance in the case of the real brain is provided by its genetics and the culture of its society. Genetics guides the cerebral embryogenesis and provides the initial wiring of the brain, or the phylogenetic initial knowledge. Formal or informal teaching supports the inheritance of the initial knowledge stored in the culture of a group of human beings. Here, the set of σ-models shared by this group of people constitutes its culture acquired in the historical process of aggregation of their society (e.g. Rocha and Rocha, 1985). Each individual of this group inherits this sociogenetic knowledge throughout formal and informal educational processes. Inheritance in the case of artificial neural nets implies programming. MPNN supports knowledge inheritance because its genetics G generates the language L(G) used to describe both the embryogenesis and programming of MPNN. L(G) also supports inductive and deductive learning.

IV.3 - Evolutive learning

Evolutive learning may be accepted (Rocha, 1982a,b) as the process of abstracting evidence from and related to a proposed question (actual knowledge) in order to develop a model about these features, which can predict (comprehension) and may be used to transmit (communication) the future behavior of the system posing the question. Thus, the evolutive σ-models M must have high predicability and must present low rates of error, thus $\delta(M)$ ---> 1.

Let h(W) be the Shannon's entropy of the discrete events W of the universe U to be modeled by a reasoning machine ERM (Shannon, 1974):

$$h(W) = - \sum^{W} p(w_i) \log p(w_i) \quad (IV.9a)$$

where $p(w_i)$ is the probability of $w_i \in W$, and log stands for the logarithm base 2. The maximal entropy of W is obtained if all $p(w_i) = p = 1/n$, n being the cardinality of W:

$$h_{Max}(W) = - \log p \quad (IV.9b)$$

Let $h_s(ERM)$ be the structural entropy of the reasoning machine ERM:

$$h_s(ERM) = = - \sum^{ERM} s_i \log s_i \quad (IV.9c)$$

where s_i is the strength of the synapsis i of ERM. The maximal entropy of ERM is obtained when all its k synapses have the same weight $s_i = 1/k$:

$$h_{Max}(ERM) = - \log k \quad (IV.9d)$$

Let $h_s(M_j)$ be the entropy of the recursive modeling M of W at the step j:

$$h_s(M_j) = = - \overset{M_j}{\Sigma} s_j \log s_j \quad (IV.9e)$$

where s_j is the strength of the synapsis of the neural net $MPNN_j$ supporting M_j.

Proposition IV.10 - The following holds for any reasoning machine ERM inheriting an initial knowledge M_0:

$$h_s(ERM) > h_s(M_0)$$

It follows from eqs. IV.9 and the definition of knowledge inheritance.

Q.E.D.

Proposition IV.11 - The reasoning machine ERM inheriting an initial knowledge M_0 may be able to model events W of U if $h_s(M_0) \geq h(W)$.

Let M_j be the refined model describing W. The learning of M_j results from changing the connectivity of ERM according to the regularities of W, thus according to the value of $h(W)$. It follows from proposition 10 and the definition of induction learning that:

$$h_s(M_j) \dashrightarrow h(W)$$

thus

$$h_s(M_0) \geq h(W)$$

Q.E.D.

Proposition IV.12 - Any reasoning machine ERM provided with an initial knowledge M_0 may learn to model W even if

$$h(W) > h_s(M_0)$$

if

$$h_{max}(ERM) \geq h_s(W)$$

If $h_{max}(ERM) \geq h(W)$ then it is possible to obtain another model M'_0 for which

$$h_s(M'_0) > h_s(M_0)$$

either by recovering some lost connections among the elements of ERM or by equalizing the connections among these elements or even by creating new connections among the elements of the ERM. The first two processes will be called here forgetting, whereas the former supports creativity. If necessary evolutive learning may obtain

$$h_s(M'_0) = h_{max}(ERM)$$

If $h_s(ERM) \geq h(W)$ then learning of W may be tried from M'_0 according to proposition IV.11.

Q.E.D.

It follows from these propositions that the role played by the evolutive learning is to adjust the entropy of the knowledge K of the ERM machine to that of the events W to be modeled. This is done by modifying the strength of the connectivity among the elements of this machine. On the one hand, inductive and deductive learning may be used to decrease $h_s(K)$ if $h_s(K) \geq h(W)$ because they can increase the strength of the connections among some of the elements of ERM while decreasing the connectivity among some other elements. They may even disconnect elements of ERM from the model being learned. This results in a decrement of $h_s(K)$. On the other hand, deductive learning and forgetting may operate in the opposite direction if $h_s(W) > h_s(K)$, because they can recreate connections among disconnected elements of ERM and can equalize the connectivity of the machine, respectively. This results in an enhancement of $h_s(K)$. The experiments described in Chapter VII, section VII.10 confirm the conclusions supported by the above propositions.

Another consequence of the above propositions is that the difficulty d(W) of modeling W is directly related to the difference of entropy of K and W

$$d(W) = f\ (h_s(K) - h(W)) \qquad (IV.13)$$

IV.4 - Inductive learning

Induction in a connectionist machine is supported by some process of modifying the relations among the elements of this machine according to the regularities of its set of inputs and the goals of the machine. In the case of neural nets, this means that the strength s_i of the synapsis is a function of both the statistics of the input and the action of the learning controllers.

The strength s_i of the synapsis is (see Chapter II, section II.3)

$$s_i = M(t_i)\ \hat{}\ M(r_i)\ *\ \mu(t_i,r_i)\ \cdot\ v_0 \qquad (IV.14a)$$

where:

IV.14c) $M(t_i)$ is the available amount of the transmitters t_i at the pre-synaptic cell;

IV.14d) $M(r_i)$ is the available amount of post-synaptic receptors r_i;

IV.14e) $\mu(t_i,r_i)$ is the binding affinity between t_i and r_i;

IV.14f) v_o is the unitary post-synaptic response triggered by one quantum (the contents of one vesicule) of released t_i, and

IV.14g) $*$, $\hat{}$ and $_1$ are t-norms.

Also, the activation of this synapsis triggers a chain of chemical events

$$T \hat{} R » C ---> action \qquad (IV.15a)$$

whose purpose is to control the very same synapsis as well as the activity of neighboring and pre-synaptic cells. These actions may control:

IV.15b) the amount and the type of the pre-synaptic transmitters;

IV.15c) the amount and the type of the post-synaptic receptors;

IV.15d) the chemical affinity between the transmitter and the receptor, and

IV.15e) the available energy at the membrane influencing the actual value of v_o.

Thus, C controls s_i.

The intensity of these actions is dependent of the amount of the released controller c_i as well as of the amount of modulators m_i produced by the learning control areas. The amount $a(c_i)$ of c_i is dependent of the statistics of the use of the synapsis and the amount $a(m_i)$ of m_i may be considered dependent of $\delta(M)$ (eq. IV.6g). It must be remembered that $\delta(M)$ measures the match between the actual performance of the net and the goals established for the learning of M. Thus:

$$s_i = f \ (a(c_i),a(m_i)) \qquad (IV.15f)$$

provides the semantic of the chemical language L(G)   defined by the genetics G of the MPNN (see Chapter III, page 21).

Rocha, 1982a,b, proposed this semantics to be that provided by a fuzzy automaton.

Let a machine M be given for which:

IV.16a) S is the set of its inputs, and

IV.16b) Q is the set of its states.

Now, let

$$\phi : Q \times S \longrightarrow Q \quad (IV.16c)$$

be a function relating the states of M to its inputs. In other words, let it be assumed that any $s \in S$ collected by M can modify its actual state q to another   state q':

$$q' = \phi (q,s) \quad (IV.16d)$$

The machine is called an automaton, and ϕ is called its next state mapping.

Given any subset $F_q \subset Q$, it is allways possible to determine the subset of inputs $E_q \subset S$ moving the actual state q of M to another state q' $\in$ F_q. In other words:

IV.16e) F_q is said to be reachable from q given E_q.

If Q_0 is the set of all possible initial states of M, then

IV.16f) F is the set of states of M reachable from Q_0 with the set of inputs of evidencs E $\subset$ S.

Conversely:

IV.16g) E is said to be accepted by M if for any e $\in$ E there exists q $\in$ Q_0 and q' $\in$ F such that

$$q' = \phi (q,e)$$

IV.16h) If the actual state q of M is required to belong either to Q_0 or F, then E is the only set of inputs or evidence accepted by M.

A fuzzy automaton is defined if ϕ is a fuzzy mapping:

$$\phi : Q \times S \longrightarrow [0,1] \quad (IV.16i)$$

In this way, the membership $\mu_F(q')$ of q' = ϕ (q,e) with F measures the degree of confidence on the acceptance of e as belonging to E.

Let it be

$$MPNN = \{ E, N, W, G, \phi, \theta, Z \} \quad (IV.17a)$$

where:

IV.17b) E is the set of inputs (evidence) of the multipurpose neural net MPNN;

IV.17c) N (of cardinality n) is the set of neurons of MPNN, so that N_n is the set of all possible synapses of MPNN;

IV.17c) W is the set of outputs of MPNN;

IV.17d) G is the set of internal Goals or final states of MPNN;

IV.17e) θ is the output function

$$\theta: N \times E \times W \longrightarrow [0,1]$$

defined in Eq. II.14a;

IV.17f) Z is the family of the matrices Z_i of state memberships $\mu(n_i,n_j)$ describing the connectivity of MPNN:

$$\mu(n_i,n_j) = s_i$$

where s_i is the strength of the synapsis between n_i, n_j. Z_i describes the state of connectivity of MPNN at the instant i. In this way, Z_0 is the set of initial connectivities of MPNN, and it encodes the initial knowledge inherited by MPNN.

IV.17g) the next state mapping ϕ modifies the connectivity of MPNN according to the goals G and the inputs E in the time continuum T:

$$\phi: Z \times G \times E \times T \longrightarrow [0,1]$$

so that

$$\mu(n_i,n_j) = \mu(T_R,t) \cdot f(n) \cdot \mu(e,e') \cdot \mu(g,g')$$

and

IV.17h) $\mu(T_R,t)$ measures the forgetting in MPNN:

$$\mu(T_R,t) \begin{cases} \text{tends to 0 if } t \geq T_R \\ \text{tends to 1 otherwise} \end{cases}$$

because T_R is the maximal time interval any information may be retained in MPNN. T_R is related to the dynamics of the synthesis, mobilization and destruction of transmitters

T and receptors R. In this way, $\mu(T_R,t)$ measures the temporal stability of the synapsis;

IV.17i) $f(n)$ measures the consistency of the statistics of MNPN because:

$$f(n) \longrightarrow \begin{cases} \text{tends 1 if } n \longrightarrow \alpha \\ \text{tends to 0 otherwise} \end{cases}$$

If α increases the number of activations necessary to modify the synaptic strength augments;

IV.17j) $\mu(e,e')$ measures the degree of the $t\hat{}r$ binding. It depends on the binding affinity $\mu(t,r)$ between t and r as well as on the amount of transmitter released by the pre-synaptic input $e \in E$ according to the pattern stored in the availabe amount M of transmitters and R of receptors (see Chapter II, section 6)

$$\mu(e,e') \longrightarrow \begin{cases} \text{tends to 1 if } e \equiv e' \\ \text{tends to 0 otherwise} \end{cases}$$

IV.17j) $\mu(g,g')$ measures the degree of performance g of MPNN concerning the goal g':

$$\mu(g,g') \longrightarrow \begin{cases} \text{tends to 1 if } g \equiv g' \\ \text{tends to 0 otherwise} \end{cases}$$

IV.17k) $\cdot$ is a t-norm.

The above properties of MPNN are supported by the physiology of the synapsis discussed in Chapters II and III showing that the synaptic strength is modified by the way it is used (conditions IV.17g,h); by the control exercised by the modulators (condition IV.17i), and by regressive phenomena recovering previous states (condition IV.17g) of the neural connectivity. These regressive phenomena put forgetting as one of the main characteristics of the real neural networks. Also, learning in the case of the real brain implies both strengthening and reducing the connectivity among neurons (Byrne, 1987; Cowan et al, 1984; Goodman et al, 1984; Hyvarinen, 1982; McConnell, 1988; Sidman and Rakic, 1973).

Let $\beta(t)$ be a string of inputs or evidences $e \in E$ distributed in the time interval $T = [t_0,t]$

$$\beta(t): E \times T \longrightarrow [0,1] \qquad (IV.18a)$$

The history from t_0 to t of the activation of MPNN

during T is the set of neuronal flow in this net elicited by $\beta(t)$. Let this be denoted by (Rocha, 1982b):

$$N_0 = \tau \ (t, t0, N_i, Z, \beta(t)), \ t \geq t_0 \quad (IV.18b)$$

where τ stands for the family of neuronal activations from a base or input set of neurons $N_i \subset N$ to another set of output neurons $N_0 \subset N$ trigered by $\beta(t)$. The actual connectivity of MPNN is a function of this history of neuronal activations. Fig. IV.3 shows the history τ of a MPNN induced by $\beta(t)$ defined in the inserted histogram and raster. The strength of the arcs or axons is correlated with the degree of activation of the corresponding neuron. In this way, disconnected neurons are cells which were not stimulated by $\beta(t)$.

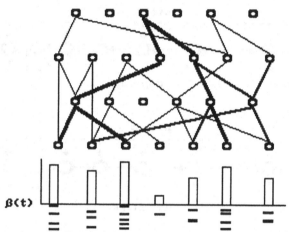

FIG. 4.3 - THE MPNN CONNECTIVITY INDUCED BY $\beta(t)$

A MPNN is considered to reach an equilibrium state if

$$N_E = \tau \ (t, t_E, N_E, Z, \beta(t)), \ t \geq t_E \quad (IV.18c)$$

Let T_E be the largest t_E for MPNN. MPNN reaches an equilibrium state because its connectivity does not change anymore (Fig. IV.4). This means that:

$$Z_E = \tau \ (t, t_E, Z_0, \beta(t)), \ t \geq t_E \quad (IV.18d)$$

either because Z_{E+1} is not a reachable set of states or because MPNN oscillates between Z_{E-1} and Z_E as a consequence of a partial forgetting. Z_E characterizes a stable learning. The actual value of T_E depends on the difficulty of the learning. From IV.13:

$$T_E = f_E \ (h_s(K) - h(W)) \quad (IV.18e)$$

The set Z_0 of the initial connectivity of MPNN is an equilibrium state for (Fig. IV.4) the empty string

$$\beta(t) = constant = \textbf{\$} \quad (IV.18f)$$

The retention time T_R is defined as the largest time interval required to recover Z_0 as an equilibrium state from any other state Z with the empty string $\beta(t) = \textbf{\$}$.

$$Z_0 = \tau \ (t,t+T_R,Z,\beta(t)) \quad (IV.18g)$$

This pathway of state transitions describes forgetting. T_R is a function of the h(MPNN)

$$T_R = f_R \ (h(MPNN)) \quad (IV.18h)$$

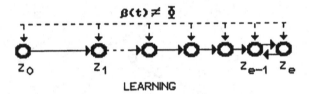

LEARNING

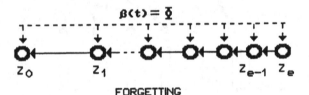

FORGETTING

FIG. IV.4 - THE STATE TRANSITIONS IN LEARNING AND FORGETTING

Let the coupling power cp(MPNN) of MPNN be defined as (Rocha et al, 1980)

$$cp(MPNN) = \overset{P}{V} \ \overset{S}{\Omega} \ \mu(n_i,n_j) \quad (IV.19)$$

the maximum connectivity of the parallel pathways P in MPNN. The strength of the connectivity of each of these pathways is the minimal $\mu(n_i,n_j)$ in these serial chains.

Let MPNN$_0$ be an ERM with no initial knowledge. This implies:

IV.20a) MPNN$_0$ to be fully connected, which means that a pathway always exists between any two neurons $n_i,n_j \in$

MPNN;

IV.20b) the strength $s_0 = 1/k$ for all connections in MPNN; where k (the cardinality of MPNN) is the number of synapses of this machine;

IV.20c) cp(MPNN$_0$) = $1/k$;

IV.20d) there is a homogeneous distribution of the parameters T$_E$, T$_R$ and α in VI.17h among the neurons of MPNN, and

IV.20e) h$_s$(MPNN$_0$) is maximal.

Theorem IV.21 (modified from Rocha, 1982b) - Let MPNN$_0$ be immersed at t$_0$ into a non-homogenous environment U. At time t $\geq$ T$_E$, the coupling power cp(M) for some subnets M of MPNN will approach 1, while cp(N) for the other subnets N of MPNN will approach 0. In other words, well learned σ-models are developed by MPNN to model a world W of U. The number of these models depends on both h$_s$(MPNN) and h(U).

If the environment U is not homogeneous:

V.21a) there exists β(t) so that the distribution of e ϵ E over the time interval [t$_0$,t] is not homogeneous (see Figs. IV.5 and 6), or

V.21b) β(t) is homogeneous and there exists a non-homogeneous conditional distribution β_c(t) of the evidences e ϵ E concerning the classes C supported by the output neurons of MPNN (see Figs. IV.7 and 8). Also

V.21c) β(t) is the world W$_t$ to be modeled by MPNN$_0$, and

$$h_s(MPNN_0) > h(\beta(t))$$

In the case of condition V.21a:

The frequency of some e$_i$ ϵ W$_t$ will be greater than the frequency of some others e$_j$ ϵ W$_t$. Thus, the subsets of very frequent, frequent and unfrequent evidences in β(t) are defined in W$_t$ according to the values of T$_R$ and α.

For those neurons activated by very frequent or unfrequent e$_j$ ϵ W$_t$:

$$\mu(T_R,t) \longrightarrow 0$$

$$f(n) \longrightarrow 0$$

Since $\cdot$ is a t-norm

$$\mu(n_0,n_j) \longrightarrow 0$$

even if

$$\mu(g,g') \text{ ---> } 1 \text{ and } \mu(e_j,e') \text{ ---> } 1$$

the consequence is

$$cp(N) \text{ ---> } 0, \ t \geq T_E$$

for the subnets N composed by these neurons.

At the same time, for those neurons activated by frequent $e_j \in W_t$:

$$\mu(T_R,t) \text{ ---> } 1$$

$$f(n) \text{ ---> } 1$$

$$\mu(e_i,e') \text{ ---> } 1$$

$$\mu(s_0,s_j) \text{ ---> } \mu(g,g')$$

the consequence is

$$cp(M) \text{ ---> } \mu(g,g'), \ t \geq T_E$$

for the subnets M composed by these neurons. Besides, for some of the subnets M it is possible to expect

$$\mu(g,g') \text{ ---> } 1$$

what means $cp(M) \text{ ---> } 1$ for a few subnets M in MPNN. Thus few well learned σ-models are generated by MPNN.

The semantics f of the linguistic qualifier FEW is related to the entropy $h(W_t)$ of W_t and that of $h_s(MPNN)$, because $h_s(M)$ must approach $h(W_t)$:

$$f = h_s(MPNN_0) \ / \ h(W) \geq 1$$

as a consequence of propositions IV.10, 11 and 12.

In the case of condition b:

Provided that T_R and α are adequated to model the world defined by $\beta(t)$, then

$$\mu(T_R,t) \cdot f(n) \text{ ---> } 1$$

but if $\beta_c(t)$ is not homogeneous, then reward

$$\mu(g_j,g_k) \text{ ---> } 1$$

and punishment

$$\mu(g_i,g_k) \text{ ---> } 0$$

may be calculated for those $c_k \in C$ represented at the output neurons of MPNN. Thus, for m models M_j

$$cp(M_j) \text{ ---> } 1$$

while for n models M_i

$$cp(M_i) \text{ ---> } 0$$

Again, the relation betwen m and n is dependent of the relation between h_s(MPNN) and $h(W_t)$.

Q.E.D.

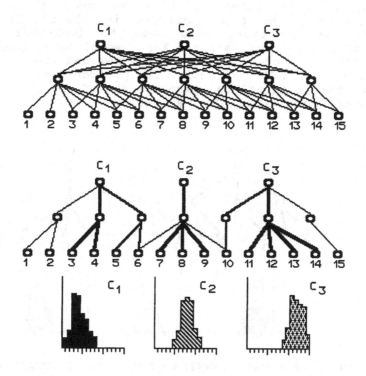

FIG. IV.5 - MODELING OF THE WORLD W_1

Figs. IV.5 to 8 shows simulations of IV.17 obtained for the same MPNN immersed in two different worlds W. Both worlds W_1 and W_2 are composed of 15 different types of evidences supporting 3 different classifications. The statistics of these worlds are showed by the histograms inserted in the figures, and they were obtained by means of controlled simulations using the Data Base Simulator of SMART KARDS(c) (Chapter IX). The main differences between these two worlds are that the distribution of the evidences in W_2 is more homogeneous than in W_1, and the conditional distribution of these evidences according to the classes C_i they must support is more homogeneous in W_1 than in W_2.

The basic structure of this MPNN is a 3 layer net whose initial connectivity is shown in Figs. IV.5a and 7a. This net was obtained using the Neural Net Generator of

Smart Kards, according to the language L(G) defined by a genetic G (Chapter III, section III.9). The differences between the simulations in Figs. IV.5 and 6 in respect to that of Fig. IV.7 are the distinct statistics of the worlds W_1 and W_2. The retention time T_{R2} used to model W_2 is twice the value of T_{R1} employed in the net immersed in W_1. Sigmoid functions were used to implement $\mu(T_R,t)$, $f(n)$ and $\mu(g,g')$. $\mu(e,e')$ was kept equal to 1, meaning that the patterns e of evidences always matched the prototypical knowledge of MPNN perfectly. The product was used as the *-norm. The simulations were performed by the Inductive Module of Smart Kards (Chapter IX).

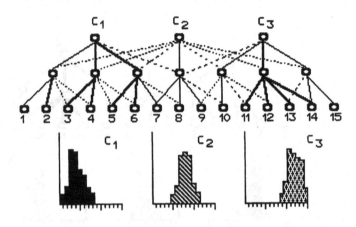

FIG. IV-6 - THE LEARNING OF W_1 AFTER 10 TRAINING CASES

Learning was processed according to the following procedure:

IV.22a) the right connections between e ϵ E and the corresponding C_i were strengthened whereas the all other connections were reduced by forgetting. Thus

$$\mu(g,g_k) \text{ ----> } 1, \ \mu(T_R,t) \text{ ---> } 1$$

for the accepted C_k and correct linkages, and

$$\mu(T_R,t) \text{ ---> } 0$$

for all the other connections in MPNN.

IV.22b) punishment was not used in the simulations presented in Figs. IV.5 to 7, but it was a key learning tool in the case of simulation shown in Fig. IV.8. In this latter case, learning was dependent of the above procedure IV.22a and of punishment

$$\mu(g, g_i) \text{ ---> } 0$$

for those wrong c_i.

The learning of the world W_1 was stabilized with a training set of 20 simulated cases (Fig. IV.5), despite the fact that c_2 was not independently supported by any evidence in W_1. All evidences associated with c_2 were also associated with either c_1 or c_3. Because of this, the learning of c_1 and c_3 was established before that of c_2. Fig. IV.6 shows the evolution of this learning after 10 training cases, when some strong ties were already established between evidences and c_1 and c_3, but not with c_2.

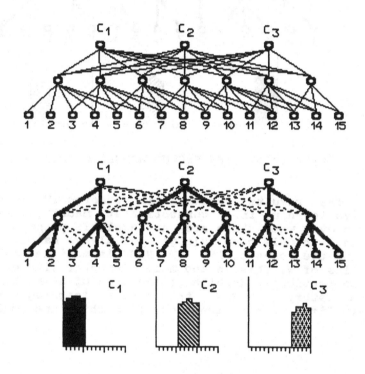

FIG. IV-7 - THE FIRST MODELING OF W_2

The learning of W_2 was much more complicated despite the fact that c_1, c_2 and c_3 were fully separable concerning the supporting evidences (Fig. IV.7). The difficulty in learning was dependent of both the fact that the distribution $\beta(t)$ of the evidences was homogeneous, and from the fact that T_R was enhanced compared to the MPNN modeling W_1. Although strong connections were already stabilized with a training set of 15 cases for C_1, C_2 and C_3, many of the initial connections were retained in the system even after the training session being increased to 40 cases (Fig. IV.7). This is a consequence of the fact that forgetting was maintained low in these simulations. The remaining initial connections were easily removed if punishment was introduced in another simulation of W_2 as shown in Fig. IV.8.

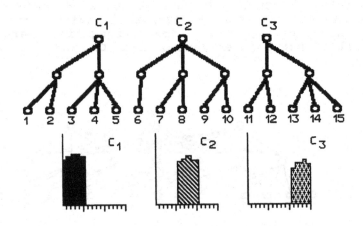

FIG. IV-8 - THE SECOND MODELING OF W_2

The results obtained in theorem IV.21 show that inductive learning in a connectionist machine like MPNN is possible even when there are no explicit purposes or goals declared. In this case, $\mu(g,g')$ is always maintained equal to 1 and the connectivity of MPNN is modified only by the statistics of W. This is the most elementar kind of learning for a connectionist machine of the type of MPNN. At the dawn of the history of a ERM, this type of learning may provide the very first initial knowledge from which the goals can be derived.

IV.5 - The role of memory

The state transitions in the fuzzy automaton defined by IV.17 are dependent of the period of time T_R The system is able to hold any information in its memory

$$\mu(T_R,t) \begin{cases} \to \text{ tends to 0 if } t \geq T_R \\ \to \text{ tends to 1 otherwise} \end{cases}$$
(IV.23a)

and on some repetition of the received information

$$f(n) \begin{cases} \rightarrow \text{tends 1 if } n \longrightarrow \alpha \\ \rightarrow \text{tends to 0 otherwise} \end{cases} \qquad \text{(IV.23b)}$$

This means that unfrequent information tends to be forgotten within T_R units of time after its introduction in the system, while its memorization depends on some relevance threshold α. By adjusting the values of T_R and α, MPNN may avoid to learn undesirable relations.

Conservative systems will operate high values of α, whereas liberal nets will use low threshold values. Besides, α may not necessarily be a fixed threshold, but it may be conceived as a function of the amount of stored or learned data:

$$\alpha = f'(1/h_s(MPNN)) \qquad \text{(IV.23c)}$$

so that at the beginning MPNN may rapidly incorporate some amount of information about U by using low values of α, and then it may eliminate irrelevant data by augmenting the value of this threshold.

Man operates different kinds of memory depending of the amount of time they hold the information: short (T_{RS}), medium (T_{RM}) and long term (T_{RL}) memory.

$$T_{RS} < T_{RM} < T_{RL} \qquad \text{(IV.24a)}$$

Many different physiological mechanisms account for the distinct types of memory implemented upon the same neural circuit:

IV.24b) short term memory (minutes to hours) can be related to modifications of the electrophysiological properties of the membrane as in the case of the sensory response attenuation (see Chapter I, section 5) or to changes in the dynamics of the transmitter release and binding (e.g. Rocha, 1980);

IV.24c) medium term memory can be related to the dynamics of the coupling of modulators at different sites of the neuron, e.g. increasing the Ca permeability at the pre-synaptic terminals (Kandel and Schwartz, 1982). This may augment the release of the transmitter during a time period ranging from hours to days, depending on the dynamics of the modulator, and

IV.24d) long term memory may be related to the modifications of the DNA reading which will result in structural modifications of both the cellular membrane and the synaptic apparatus.

Besides,

IV.24c) the chemical processing supported by L(G) defined in III.14 and 15 controls the dynamics of these memories and provides a mechanism for moving the information from one to another memory.

Proposition IV.25 (Rocha, 1982)- The input frequency necessary to learn and to maintain σ-Models is inversely related to T_R, so that low frequencies are expected to stabilize long term better than short term memories, unless changes in α overcome such effects.

It follows from IV.24a,b and IV.17f.

Q.E.D.

The movement of information among the different memories of a MPNN may be assumed as dependent of the relevance of the information. It may be accepted that the most relevant findings supporting a model or hypothesis M_j are to be moved to the long term memory, while those less significant pieces of information are to be maintained in the short term memory.

Proposition IV.26a - Let α_S, α_M and α_L be the short, medium and long term thresholds of a ERM, respectively. If

$$\alpha_K = f''(T_{RK}), \quad K = \{S, L, M\}$$

then very frequent evidences can be stored in both short, medium and long term memory, frequent events can be stored in both short and medium term memories and less frequent data are stored in short term memory. The semantics of very frequent, frequent and less frequent is dependent of the slope of the function f''.

From IV.24a and the definition of f''

$$\alpha_S < \alpha_M < \alpha_L$$

Therefore, there exists a semantic for the frequency of e in $\beta(t)$ or W_t so that the value of f in IV.17i:

a) for very frequent events $e_V \in E$

$$f_S(n) \dashrightarrow 1, \quad f_M(n) \dashrightarrow 1, \quad f_L(n) \dashrightarrow 1$$

b) for frequent events $e_F \in E$

$$f_S(n) \dashrightarrow 1, \quad f_M(n) \dashrightarrow 1, \quad f_L(n) \dashrightarrow 0$$

c) for less frequent events $e_L \in E$

$$f_S(n) \dashrightarrow 1, \; f_M(n) \dashrightarrow 0, \; f_L(n) \dashrightarrow 0$$

and for all of them

$$\mu(T_R, t) \dashrightarrow 1$$

so that provided

$$\mu(e, e') \dashrightarrow 1 \text{ and } \mu(g, g') \dashrightarrow 1$$

the memorization of $e \in E$ can be done as proposed above.

The results are dependent of the actual values of α defined by f''. Thus, the semantic of the fuzzy probability of the evidences $e \in E$ depends on the slop of f''.

Q.E.D.

Proposition IV.26b - The results of proposition IV.26a can be modified in order to avoid some information to be stored in specified memory.

It follows from $\mu(n_i, n_j)$ being dependent also of $\mu(g, g')$.

Q.E.D.

Let it be

$$0 \leq \Omega_I < \alpha_S < \alpha_M < \alpha_L < \Omega_U \leq 1 \qquad (IV.26b)$$

where Ω_I and Ω_U are, respectively, the lower and upper limits of the variation of these thresholds α.

Proposition IV.26d - If

$$\Omega_U \dashrightarrow 0$$

then

$$h_S(M_j) \dashrightarrow h_S(MPNN)$$

while if

$$\Omega_I \dashrightarrow 1$$

then

$$h_S(M_j) \dashrightarrow 0$$

If $\Omega_U \dashrightarrow 0$ then $\alpha_L \dashrightarrow 0$ guaranteeing a value of α_L which allows any information to be stored at least in the short term memory of MPNN. In this condition:

$$h_S(M_j) \dashrightarrow h_S(MPNN)$$

meaning that MPNN may use all its entropy to model $\beta(t)$.

If $\Omega_i \dashrightarrow 1$ then $\alpha_S \dashrightarrow 1$ and no information about $\beta(t)$ can be stored even in the short term memory.

Proposition IV.26e - If

$$\Omega_k = f'''(h_s(M_j)), \quad k=i,u$$

then MPNN is able to learn the most relevant features of W_j being modeled by M_j.

The proof is a consequence of:

1) if $\Omega_u \dashrightarrow 0$ then $h_s(M_j) \dashrightarrow h_s(MPNN)$ forcing $\Omega_I \dashrightarrow 1$, and

2) if $\Omega_u \dashrightarrow 1$ then $h_s(M_j) \dashrightarrow 0$, forcing $\Omega_i \dashrightarrow 0$.

$$Q.E.D.$$

IV.6 - The labelling of MPNNs

Any MPNN is composed of the following:

IV.27a) input layer: formed by a set N_i of input neurons collecting evidences from the external world or from some other MPNN;

IV.27b) hidden or reasoning layers: formed by a set N_H of neurons whose pre-synaptic terminals come from any other layer of the same MPNN, and

IV.27c) output layers: formed by a set of neurons N_O whose axons send terminal branches to other MPNNs or to some actuators A exerting some action $a \in A$ over U, so that:

IV.27d) the intersection $N_H \cap N_O$ is not necessarily empty

$$N_H \cap N_O <> \emptyset$$

because neurons $n \in N_H$ may send axonic branches to both neurons in N_O and N_H in other MPNN or some actuators in A, and

IV.27f) the set N of neurons of any MPNN is

$$N = N_I \cup N_H \cup N_O$$

In this way:

IV.27e) a set of MPNNs may be hierachically organized into a MPNN circuit or system (NPNS) if neurons of N_H and N_O of some of these MPNNs send axonic branches to N_i of some other nets in the MPNNS;

Let V be a set of fuzzy labels (e.g. temperature, pressure, velocity, ..., or A, B, C, ..., etc.) and δ a function assigning these labels to the neurons N of a MPNN

$$\delta: V \times N \longrightarrow [0,1] \qquad (IV.28a)$$

or

$$v_i = \delta(n_i), \; v_i \in V, \; n_i \in N \qquad (IV.28b)$$

The label v_i assigned to n_i is assumed to represent either the function of this neuron or the concept it represents.

In this context:

IV.28c) a subset V_I of these labels is used to name the set I of instruments (matching functions) associated with the neurons N_I. In this way, a prototypical knowledge K_p is encoded in N_I by means of V_I; the matching functions σ_i defining the measurements to be done by the input layer I, and the W_i (axonic output) semantics of these measures

$$K_p = \{ \; V_I, \; \{\sigma_i\}_I, \; \{W_i\}_I\}$$

The matching function σ_i associated with the neuron n_i is either the encoding function I.18c in the case of the sensory neuron (Chapter I, section I.5) or the matching function II.13d (Chapter II, section II.4) in the case of the input neurons receiving information from other MPNNs. The semantics W_i is provided by the axonic encoding functions I.18e and II.14a. Since different filtering properties can be assigned to the distinct branches of n_i, then knowledge can be encoded in N_i either as fuzzy variables like

Temperature is Fever = Temperature > 37º Celsius

or as linguistic variables like

Fever is (Absent, Low, Moderate, High)

IV.28d) a subset V_R of these labels is assigned to N_H as an auxiliary set of labels to describe the reasoning R in the hidden layers R. In this way

$$R = \{ \; V_R, \; \{\sigma_p\}_H, \; \{W_r\}_H \; \}$$

where σ_p and W_r are supported by II.13c and II.14c, respectively. Examples of labels assigned to the neurons at the reasoning or associative layer are: X is B; Y is C ..., etc., where B, C, etc., are fuzzy sets, or Ask Information About Velocity, about Temperature, etc.

IV.28e) a subset V_0 of these labels is assigned to N_0 to describe the output of the MPNN according to the decision process D

$$D = \{ V_0, \{\sigma_0\}_0, \{W_0\}_0 \}$$

The labels assigned to the output neurons are either fuzzy sets like X is A, Y is B, ..., etc. if MPNN is in charge of a numeric control (e.g. motor control) or are linguistic variables like

Diagnosis X is {Impossible, Probable, Compatible, etc.}
or
Procedure Y is {Forbidden, Recommended, Accepted, etc. }

In this way:

IV.28f) The fuzzy vocabulary V is the set of fuzzy labels associated with N, so that for each $n_n \in N$

$$n_n \text{ is } v_n \in V$$

where $v_n \in V$ is the fuzzy set defined by σ_n, whose semantic is provided by W_n.

Let
$$T = \{t_1, \ldots , t_n\} \qquad \text{(IV.29a)}$$

be a set of terms like small, medium, high, etc;

$$Q = \{q_1, \ldots , q_n\} \qquad \text{(IV.29b)}$$

be a set of quantifiers like very, more or less, less, etc. Let the grammar G

$$G = \{ D, Q, P, \Theta \}$$
generate

IV.29c) the set P of productions of the type dq^i, $d \in D$, $q \in Q$, i is a finite integer, according to

IV.29d) the sintax Θ

$$\Theta: D \times Q \times P \dashrightarrow [0,1]$$

Examples of productions in P are: very small, more or less high, etc.

In this context, the labelling $\pounds$ of MPNN is:

$$\pounds = \{ N, V, P, U, W, S, \delta \} \qquad \text{(IV.30a)}$$
where

IV.30b) N is the set of neurons of MPNN;

IV.30c) V is the set of names or labels (IV.28f) of the variables encoded by the neurons N of MPNN;

IV.30d) P is the set of productions in IV.29;

IV.30e) U is the universe of discourse;

IV.30f) W C U is the base set of the world to be modeled;

IV.30g) the semantic S of L is provided by the filtering properties of the axonic branches of n ϵ N, thus it is supported by eq. II.17c, and

IV.30h) δ as in IV.28:

$$\delta: V \times N ---> [0,1]$$

so that

$$v_i = \delta(n_i), \quad v_i \epsilon V, \quad n_i \epsilon N$$

As the result of the labelling process, MPNN is easily readable as a structured set of fuzzy productions of the type

$$\text{if } (X_i \text{ is } A_i \dots) \text{ then } Y \text{ is } B \qquad (IV.30g)$$

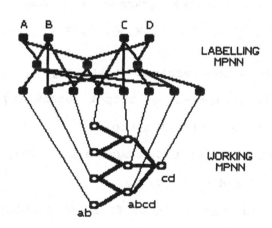

FIG. IV.9 - THE LABELLING OF MPNN

In the case of the real brain, some MPNNs specialize to provide both the grammar and the labelling of the other MPNNs involved in the human reasoning process (see Fig. IV.9). These specialized MPNNs support the human natural languages or some other symbolic language (e.g. visual symbols like in fuzzy graphs), and they constitute a hierachic set of MPNNs. Artificial Evolutive Reasoning Machines must also include this type of language system, because it provides an easy description of the knowledge encoded in the net, and it can support the inquiring required to obtain the input data to be used in the reasoning process R defined in VI.28d.

In the case of classic neural nets, labels are assigned to the input and output layers, but not to the hidden layers. Because of this it is very difficult to get a formal description of the knowledge wired in these layers after the training of the net. This is the reason why these layers are called hidden layers. Also, no natural linguistic label is in general assigned by the expert to the non-terminal nodes of his knowledge graphs (see Chapter V). This keeps the complexity of his jargon as low as possible and explains why knowledge acquisition by means of production rules is a very hard task. However, provided another labelling tool has a high semantic complexity, e.g. fuzzy graphs, the expert easily expresses the knowledge wired in his brain (see Chapter V)

IV.7 - Other properties of the inductive σ-models

Let the fuzzy grammar FG (Negoita and Ralescu, 1975) be:
$$FG = \{ V_n, V_t, V_i, M, Q, \phi, P \} \qquad (VI.31a)$$
where

IV.31b) V_n is a non-terminal vocabulary;

IV.31c) V_t is a terminal or output vocabulary;

IV.31d) V_i is the initial symbol;

IV.31e) M is a set of matchings in the closed interval [0,1]

IV. 31f) $\phi: M \times M \longrightarrow [0,1]$;

IV.31g) P is a finite set of fuzzy productions

$$v_j(m_j) \longrightarrow v_k(\phi(m_j))$$

$$v_j, v_k \in V, \quad V = V_n \cup V_t \cup V_i,$$

$v_k(\phi(m_j))$ representing that the actual production matches v_k with degree $m_k = \phi(m_j) \in [0,1]$. In other words, the productions $p \in P$ are fuzzy relations between terms of V, and v_k is said to be derived from v_j with confidence $m_k = \phi(m_j)$;

IV.31h) If r_j is the relevance of v_j to support the

derivation of vk then

$$r_j \cdot v_j(m_j) \dashrightarrow v_k(\phi(m_j \cdot r_j))$$

where $\cdot$ is a $\top$-norm. In other words, the confidence m_k on v_k is dependent of both r_j and m_j. Finally

IV. 32i) A derivation chain from $v_j \epsilon$ V_i to $v_p \epsilon$ V_t is (Negoita and Ralescu, 1975):

$$r_j \cdot v_j(m_j) \dashrightarrow \ldots \dashrightarrow v_p(\phi(m_{p-1} \cdot r_{p-1}))$$

The fuzzy language L(FG) generated by FG is (Negoita and Ralescu, 1975):

$$L(FG) : (V_t)_s \dashrightarrow [0,1] \qquad (IV.32a)$$

is the set of the strings s_i of size s or less produced by the derivation chains:

$$r_j \cdot v_j(m_j) \dashrightarrow \ldots \dashrightarrow v_p(\phi(m_{p-1} \cdot r_{p-1})) \quad (IV.32b)$$

for all $v_j \epsilon V_i$, so that
$$L(FG)(s_i) = \overset{C}{\max} \ (r_j \cdot m_j \ast \ldots \ast r_{p-1} \cdot m_{p-1}) \ (IV.33c)$$

measures the confidence in s_i, and it is the maximum over the set C of all derivation chains of s_i. $\cdot$ and $\ast$ are, in general, t-norms combining information in each of these derivation chains.

FG may be implemented by an MPNN (Fig. IV.9) since:

$$MPNN = \{ \ V_I, \ V_R, \ V_O, \ W, \ \{\Theta_n\}_n, \ P \ \} \ (IV.33a)$$

IV.33b) $M \equiv W$, where W is the axonic encoding;

IV.33c) $\phi \equiv \{\Theta_n\}_N$, where Θ_n is the encoding function defined in III.14, and

IV.33d) P is the set of productions at the synapsis between the pre-synaptic neurons n_i and post-synaptic neurons n_j, $n_i, n_j \epsilon$ N

$$l_i(w_i) \cdot s_i \dashrightarrow l_j(\Theta_j(w_i \cdot s_i))$$

$$s_i = \mu(n_i, n_j)$$

so that it generates a fuzzy language L(MPNN)

$$L(MPNN) : V_O \times V_O \times .. \dashrightarrow [0,1] \qquad (IV.33e) \text{ and}$$

$$L(MPNN)(c_i) = \max \ (\ s_1 \cdot w_1 \cap \ldots \cap s_t \cdot w_t \ ) \ (IV.31c)$$

• and ∩ are, in general, the t-norms over all derivations from $s_i \in V_I$ to $c_i \in V_0$.

L(MPNN) is used to describe the behavior of MPNN nets and it represents a level of description higher that the labelling £ described in the previous section. This is one of the roles played by natural language. L(MPNN) uses the labels provided by £ as its input vocabulary, but its syntax is used to relate the internal description of a given MPNN with an external representation (Negoita and Ralescu, 1975) of this very same MPNN. This abstraction is crucial to make possible the external observation of the MPNN.

Theorem IV.32 (Rocha, 1982b) - Encoding in well- learned σ-models M_j tends to be context-free.

Let $MPNN_j$ be the neural net supporting M_j. Since

$$\delta(M_j) \text{ ---> } 1$$

then

$$cp(MPNN_j) \text{ ---> } 1$$

and

$$\mu(n_i, n_j) \text{ ---> } 1 \ , \ n_i, n_j \in M_j$$

As a consequence of VI.33d

$$L(M_j)(c_i) = (\ w_1 \cap \ldots \cap \ w_t \ )$$

independently of the previous history of use of M_j.

Q.E.D.

Theorem IV.33 - Well learned σ-models tend to be trees or cyclic graphs composed by a small number of nodes.

σ-models are recursive because their output may be direct to the external world W they are modeling. As a consequence:

IV.33b) cyclic pathways are allowed between the output and input neurons of the MPNN supporting these models, provided that the statistics of the new W' is supported by MPNN, otherwise

IV.33c) the strength of these recursive connections is reduced to maintain the statistics of W supported by MPNN. These circuits tend to be trees.

The small number of neurons in such circuits is a consequence from theorem IV.21.

Q.E.D.

Theorem IV.34 - The productions uv_ix_iz of the well

learned σ-models are:

IV.34a) dynamic or phasic productions, for which i has a finite and small limit, and

IV.34b) static or tonic productions, for which vx are elements periodically repeated with a frequency dependent of the retention time T_R of M_j.

It follows from theorems IV.32 and IV.33.

Q.E.D.

Theorem IV.35 - The processing time at well-learned phasic σ-models is small.

It follows from theorem IV.33 and IV.34.

Q.E.D.

The functional entropy of the neuron n_i is:

$$h(C) = \overset{C}{\Sigma} h(c_i) \qquad (IV.35a)$$

where c_i is a basin in the phase space of the neuron n_i (see Chapter I, section I.4). The functional entropy $h_f(W_i)$ of the output code W_i of n_i is (Rocha, 1985):

$$h_f(W_i) = \overset{W}{\Sigma} p(w_i) h(c_i) \qquad (IV.35b)$$

where $p(w_i)$ is the probality of occurence of $w_i \in W$. If M_j models the universe U, then $p(w_i)$ at the terminal or sensory neurons is dependent of the probability of $p(u_j)$, $u_j \in U$ encoded by w_j:

$$p(w_i) = \Sigma\, p(u_j) * \mu(u_j, w_i) \qquad (IV.35c)$$

where $\mu(u_j, w_i)$ is the possibility that u_j is encoded by w_i, provided that:

$$\overset{}{\underset{U}{V}}\, \mu(u_j, w_i) = 1 \qquad (IV.35d)$$

where V is the maximum. In the case of non-terminal neurons,

$U = \{ W_{i-k} \mid n_{i-k}$ is a pre-synaptic neuron to $n_i \}$

(IV.35g)

$$\mu(u_j, w_i) = \mu(n_j, n_i), \; j < i \qquad (IV.35h)$$

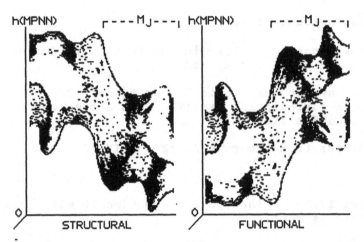

h(MPNN)   ⌐- - -M_J- - -¬   h(MPNN)   ⌐- - -M_J- - -¬

STRUCTURAL   FUNCTIONAL

FIG. IV.10 - THE ENTROPY TRADEOF

Theorem IV.36 - The learning of well-learned σ-models M_j decreases $h_s(M_j)$ and increases $h_f(M_j)$.

From theorem IV.35 and the above definitions of $h_f(M_j)$. It may be said that there is a trade-off of entropies during the learning of M_j.

$$Q.E.D.$$

The meaning of theorem IV.36 is that the descriptive complexity of σ-M_j is low, and its functional (semantic) complexity is high (Fig IV.10). This result agrees with the proposition of Lofgren, 1977, about the complexity of systems, and it is obtained because the output W_i of the neurons of M_j is shared by a small number of pre-synaptic neurons. Because $h_s(M_j)$ decreases, the description of M_j

$$L(M_j)(c_i) = (\ W_1 \cap \ldots \ldots \cap\ W_t\ )$$

becomes simple. But at the same time, the entire set W_i of outputs is used to measure the matching between the world being modeled by M_j and M_j itself. Thus, the functional complexity increases.

IV.8 - Inductive and deductive learning

The representation of the knowledge about the observed world $W \subset U$ in a MPNN is a process founded on the recognition and interpretation of messages channeled by the sensory neurons (Rocha, 1982a). Reality R is, therefore, a model of W expressing the relations between events $e \in E$

sensed by the set I of these neurons. These relations are learned as associations between the neurons N of MPNN. This knowledge is used to handle the actuators A under control of MPNN. This permits the MPNN to manipulate the universe U according to R in order to generate its own actual world W_R perhaps real, perhaps imaginary. This process is a partially closed process (Rocha, 1982a), what means that part of the evidences (E_R) composing E may be generated by MPNN itself and part of them (E_O) is provided by the complement $\Gamma_U W_R$ of the reality created by the MPNN:

$$E = E_R \cap E_O, \quad E_R \subset W_R, \quad E_O \subset \Gamma_U W_R \quad (IV.37b)$$

In other words, E is the union of E_R and E_O.

The partial closure of IV.37 makes MPNN self-observable and supports a controlled observation $\Theta(M_i)$ of U according to the knowledge M_i:

$$\Theta(M_i): R \times A \longrightarrow W_i \quad (IV.38a)$$

$$\Theta(M_i): W_{Ri} \times I \longrightarrow E_i \quad (IV.38b)$$

$$\Theta(M_i): E_{Ri} \times N \longrightarrow M_j \quad (IV.38c)$$

$\Theta(M_i)$ focus the attention of MPNN over U according to M_i.

$\Theta(M_i)$ allows MPNN to modify the observable world W by:

IV.39a) manipulating the surrounding universe according to its set of actuators A and the knowledge M_i;

IV.39b) modifying the set I of instruments used to sense this universe. This may be obtained by adjusting the prototypical knowledge encoded by I or even by creating new instruments (matching functions) as the result of a manipulation over both the MPNN and/or U, and

IV.39c) looking to data on U other than those evidences E supporting W.

In this way:

IV.39d) The world $W_i \subset U$ modeled by M_i is said to be opened under the point of view $\Theta(M_i)$ if IV.38 holds to modify W_i itself, otherwise it is said to be a closed world.

IV.39e) The entire observable world W is

$$W = \bigcup_{j=1}^{n} W_j$$

the union of all observed worlds W_i of U. The observable world W may contain both closed and opened worlds W_j.

IV.39f) The observable world W may be expanded either because some opened W_i may be modified, or because $\Theta(M_i)$ may create new different observable closed or opened worlds W_j.

IV.39g) The models M_j are developed by MPNN according to a set G of goals orienting the process of observation of U:

$$\Theta(G) : U \times R \longrightarrow M_j$$

As a consequence, W is organized according to these goals.

The performance of M_i is measured by the degree $\mu(M_i,G)$ it matches the goals of MPNN. The degree of matching $\mu(M_j,M_i)$ between the new observation M_j and the previous knowledge M_i may be calculated as a function of how much these models fulfill the goals of MPNN:

$$\mu(M_j,M_i) \quad \begin{cases} \longrightarrow \text{ tends to 1 if } \mu(M_j,G) \geq \mu(M_i,G) \\ \longrightarrow \text{ tends to 0 otherwise} \qquad \text{(IV.40a)} \end{cases}$$

The matching between models is not necessarily reflexive because it is not a measure of equivalence between models, but a relative measure of the performance of these models. Models are equivalent if

$$\mu(M_j,M_i) = \mu(M_i,M_j) \qquad \text{(IV.40b)}$$

Let M_i and M_j be as in IV.38. Whenever

$$\mu(M_j,G) \longrightarrow .5 \qquad \text{(IV.41a)}$$

questions can be raised about the choice of M_i to model W_i:

IV.41b) if an evidence $e \in E$ supports a model M_i: if the removal of $e \in E_{Ri}$ reduces $\mu(M_i,G)$; or

IV.41c) a new model M_j may be generated from M_i: if the modification of E_{Ri} results in

$$\mu(M_j,G) > \mu(M_i,G), \text{ or}$$

IV.41d) a new model M_j having a performance better than M_i may be generated from model $M_k \in \Gamma_{MPNN} M_i$: if IV.41c does not hold and the modification of E_{Ri} results in

$$\mu(M_j,G) > \mu(M_i,G), \ \mu(M_j,G) > \mu(M_k,G)$$

On this condition, a new model M_j can be derived from the complementary knowledge Γ_{MPNN} M_i of M_i.

Let $MPNN_i$ and $MPNN_j$ be the neural nets supporting M_i and M_j, respectively. If

$$W_i \cap W_j = W_c <> \emptyset \qquad (IV.42a)$$

from the reasoning in IV.41, it is possible to evaluate whether a model M_i of $MPNN_i$ can be inserted by this net into $MPNN_j$. In other words, if two MPNNs share some common model or knowledge W_c, then the exchange of the not-shared models is possible.

Here, IV.42a characterizes an instructional process between $MPNN_i$ and $MPNN_j$., with W_c being called the culture shared by $MPNN_i$ and $MPNN_j$. An example of this inheritance is the process of "learning by being told" used by JARGON (Chapter VIII) to learn about natural language syntax. The existence of a common labelling language L(MPNN) for both $MPNN_i$ and $MPNN_j$

$$L(MPNN) \subset W_c \qquad (IV.42b)$$

permits $MPNN_i$ to insert its models M_i into W_c, and $MPNN_j$ to inherit the models of W_c. This is because L(MPNN) guarantees the noteworthiness of M_i. In this way, L(MPNN) is a very important cognitive tool. This is one of the roles played by human natural languages. Also, this is the role Fuzzy Logic may play in the case of MPNNs.

Let the following properties of a model M_i (Rocha, 1982a) be considered:

IV.43a) plasticity: the capacity $p(M_i)$ of M_i to generate other models M_j to expand the observable world W:

$$p(M_i) = (W_i \cap W_j)/ W_i$$

This means that the plasticity of M_i is related to how much of W_i is preserved in W_j supported by the new models M_j. Plasticity provides the substrate for knowledge evolution.

IV.43b) autonomy or self-reproduction: the capacity $a(M_i)$ of M_i to be maintained or recreated as a similar model M_j in $MPNN_i$:

$$a(M_i) = (W_i \cap W_j)/ W_j$$

Autonomy provides the substrate for comprehension.

IV.43c) mobility or exogenous reproduction: the capacity m(Mi) of Mi to move from MPNNi to other MPNNs. Thus

$$m(M_i) = (W_i \cap W_c)/ W_c$$

where W_c is the common culture (models) shared by a population of MPNNs. Mobility provides the substrate for communication. The mobility of Mi increases as more of Mi is shared by W_c. In other words, the mobility of Mi depends on how much Mi is close to the culture W_c shared by different MPNNs.

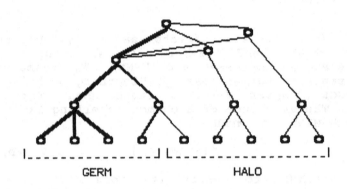

GERM   HALO

FIG. IV.11 - GERM AND HALO

Plasticity and autonomy may initially be considered antagonistic properties because strong autonomy implies

$$a(M_i) ---> 1 \qquad (IV.44a)$$

$$\mu(M_j,M_i) ---> 1 \qquad (IV.44b)$$

whereas the enhancement of plasticity requires

$$p(M_i) ---> 1 \qquad (IV.44c)$$

$$\mu(M_j,M_i) ---) 0 \qquad (IV.44d)$$

However:

Proposition IV.45 - Well Learned σ-Models Mi may enjoy strong autonomy and high plasticity.

Let Mi be composed (Fig. IV.11):

IV.45b) a subnet of strongly connected neurons called the germ $g(M_i)$ of M_i, and

IV.45c) a set of subnets of not so strongly connected neurons; each of these subnets connected to $g(M_i)$. These subnets will be called the halo $h(M_i)$ of M_i.

In this case, the germ provides a strong autonomy for M_i while the halo supports the high plasticity of M_i. This means that M_i has a very strong nucleus of knowledge which may be associated with other different pieces of knowledge to support a set of (new) related ideas.

Well Learned σ-Models M_i have

$$cp(M_i) \dashrightarrow 1$$

what means that they can be composed by a nucleus of well connected neurons compared with some other neural chains of lower connectivity, since

$$cp(M_i) = \overset{P}{V} \overset{S}{\Omega} \mu(n_i, n_j)$$

is a max-min operation over the connections of M_i. Thus, well-learned σ-models may enjoy high autonomy and plasticity.

Q.E.D..

Proposition IV.46 - Given a set of MPNNs sharing a culture W_c of well-formed σ-models, the mobility of these models among the nets is guaranteed.

It follows from IV.42 and proposition IV.45.

Q.E.D.

Theorem IV.47a - If the observable world W_i of U is not a closed world, then there exists a σ-M_j modeling of W_i with strong autonomy and high plasticity.

If W_i is not closed, there exists

$$\Theta(M_i): R \times A \dashrightarrow W_i$$

$$\Theta(M_i): W_i \times I \dashrightarrow E_i$$

$$\Theta(M_i): E_i \times N \dashrightarrow M_j$$

so that

$$W_i \subset W_j \text{ and } M_j = M_i \cup M_h \text{ where}$$
$$cp(M_h) < cp(M_i)$$

Thus M_j contains a germ M_i and a halo M_h. Plasticity ensures the opening of Wᵢ because it guarantees the knowledge evolution.

Theorem IV.47b - If the observable world W_j of U is a closed world under a set of goals G, the σ-modeling of W_j results in models M_j enjoying strong autonomy and low plasticity.

This is a consequence from the fact that $cp(M_j)$ is a max-min operation, and that the continuous observation of a closed world W_j may result into a complete disconnection of the poor linked neurons of M_j.

Q.E.D.

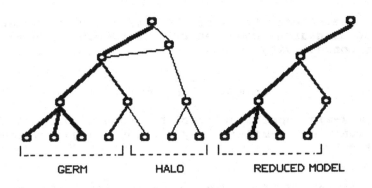

GERM   HALO   REDUCED MODEL

FIG. IV.12 - THE REDUCED MODEL

Theorem IV.48 - Whenever

$$\mu(M_i,G) \;\text{---}\!\!> .5$$

there is a procedure to evaluate the possibility of a new model M_j to be developed from M_i or from M_c belonging to the complementar space $\Gamma_{MPNN} M_i$ of M_i. Well learned σ-models provide good germs for developing this new M_j.

Let it be

IV.48b) the reduced model $\Phi(M_i)$ of M_i (Fig. IV.12):

$$\Phi_k(M_i) = \{ \; n_i, n_j \in M \mid \mu(n_i,n_j) > \alpha_k \; \}$$

$$\Phi_k(M_i) \quad C \quad \Phi_{k+1}(M_i) \text{ provided}$$
$$\alpha_{k+1} > \alpha_k$$

The reduced model is, therefore, a p-level set of M_i

(Negoita and Ralescu, 1975; Rocha, 1982a,b). The reduced model $\Phi_k(M_i)$ is obtained by pruning on the level k the MPNN supporting M_i. This means that all neurons connected to this net with a synaptic strength less than α_k are disconnected from the MPNN. The germ M_i is a reduced model of its MPNN for which $\alpha_k \longrightarrow 1$.

Now if there exists

IV.48c) $g(M_i) = \lim_{\alpha \to 1} \Phi_k(M_i)$ so that $\mu(g(M_i),G) \longrightarrow 1$

then it is possible to try to use M_i' as the germ for the development of a new M_j to support a changing W_i or a set of changing goals, otherwise it is necessary to search for

IV.48d) M_c in Γ_{MPNN} M_i for which $\mu(M_c,G) > .5$, and

$g(M_c) = \lim_{\alpha \to 1} \Phi_k(M_c)$ such that $\mu(g(M_c),G) \longrightarrow 1$

to be used as the germ for the development of the new M_j.

The meaning of IV.48c,d is that a new model may be derived from the knowledge about M_i if M_i has a strong germ or if it points to a germ in its complementary space. The new model M_j is obtained by joining haloes M_h to the germ $g(M_i)$ and/or $g(M_c)$. In this case, $M_h = h(M_j)$.

Whenever IV.48c and d do not hold, then no deductive learning can be done from M_i. In this case, inductive learning may be tried in order to model the surrounding environment.

Well learned σ-models M_i supply good germs for the development of new models M_j because in this case there always exists

IV.48e) $g(M_i) = \lim_{\alpha \to 1} \Phi_k(M_i)$ such that $\mu(g(M_i),G) \longrightarrow 1$

$$Q.E.D.$$

The condition
$$\mu(M_i,G) \longrightarrow .5$$

may result either:

IV.49a) from a changing observable world W, or

IV.49b) by modification of the set G of goals of the MPNN.

A changing world W may be the result of the action of

IV.49c) this very same MPNN or

IV.49d) some other system.

Therefore:

IV.481) the change of $\mu(M_i,G)$ is related to $\Theta(M_i)$, at least in the cases of IV.49b and IV.49c,d.

      If M_j may be obtained from M_i according to IV.48, then M_j is said to be deductible from M_i. This will be denoted here by

$$g(M_i) \;\vdash\; M_j \qquad (IV.50a)$$

Whenever the deduction of M_j is induced by $\Theta(M_i)$

$$g(M_i) \;\underset{\Theta(M_i)}{\vdash\!\!\!\!\!\!-\!\!\!-\!\!\!-}\; M_j \qquad (IV.50b)$$

M_j is said to be deductible from M_i under the theory $\Theta(M_i)$. The word theory is used here in its broad sense, meaning either formal or informal, scientific or religious theories, etc.

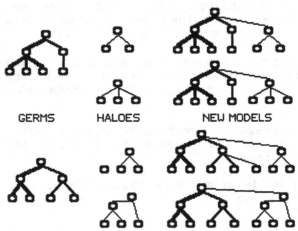

FIG. IV.13 - GENERATING NEW MODELS

      Different models may be deductible both from the same germ by addition of different haloes and by the utilization of different germs to try to model the new observable world W (Fig.IV 13). This approach provides the variability (IV.8a) required by the Evolutive Learning, and modifies the knowledge of MPNN.

The new models M_j will be used to new observations of U, such that the connectivity of the subnets N_j supporting them will be changed according to eqs. IV.17. This will modify the values of $\mu(M_j,G)$, so that (Fig. IV.14):

IV.50c) if $\mu(M_j,G)$ ---> 1 then M_j may be accepted in MPNN and inserted into W_{Rc}, or

IV.50d) if $\mu(M_j,G)$ ---> 0 then M_j is eliminated from MPNN.

This provides the selective process (IV.8b) required by the Evolutive Learning and consolidates the knowledge of MPNN.

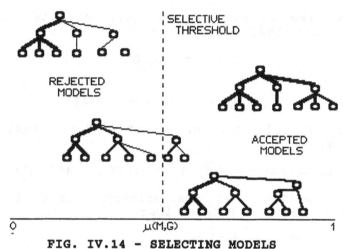

FIG. IV.14 - SELECTING MODELS

IV.9 - Evolution of learning

Let the Evolutive Reasoning Machine ERM be the following fuzzy automaton:

$$ERM = \{ W, G, \{ MPNN \}_{NS}, M_0, M_t(M_0), \beta_a, \beta_r \} \quad (IV.51a)$$

where

IV.51b) W is the observable world of ERM;

IV.51c) G is the set of goals of MPNN;

IV.51d) { MPNN }$_{NS}$ is the family of MPNNs supporting ERM in the natural or artificial nervous system NS;

IV.51e) Mo is the initial knowledge or germs of ERM:

$$Mo = \{\text{\$}_k(M_i)\}_{NS}$$

IV.51f) $M_t(Mo)$ is the actual knowledge of ERM at the time t, so that given $M_i \in Mo$ and $M_j \in M_t(Mo)$ then

$$M_i \vdash\!\!\!\underset{\Theta(Mo)}{\rule{3cm}{0pt}}\!\!\! M_j$$

M_j is deductible from M_i in the theory $\Theta(Mo)$, according to the procedures of theorems IV.21 and IV.48;

IV.51g) $\qquad \beta_a\colon M_t(Mo) \times G \dashrightarrow [0,1]$

measures the acceptance of M_j by its capacity of supporting G. Thus $\beta_a(M_j)$ is

$$\beta_a(M_j) = \mu(M_j, G)$$

IV.51h) $\qquad \beta_r\colon M_t(Mo) \times Mo \dashrightarrow [0,1]$

measures the rejection of M_j by its membership with Mo. Let it be

$$\beta_r = 1 - [\; 0 \; V \; (\mu(M_j,G) - \mu(Mo,G) / (1 - \mu(Mo,G)]$$

In this condition, if the performance of M_j is not better than Mo its rejection is maximum.

In this condition:

IV.52a) M_j is accepted as a deduction from Mo if

$$\beta_a(M_j) > \beta_r(M_j)$$

IV.52b) otherwise, M_j is rejected as a deduction from Mo.

Theorem IV.53 - At the moment t_0 of the genesis of ERM

$$0 < \mu(Mo,G) < 1.$$

If $\mu(Mo,G) = 0$ then there is no possibility to apply the procedure of theorem IV.48 because no germ is provided. On the contrary, if $\mu(Mo,G) = 1$ all new models M_j will be rejected.

$$\text{Q.E.D.}$$

Theorem IV.54 - Whenever

$$\mu(M_t(M_0),G) \longrightarrow 1$$

the observable world W tends to become closed under the theory $\theta(M_0)$.

If
$$\mu(M_t(M_0),G) = 1$$

it is impossible to promote the modification required by IV.39.

Q.E.D.

The key issues of the evolutive learning supported by the above ERM are:

IV.54a) to characterize its initial knowledge M_0 or germs: as discussed before, part of this initial knowledge is genetically inherited by the topologies of { MPNN }$_{NS}$, and part of then is socially acquired from W_c by means of formal and informal instructive processes;

IV.54b) to obtain the haloes to create offspring of models from these germs, and

IV.54c) to select the best elements in these offsprings according to the goals of ERM.

The evolutive learning paradigm may also be used to modify the genetic G of LG according to the requirements of W. This enhances the learning capabilities of the MPNN.

The basic requirements of IV.55c were discussed in section IV.4 since natural selection is mainly provided by inductive learning, although deduction may also be used to select models. The requirements of IV.55a were analyzed in section IV.8. The next section devotes its attention to the requirements of IV.55b)

IV.10 - Creativity

The selection of neurons or pathways to compose the halo of new models is a central point in Evolutive Learning. Many mechanisms account for this selection. They may be ordered by the amount of knowledge required to select the candidates. At one extreme is the random choice of a set of possible circuits, at the other extreme is the formal process of logically deducing the best elements from a well structured theory. In the middle is the associative knowledge supported by the modulator learning paradigm introduced in Chapters III, section III.4.

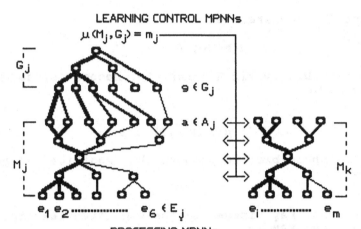

LEARNING CONTROL MPNNs

PROCESSING MPNNs

FIG. IV.15 - THE MODULATOR PARADIGM

The modulator paradigm of learning control MLC implies (Fig. IV.15):

IV.55a) the output A_j of the model M_j is to be compared with the goal G_j of the ERM machine, represented by the net MPNNc controlling the learning of the ERM;

IV.55b) the degree $\mu(M_j,G_j)$ of this matching determines the amount of modulator m_j to be released upon a area of processing nets MPNNp under control of MPNNc, so that

IV.55c) $\beta_a(M_j)$ determines the amount of reinforcement provided by m_j;

IV.55d) $\beta_r(M_j)$ determines the amount of punishment provided by m_j, and

IV.55e) The modulator m_j acts upon those MPNNp activated within the period of time $t \pm T_R$, where T_R is the retention time in these MPNNp and t is the moment the corresponding MPNNj is activated. This is because chemical signals are retained in those MPNNp activated at

$$t_0 < t - T_R$$

such that they may be used to generate the haloes to be combined with $g(M_j)$. In the same way, chemicals are retained by $g(M_j)$, such that also those MPNNp activated at

$$t_p > t + T_R$$

may be used with the same purpose.

V.55f) As a consequence of the above, MLC favours modifications of the links between different models M_i stored in the $MPNN_p$ under the control of the same $MPNN_c$ and activated during the period of time t ± T_R (Fig. IV.15). In other words, it supports associative learning.

Let $g(M_j)$ be the set of germs to be used to generate a set O_{Mj} of new models M_k. O_{Mj} is said to be the offspring of M_j. The success of obtaining surviving models M_s in O_{Mj} is

IV.56a) inversely related to the amount of knowledge used in the selection of the haloes, and

IV.56b) directly dependent of the number of putative generated new models in O_{Mi}.

In this way:

IV.56c) Random selection requires the creation of a large O_{Mj} in order to enhance the probability of obtaining a final successful modeling of U. This process is very similar to that proposed in Genetic Algorithm as the source of variability of the new strings (Booker et al, 1989). Like in Genetic Algorithm, heuristic strategies may reduce the size of the offspring and may enhance the probability of its success. But this is to increase the knowledge used in the selection.

IV.56d) Strong theories may support successful decisions in creating a very selective and successful offspring O_{Mj} because previous knowledge can be used to analyze the most adequate and the most inadequate haloes to be used.

IV.56e) Associative learning supported by the modulator paradigm will favor those circuits or neurons activated during a period of time

$$T_k = t \pm T_R$$

to compose the halo $h(M_k)$ and the germ $g(M_k)$ of M_k to create O_{Mj} at the moment t. The modulator paradigm is a very potent tool for discovering temporal relations in U. Besides, the size of $h(M_i)$ tends to be intermediate between the cardinality of the haloes generated by the random and deductive processes.

The other point to analyze in the generation of an offspring of models is the motivation to create new models M_j to try to understand the observable world W. There are two main sources for this motivation:

IV.57a) disagreement between evidences $e \in E$ and the

expected   behavior   of   W$_j$ supported   by   M$_j$,   or

IV.57b) the modification of the set G of goals of a MPNN.

Disagreement results because some of the observed
evidences e ϵ E are a source of error for M$_i$. These
evidences:

IV.58a) must be linked to the germs g(M$_j$) in order to be
incorporated into the new model M$_k$; but they

IV.58b) may have activated other models M$_L$, too. These
models M$_L$ can provide a better support to create the
required new knowledge M$_k$. The reduced models $\#$(M$_L$) must
compose g(M$_L$).

As a consequence:

IV.58c) disagreement automatically defines the haloes and
the germs to be used to compose the offspring O(M$_j$) of the
model M$_j$.

The modification of G is obtained by altering the
correspondent MPNN$_p$ by the procedure described in IV.55.
Naturally, the modification of G results in disagreement for
some MPNN$_p$ and induces some learning of the type discussed
above. Here, the interesting point to question is the
motivation to change G, which may be:

IV.59a) no successful offspring of models are obtainable
with the actual set G to accomodate disagreements observed
in W concerning the knowledge of MPNN, or

IV.59c) to avoid the closure of the actual observable world
W under the actual theory Θ(G). In this condition, the set
of germs provided by M$_t$(M$_0$) will provide the initial
knowledge for the new ERM defined by this new set G' of
goals. The present book is an example of this approach. The
goals of neural nets theory were changed to open this theory
to symbolic reasoning.

IV.11 - An example for the use of ERM

Let an example illustrate what was discussed so far.
Let this example show how to use an ERM to solve a problem
defined in U.

Scheduling nursing personnel in hospitals is a very
complex task due to a variety of conflicting interests or
objectives between hospitals and nurses (Ozkarahan and
Bailey, 1988). This process may be viewed as one generating

a configuration of nurse schedules that specify the number
of identities of the nurses working each day of the
scheduled period (Miller et al., 1976). The purpose of the
process is to generate patterns of scheduled days on and
days off which minimize the institutional costs c_i and the
personnel dissatisfaction d_i (Miller et al., 1976; Musa
and Saxena, 1984; Ozkarahan and Bailey, 1988). In other
words, the major goal is to minimize the objective function
measuring the costs c_i and dissatisfaction d_i:

$$\min (z) = \Sigma\, c_i + \Sigma\, d_i \quad (IV.60)$$

subject to institutional and individual constraints.

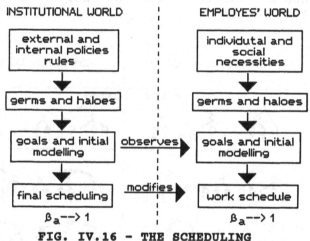

FIG. IV.16 - THE SCHEDULING

The problem may be formalized as (Fig. IV.16):

IV.60a) there are two worlds to be modeled: the employee's
W_E and the institutional W_I worlds;

IV.60b) labor policies inside and outside the institution
and the required labor tasks result in the constraints and
define the goals to model in W_I;

IV.60c) individual and social demands are the constraints in
W_E to define the goals and models in this world;

IV.60d) different constraints are more likely to have
different relevances in the solution of the problem, so that
it is possible to speak og strong (germs) or weak (haloes)
constraints in both worlds;

IV.60e) the scheduler, here the ERM scheduler or sERM, uses
the goals and the initial modeling provided by the

institutional world in order to observe W_E, defined here as the family of individual goals and desired schedules;

IV.60f) this interaction must result into a final schedule S_f which tries to maximize the interests of both worlds

$$\beta_a(S_j) \text{ ---> } 1, \ S_j \in W_I \cup W_E$$

subject to the constraints imposed by W_i and W_E;

IV.60g) this final schedule S_f modifies W_E and triggers individual responses which may be taken as the main motif for natural selection of these S_f, and finally

IV.60f) the successful S_fs may be used to refine the germs and haloes of the sERM.

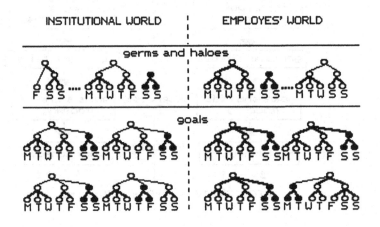

FIG. V.17 - THE GOALS

Let (Fig. IV.17):

IV.61a) $g(M)_I$ and $h(M)_I$ the set of germs and haloes in the W_I, and

IV.61b) $g(M)_E$ and $h(M)_E$ the set of germs and haloes in the W_E;

be composed by input neurons representing the week days of days on (unfilled neurons) and off (filled neurons in the Fig. IV.17) duty.

These germs and haloes correspond:

IV.61c) to the constraints imposed by the internal and

external labor policies in W_I: specifying e.g. the patterns of work stretches, weekend policies, etc. (Miller et.al., 1976; Musa and Saxena, 1984; Ozkarahan and Bailey, 1988). For instance, germs and haloes of adequate size must be produced to accomplish both constraints about the maximal and minimal stretches of working days on and off. Also, they must support the work load required in the different days of the week, so that germs $g(M)_B$ and haloes $h(M)_B$ combining working days on in the beginning must be produced in a number higher than those $g(M)_W$ and $h(M)_W$ composed by days on at the end of the week, since the activities in hospitals tend to be reduced at the weekends.

IV.61d) to the social and individual necessities in W_E: specifying the individual preferences of patterns of working days on and off duty.

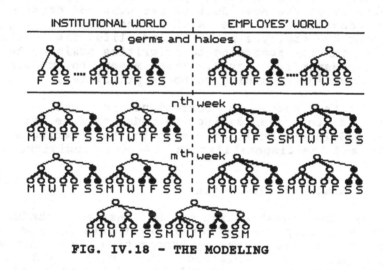

FIG. IV.18 - THE MODELING

In this way:

IV.61e) the strength of the arcs (synapsis) of these germs and haloes is determined by the relevances of the constraints in W_I and W_E. For example, the strength of the connections of $g(M)_W$ and $h(M)_W$ must be related to the relevance given to adequate the work load with the available staff, and

IV.61f) strong incompabilities between constraints such as the length of the consecutive days on duty and weekend policies, or days off policy, etc., may be encoded by means of different transmitters and receptors allocated to the correspondent incompatible germs and haloes. For example, if the work stretch is restrained to 5 consecutive days on and 2 consecutive days off, no clustering of germs and haloes of size 3 or greater must be allowed by providing them with incompatible transmitters and receptors, while high t^r affinity must be used to favor the combination among germs and haloes of size 3 and 2. In the same line of reasoning, the t^r coupling must be programmed to provide the best combinatorial composition of days on and days off.

Both $g(M)$ and $h(M)$ are used to build the goals and models of ERM concerning both worlds W_I and W_E:

IV.61g) the goals G correspond to those models which are optimal solutions to minimize each of the constraints defined in the corresponding worlds (Fig. IV.17). Thus, G_I in W_I is composed by those models optimizing the labor force concerning the labor policies such as the length of the work stretch, weekend policies, etc., whereas G_E is the family of the most desired patterns of days on and off duty according to the social and familiar requirements defining the employee's style of life. The goals $g_i \in G$ may be ranked according to priorities assigned to them by the corresponding world. This means to modulate their connectivities according to these priorities,

IV.61f) the models M correspond to set of schedules generated to cover defined periods of time (Fig. IV.18). These models are built according to the covering necessities in W_I and the disponibilities of the individuals in W_E.

Now, let sERM begin:

IV.62a) the construction of the desired schedule S_d in W_I as a family of patterns p_i of days on and off duty according to the covering necessities. Each pattern is an individual graph obtained from $g(M)_I$ and $h(M)_I$ to describe the days on and off duty of the required number of employees:

$$S_d = \{ p_i \mid p_i = \overset{D}{U} M_i , M_i \in g(m)_I \ U \ h(M)_I \}_C$$

where D is the period of time to be covered and C is the required number of employees. Different S_ds may be obtained for min and max values of C, respectively.

At the same time, W_E provides

IV.62b) the construction of the desired patterns p_j of each employee j of W_E.

In the next step:

IV.62c) sERM observes W_E according to $\Theta(S_d)$, in order to rank the patterns $p_j \in W_E$ in respect to their matchings with $p_i \in S_d$. These matchings are calculated by the degree of satisfaction they provide to reach the goals of W_i.

IV.62d) These matchings are calculated taking into consideration the priorities assigned to the goals of W_I, and they guide the changes in $p_i \in S_d$ necessary to reduce the disagreement between W_I and W_E.

But

IV.62e) the adequation of S_d to the patterns $p_j \in W_E$ is required to minimize the dissatisfaction in W_E, and

IV.62f) the modification of S_d may enhance the costs in W_I.

Thus, sERM

IV.62g) modifies the patterns $p_i \in S_d$ with the lowest matchings with $p_E \in W_E$,

IV.62h) to obtain the germs and haloes of both p_i and p_j according to pruning thresholds α_I and α_E. This is done in order

IV.62i) to build new patterns p_n of days on and off duty by the combination of these germs and haloes.

These new patterns

IV.62j) must provide the best matching between germ and halo subgraphs. The degree of matching may be guided by the t^r compatibility, since the released transmitter may be a function of the filtering properties of the terminal branch which are, in turn, dependent of the strength of the correspondent synapsis. In this way, the t^r affinity can be used to encode the incompatibility of constraints in W_I.

In the sequence:

IV.62k) the new patterns must be matched with the goals of W_I according to the priorities defined in W_i by the internal and external policies;

IV.62l) the best matchings p_n replace the undesirable p_i in S_j. By this process a new S_r is obtained;

IV.62l) the value of the objetive function to S_r is calculated by an specific MPNN(c) composed by an output min-neuron (see Chapter II, section II.6).

IV.62m) If the value of the objective function falls bellow some threshold α_a of acceptable cost then S_r is accepted as the putative final schedule S_f, otherwise

IV.62n) the values of α_I and α_E are modified accordingly, if the institutional cost or the individual dissatisfaction is the most proeminent factor in determining the actual value of the objective function, and the steps from IV.62d on are recursively repeated.

Optionally, at the end of the cycle in IV.62m:

IV.62o) α_a may be decreased and the entire process repeated in order to test the existence of another $S_{f'}$ better than S_f in satisfying the objective function.

Otherwise

IV.62p) the scheduling may be considered finished with a degree of satisfaction equal to α_a.

The matching required by IV.62d may involve

IV.62q) a non-monotonic reasoning of the type "X unless Y", because the rules proposed in the schedulling literature maximize the coverage unless disatisfaction increases; minimize weekend splitting unless working stretch increases; etc.

Thus:

IV.62r) neurons of the MPNN(c) controlling the processing in ERM must be non-monotonic devices. This can be guaranteed by adequate filtering properties of their axon and by the adequate t^r coupling.

The process defined in IV.62 supports the goal programming of the schedule processing by sERM.

The following characteristics of the scheduling systems are important:

IV.63a) Coverage: the number of nurses (by skill class) assigned to be on duty in relation to some minimum number of nurses required.

The sERM defined here can work different desired schedules specified for the minimum and maximum coverage, so that recursive processing can be used to obtain an optimun S_o schedule between these minimum and maximum schedules.

IV.63b) Quality: a measure of a schedule desirability as judged by the nurses who will have to work according to it.

If the employees are allowed to match the schedules S_f supplied by sERM with their goals, this degree of matching may be used as the measure of the quality of the output of sERM. If this information is used to modify the strength of the MPNNs supporting these schedules, then eq. IV.17f provides the way of adapting W_I to W_E. It must be remarked that $\mu(g,g')$ in IV.17f,j allows this adaptation to obey the most important constraints in W_I.

IV.63c) Stability: a measure of the extent to which nurses know their future days off and on duty according to some consistent and stable set of policies.

IV.63d) Flexibility: the ability of the scheduling system to handle changes on both the institutional and employee's worlds.

The stability and flexibility of the scheduling process is determined by the autonomy and plasticity of the germs and haloes of sERM. Since evolutive learning may result into well σ-models, sERM may learn stable scheduling policies which can be flexible enough to acommodate the variability of W_E, and stable enough to support other plannings in both W_E and W_i.

IV.63e) Fairness: a measure of the balance of the influences of the constraints in both worlds in the final schedule S_f.

The fairness of sERM is determined by the relation between the pruning thresholds α_I and α_E in IV.62h. These thresholds may be programmed and/or learned. This provides the control of the fairness of sERM.

IV.63d) Cost: the resources consumed in making the scheduling decision.

The computational cost of sERM can be reduced by the evolutive learning since its knowledge can be adapted by the quality measures provided by W_E.

SCHEDULLER is a MPNNS being now developed following the above guidelines with the purpose of schedulling the nursing staff in a medium size hospital. The initial results seems to confirm MPNN as an suitable tool for implementing this kind of decision making.

IV.12 - Some related theories

The ideas discussed so far are tightly related to the concepts of Classifier Systems (CS) described by Booker et al, 1989. CS are correlated with the notion of modular

MPNN in the sense that both structures represent rule-based pieces of knowledge combined into building-blocks (germs and haloes in the case of MPNN), which are the basic units for constructing knowledge nets. The major differences between MPNN and CS are:

IV.64a) the structure of knowledge representation in MPNN is stronger than the structure provided by CS, because

IV.64b) MPNN uses both mail and broadcasting strategies to spread information in the knowledge net, whereas CS makes use only of the broadcasting system.

IV.64c) CS are supported by a binary crisp logic, whereas MPNN handles both crisp and fuzzy operations, and

IV.64d) the capacity of MPNN to handle uncertainty is greater that that of CS. The neuronal activation and connectivity encode confidence (uncertainty of matching) and relevance (uncertainty of frequency). Credit assignment in CS encodes relevance, but tag matching is a crisp operation.

Both MPNN and CS take advantage of combining or modifying building-blocks to provide the augmentation of the modeling entropy required to cope with novelty in the surrounding environment. Both theories make use of natural selection to obtain the best solution of the problem and to guarantee knowledge consistency. However:

IV.65a) CS uses heuristic knowledge to define and control the genetic operators, but the combinatorial procedure of generating the model's offspring is essentialy a random process and

IV.65b) production of variability in MPNN is supported by an unlimited number of strategies ranging from randomness to associative learning, causal reasoning, etc. Genetic algorithms are among the allowed strategies in MPNN. The enhancement of the entropy in MPNN may be the result of both numerical (strength) as well as symbolic changes (t^r coupling) in the connectivity among the elements of the net. The production of symbolic variability is supported by a very rich and strong formal language.

Booker et al., 1989 , pointed out that although CS and classic neural nets share many interesting properties, knowledge representation in CS is not restrained to a set of connection strengths. Because of this, they proposed CS to occupy an important middle ground between symbolic and connectionist paradigms. Knowledge representation in MPNN is a process richer and stronger than in both classic neural nets and CS, because

IV.66a) it takes advantage not only of the power of the

numeric encoding provided by the synaptic strength or credit assignment, but also of the fuzzy symbolic processing capacity of $L(G)$, and

IV.66b) it combines these two approaches to support many kinds of knowledge representation such as rule-based, mathematical programming, and goal programming strategies.

MPNN shares also many basic concepts and properties with the theory of Neuronal Group Selection (NGS) of Edelman, 1987. MPNN and NGS assume that:

IV.67a) while the world is not amorphous and the properties of objects are describable in terms of chemistry and physics, it is clear that objects do not come in predefined categories, are variable in time, occur as novelties, and are responded to in terms of relative adaptative value to the organism rather than of veridical descriptions, so that

IV.67b) perceptual categorization occurs without assuming that the world is prearranged in an informational fashion or that the brain contains a homunculus.

But:

IV.68a) NGS proposes that all variability required to model the environment must be randomly prewired in the brain during the embryogenesis. This process provides the primary repertory of neuronal ensembles. Epigenetic postnatal factors select the best neuronal groups to form the secondary repertory of the animal to cope with the surrounding environment. The necessity to assume all variability to be provided during the prenatal period is a consequence of the dogma that neurons cannot reproduce after the birth. However, a new synapsis may grow in the post-embryonary period between neurons and axons nearby. In this way, systems of parallel fibers may provide a suitable neuroanatomic substrate for potential synapsis, without paying the price of maintaining their physiology (Chapter VI, section VI.7) Because of this,

IV.68b) MPNN assumes the importance of both genetic and sociogenetic knowledge inheritance to avoid the hard task of the creation of a new observable world W from the scratch by each newborn brain. As a matter of fact, MPNN takes advantage of these knowledge inheritance to guarantee communication among different individuals sharing this common social and biological culture and to support the temporal and spatial spread of knowledge among brains.

In sum, MPNN not only combines numeric and symbolic reasoning to make them complementary tools for human thinking, but it also takes advantage of inheritance and evolution to make progress in learning. Evolution implies guided and random modification of the inherited knowledge, thus it avoid a closed genetics. MPNNs are σ-systems: partially opened and partially closed structures. According to Wah et al., 1989, Intelligent Systems are non-deterministic machines, thus at least partially opened systems. MPNN are intelligent systems. Proofs of consistency cannot be performed in opened worlds, but (natural) selection may filter the best models to handle these opened worlds.

ACKNOWLEDGEMENTS

I am in debt with Pietro Torasso, Luca Console and Ricardo J. Machado because some of the ideas presented in this chapter were inspired on the results of the many discussions we had together. Also, the student Clezio Gelli developed the MPNNS used the initial testing of the ideas discussed in section IV.11. The success of this initial test encouraged us to begin the development of SCHEDULLER.

CHAPTER V

INVESTIGATING EXPERTISE

V.1 - The purpose

Our culture has privileged logic as the tool for treating and explaining the human reasoning for almost two thousand years. Only a few decades ago, Biology discovered conditioning as another way of learning and Neuroscience started to furnish knowledge about information processing in neural circuits. The last decades have been characterized by a continuous competition between connectionism and logic in the AI field, which has resulted in an alternating predomination of one of these approaches as the most popular tool for modeling the human mind (for some of the discussion on this topic, see Anderson and Rosenfeld, 1989; Chandrasekaran et al, 1988; Gallant, 1988; Hinton, 1989; Minsky and Papert, 1969; Rosenblatt, 1958; Shastri, 1988). The inflection point in this alternation of popularity has been determined by the unsuccess of the predominating theory in solving particular problems. The main drawbacks explaining this oscillation are:

a) connectionism is blamed for not being able to handle symbolic reasoning adequately, whereas

b) logic has to rely on heuristic because it cannot handle learning and uncertainty adequately.

The simplification of the behavior of the neuron apparently to favor the mathematical analysis of the proposed models (e.g. Hinton, 1989; Hopfield, 1982; Kohonen, 1982; Rumelhart et al, 1986) is the main source limiting the symbolic power of connectionism as practiced nowadays. This approach is misleading because the reduction of the complexity of the real neuron does not contribute to the real understanding of the entire behavior of the net. On the contrary, the simplification of the neuron eliminates important elements controlling the activity in the neural circuit and generates unrealistic modelings of the brain. The updating of the biology of the artificial neuron (Chapters II and III) greatly increases the computational power of neural nets and eliminates the actual conflict in combining numerical and symbolic processing into intelligent neural models.

The binary encoding of uncertainty is unrealistic if the human thinking is concerned or even if the relations between man and the surrounding environment are considered. Most frequently, people gather information to gain confidence on fuzzy rather than crisp decisions (Rocha et

al, 1990; Zadeh, 1965). Gain of confidence is prohibited by
max-min procedures used in general to handle uncertainty in
most of the conventional systems used to simulate the human
reasoning. In the worst situation, man at least averages
uncertainty (Booker, 1989; Greco and Rocha, 1987; Theoto and
Rocha, 1989; Yager, 1988b; Zimmermann and Zysno, 1980) to
evaluate the success of his decision making. If he is
allowed, man tries to make profit in gathering information
to reduce uncertainty (Maeda and Murakami, 1988; Maeda and
Theoto, 1990; Rocha et al, 1989; Theoto and Rocha, 1989;
Zimmerman and Zysno, 1980), what means a non-linear
processing of the confidence in the collected data (Rocha et
al, 1989; Theoto and Rocha, 1989).

Man has progressed quickly from using only
probability as the tool for formalizing uncertainty, to
developing other concepts as possibility, belief,
plausibility, etc. (e.g. Klir, 1989; Shaffer, 1976;
Smithson, 1987; Zadeh, 1978). It could be interesting to
take advantage of the numerical processing capabilities of
neural nets to get a better treatment of uncertainty in
symbolic reasoning. The results discussed in the previous
chapters show that MPNNs handle well both numeric and
symbolic processing and they could be the adequate candidate
to process the various dimensions of uncertainty state space
of the human reasoning.

The relevance of the study of the human brain to the
study of artificial intelligence has long been an issue of
debate both in the AI community (Smoliar, 1989) and in the
Neuroscience field (Edelman, 1987). On the one hand, it has
been claimed that the study of the brain is too complex and
mysterious to yield useful guides for the construction of
intelligent machines and that these machines may be
developed under different guidelines and hardware. On the
other hand, it is claimed that the behavior of the brain is
the very inspiration for the study of artificial
intelligence (Smoliar, 1989).

Of course, when neurons agglutinate into brains, new
(emergent) properties arise, not possessed by each of these
neurons themselves, which derive from their association
(Bunge, 1977; Eccles, 1981; Popper and Eccles, 1985; Rocha
and Rocha, 1985). This could be the argument to support the
point: Neuroscience is irrelevant to AI. But besides these
emergent properties, the high order system (brain) also
inherits properties from the unitary elements (neurons)
composing it (Bunge, 1977; Rocha and Rocha, 1985). This
could be the justification for the attempt to reduce
reasoning to the physiology of the neuron. However, as the
new connectionism Neuroscience is discovering, it stresses
the mutual influence between mind and brain, and it starts
to provide the ways to approach experimentally this mutual
correlation under the optics of science, that is, putting it
as a workable hypothesis which may be falsified by empirical

data. This new approach means that reasoning must be put under experimental investigation instead of remaining only a matter of theory, even if supported by a strong logic.

The development of new techniques to knowledge acquisition and analysis was very much encouraged by the success of the Expert System technology. They were initially developed with the purpose of obtaining the contents of the knowledge base of the expert system. However, these methodologies may be applied to investigate the human thinking (Rocha et al, 1990), the differences between expert and non-expert reasoning (Leão and Rocha, 1990), the properties of the calculus of uncertainty (Greco and Rocha, 1987; Maeda and Theoto, 1990; Theoto et al, 1987;1990), etc.

In this line of work, the use of empirical data gathered by Neuroscience about neuron and brains, and by AI about symbolic reasoning, can be combined to derive and test theories and models. This is a strong scientific commitment, because it renders the mind-brain dualism a hypothesis that can be falsified, therefore, a true scientific theory (Popper, 1967). This is the line of investigation followed in this book. In this chapter, attention is devoted to the experimental study of expertise.

V.2 - Knowledge elicitation

The approach used here for knowledge elicitation (Leão and Rocha, 1990; Machado, Rocha and Leão, 1990; Rocha et al, 1990; Theoto and Rocha, 1989, 1990; Theoto et al, 1989) takes advantage of the fact that the human expert uses a jargon to speak about his field of specialization (Rocha and Rocha, 1985; Sager et al, 1987). The jargon is a subset of the natural language having its semantic restricted by the models defining this specialization.

The first step in this process of knowledge elicitation is to characterize the jargon, by asking the experts to supply the list of the signs, symptoms, laboratory test results, etc. they assume necessary to speak about the reasoning models in their field of specialization (Fig. V.1A). This jargon can also be obtained from any available specialized data base (Chapter VII, section VII.9). The second step of knowledge elicitation is to use the jargon list as terminal nodes in order to obtain the graphs (Fig. V.1,2) of the reasoning models supporting decision in the area of expertise under investigation (Leão and Rocha, 1990; Machado, Rocha and Leao, 1990; Rocha et al, 1990). Here these graphs are called knowledge graphs and denoted KG.

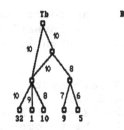

SUPPORTING DATA   A

1 - presence of the bacillus in the sputum
5 - weight loss
9 - fever
10 - cough and sputum
15 - culture of the bacillus
20 - biopsy
32 - compatible XR

IMPORTANCE ORDER   TEMPORAL ORDER

```
□ □ □   □ □     □ □ □ □   □  □
32 1 10  9  5   10 9 32 1  15 20
```

DECLARATIVE   PROCEDURAL
KNOWLEDGE

DECLARATIVE KNOWLEDGE

**FIG. V.1 - THE JARGON LIST AND THE DECLARATIVE KNOWLEDGE
GRAPH ABOUT TUBERCULOSIS**

The experimental data discussed in this chapter were obtained by interviewing 48 physicians working on different fields of medicine including: Cardiology, Pulmonary Diseases, Nephrology, Urology, Dermatology, Gynecology, Pathology, Ophthalmology and Dentistry. All of them have at least 5 years of practice in their field of specialization and most of them obtained the title of specialist from the corresponding medical association in Brazil. Some were also involved in teaching at several medical schools in different parts of the country, but others are merely country doctors. Despite this diversity of the studied population, the results are very consistent and some general rules of the expert reasoning can be disclosed from this analysis. No serious problem occurred while eliciting the expert knowledge with the technique described below. As a matter of fact, in some cases, after having worked with one of the researchers in one or two diagnoses, the volunteer decided to build the other KGs by himself at home or in the office.

V.2a - The jargon list

The structure of the data in this jargon list is of the type

$$X \text{ is } A \qquad (V.1a)$$

where X is a fuzzy or a linguistic variable (Zadeh, 1975) defined over a universe of discourse U. Thus, A is either a fuzzy set of U or a set of linguistic labels (Fig. V.3) in a given language L, either a natural or a formal language.

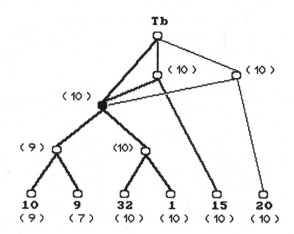

FIG. V.2 - THE PROCEDURAL KNOWLEDGE ABOUT TUBERCULOSIS

A fuzzy variable X is A over a universe of discourse U (base set) is defined by an assignment equation (Zadeh, 1975):

$$x = u:R(X) \qquad (V.1b)$$

which represents the assignment of a value u to x subject to the restriction R(X). The membership function $\mu_A(u)$ defining the fuzzy set A measures the compatibility between the actual value of u and the concept represented by A. In this sense, A represents the prototypical knowledge encoded by R(X).

For example, Fever may be a fuzzy variable (Fig. V.3a) defined by the fuzzy relation:

temperature above 37º Celsius

In this case, A is the fuzzy set of all temperatures higher than 37º C in the set T of human temperatures, and fever is defined by the following assignment relation (Zadeh, 1975):

$$F = T:R(F) \qquad (V.2a)$$

$$R(F) \equiv t > 37ºC \qquad (V.2b)$$

or

$$\mu_F(T) : T ---> [0,1] \qquad (V.2c)$$

Here, the symbol ≡ is used with the meaning of "denotes" or "equal by definition". $\mu_F(t)$ measures the compatibility of the actual temperature t ∈ T with the prototypical knowledge about Fever represented by R(F).

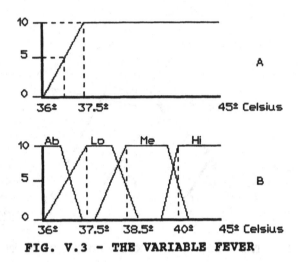

FIG. V.3 - THE VARIABLE FEVER

But Fever may also be, and usually is used as a linguistic variable (Fig. V.3b) if A is assumed to be the set of labels

$$A = \{ \text{high, medium, low, absent} \} \qquad (V.3a)$$

which may be expanded with the following set of modifiers

$$M = \{ \text{very, more or less, etc.} \} \qquad (V.3b)$$

according to some syntax θ restricting the use of a combination of terms which are not of common use in the medical jargon, e.g. more or less medium.

The allowed productions of this language:

$$\theta : M \times A \longrightarrow [0,1] \qquad (V.3c)$$

are fuzzy sets in the base set T. Let L be the set of all these allowed productions l_i. The semantic of each $l_i \in$ L is provided by the corresponding membership function

$$\mu_{Li} : L \times T \longrightarrow [0,1] \qquad (IV.3d)$$

which measures the compatibility of t ∈ T with the concept represented by l_i (Zadeh, 1975).

The expert is asked to provide the information about the compatibility (membership) functions used to define the semantics of both the fuzzy and linguistic variables in the jargon list. He provides this information without difficulty either as a graphic or a table of values of confidence (Rocha et al, 1991d). Figs. V.3 and 4 show examples of real data obtained from the experts.

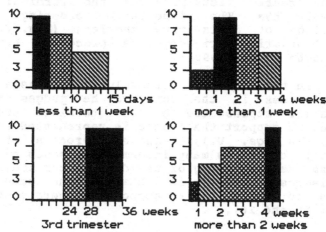

FIG. V.4 - COMPATIBILITY OF FUZZY VARIABLES

Many of the variables in the jargon list are composite interactive variables in the sense defined by Zadeh, 1975. Thus the n-ary composite variable

$$X \equiv (X_1, \ldots, X_n) \qquad (V.4a)$$

is defined by the n-ary assignment equation

$$(X_1, \ldots, X_n) = (u_i, \ldots, u_n) : R(X_1, \ldots, X_n) \qquad (V.4b)$$

representing the action of putting u_i in $X_i, \ldots,$ and u_n in X_n simultaneously, under the restriction that the n-tuple of objects $(u_1, \ldots u_n)$ must be in the $R(X_1, \ldots, X_n)$ list.

Because of the complex interactivity among the atomic terms x_i of X disclosed by the experimentation, Rocha et al, 1991d, proposed to use the same graph approach described below to handle this type of variable.

V.2b - Knowledge graph

After providing the jargon list, the expert is asked to consider this list as the terminal nodes of a graph (e.g.

Figs. V.1,2) which may represent the knowledge about the diagnosis being investigated.

The first step in the process of building KG is to order the terminal nodes by importance and by the temporal sequence each datum is obtained in practice (Fig. V.1). Each of these ordered lists provides the terminal nodes of two different KGs. The importance ordering induces the elicitation of declarative knowledge (Fig. V.1B) and the temporal ordering triggers the procedural knowledge (Fig. V.2) (Rocha et al, 1990a).

In the sequence, the volunteer is asked to join the terminal nodes in the same way he judges the pieces of information represented by the terminal nodes must be combined to support the diagnosis represented at the root of KG (TB in Figs. V.1,2). In this process, the expert is allowed to create as many intermediate nodes as necessary to organize the clustering of the data. After the KG describing the diagnosis process is obtained, the same procedure is applied to build the KG supporting therapy decision (see Fig.s V.15 and 16).

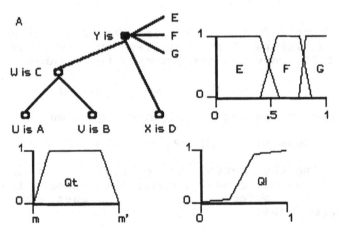

FIG. V.5 - THE KNOWLEDGE GRAPH AS A FUZZY DEDUCTION

KG may be read as a set of propositions of the type (Fig. V.5):

if X is A and Y is B .... then Z is C   (V.5a)

which are propositions in fuzzy logic (Zadeh, 1979, 1983a)

The solution of V.5a is proposed (Zadeh, 1983a) to

be provided by the extended version of modus ponens (EMP):

$$\text{if X is A then Y is B}$$

$$\text{X is A'}$$

$$\text{Y is B'} \qquad\qquad\qquad \text{(V.5b)}$$

which means to find the fuzzy set B' given the fuzzy set A' and the implication function f relating the fuzzy set A to the fuzzy set B (Godo et al, 1991; Katai et al, 1990a,b; Trillas and Valverde, 1987; Yager, 1990d; Zadeh, 1983a).

       The solution of EMP is obtained in 4 steps (Zadeh, 1983a):

V.5c) Matching: the compatibility σ between A and A' is the measure of the equality [A≡A'] between the fuzzy sets A and A' (Pedrycz; 1990a,b), so that the matching between (X is A') and (X is A) (Godo et al, 1991) is calculated as:

$$(\text{X is A'}) \equiv (\text{X is A}) \text{ is } \sigma$$

Since, $A \equiv A'$ implies

$$\text{A C A'} \quad \text{ and } \quad \text{A' C A}$$

$$\mu_A(x) \;\le\; \mu_{A'}(x) \qquad \text{and} \qquad \mu_{A'}(x) \;\le\; \mu_A(x)$$

the assessment of the value of σ means to evaluate how equal are these two fuzzy sets taking into account their elements (see Chapter X, section X.7).

      The information provided by the variables in the jargon list   is considered as pieces   of prototypical knowledge about the disease. In this way, the actual datum is matched against this knowledge by the expert to provide the degree of compatibility in which the patient fulfills the requirements to be classified under a given diagnosis. This compatibility measures the confidence the physician has that the patient exhibits the sympton, sign, etc. required by the prototypical knowledge encoded in the corresponding jargon variable. For example

    if Temperature is Fever then Infection is Present

    given Temperature is High

    Infection is Compatible

becomes

> if Temperature is Fever then Infection is Present
>
> (Temperature is Fever) is σ_a
> _____
>
> (Infection is Present) is σ_c

which is very much related to the natural way of reasoning of the expert.

V.5d) Aggregation: all compatibilities σ_i assigned to the atomic propositions in the antecedent part of the implication are aggregated into a unique value representing the compatibility σ_a of the antecedent

$$\sigma_a = \underset{i=1}{\overset{n}{\Theta}} \, (\sigma_i)$$

Aggregation is a very important issue in the expert reasoning. As a matter of fact the non-terminal nodes of KG are classified according to the type of aggregation they carry out. This will be discussed in detail in section V.4 and 5.

V.5e) Projection: the compatibility σ_c of the consequent is obtained as function of the aggregated value σ_a (Delgado et al, 1990b; Diamond et al, 1989; Godo et al 1991; Katai et al, 1990a,b)

$$\sigma_c = f(\sigma_a)$$

σ_c measures the compatibility of (Y is C') to (Y is C):

> if X is A ... then Y is C
>
> (X is A) is σ_1 and ...
> _____
>
> (Y is C) is σ_c

The slope of f is one of the parameters affected by learning resulting in expertise. As the expert knowledge consolidates the slope of f increases and it may become greater than 1. The result of this learning is a net gain of confidence in the solution of the implication because:

$$\sigma_c > \sigma_a$$

This type of reasoning is here called additive reasoning. It means that the expert learns to recognize some pieces of information as the fundamental data supporting the decision represented by the consequent. Whenever these data are fully observed they guarantee the decision by themselves. Whenever $\sigma_i \ll 1$ is assigned to these relevant data, other pieces of information must be inspected to add, if possible, confidence in the final decision. The final result of the aggregation and projection steps is very dependent on the relevance of the observed data. The value of this relevance δ is the strength of the arcs (Fig. V.6) of KG (numbers assigned to KG in Figs. V.1 and 4). The role played by relevance will be further discussed in the next sections.

V.5f) Inverse-Matching and Defuzzification: given σ_c it is necessary to obtain C' or a singleton $c \in C'$ to represent the final result of the calculation.

Many approaches have been proposed to handle both the inverse-matching and the defuzzification (see Chapter X, section X.7) However, defuzzyfication is unnecessary in the case of the expert reasoning involved in classification tasks, since the purpose of this kind of thinking is to obtain a degree of matching between the actual pattern and the prototypical description of the possible classes or diagnoses. This degree of matching is σ_c obtained in V.5e. Whenever the confidence in the fuzzy decison has to be transformed into a measure of probability, the conversion proposed by Klir, 1989, may be used.

V.2c - Relevance

After KG is built, the expert values its arcs according to the relevance (numbers at the side of arcs in Figs. V.1 and V.2) of the information represented at its leaving node, to support the decision about the diagnosis.

Relevance is the measure of the uncertainty about the frequency of association between the supporting datum and the decision making (Greco and Rocha, 1987; Rocha et al, 1989; Theoto et al, 1987). Confidence (σ_i) is the measure of the uncertainty of matching discussed above. Relevance and confidence are two different psychologic constructs (Greco and Rocha, 1988), which are combined to express the fact that:

V.6a) the greater the relevance, more influent is the datum, and zero relevance means no influence (Bartolin et al, 1988; Sanchez and Bartolin, 1989; Sanchez, 1989; Soula and Sanches, 1982; Yager, 1990c), while

V.6b) the greater the confidence, bigger is the support

provided by the datum for the decision making.

Kacprzyk, 1988, Kacprzyk et al, 1990, and Rocha et al, 1989, proposed that relevance and confidence must be ANDed to express their influences upon the decision making; the first author proposed the min operator for the accomplishment of this aggregation, while the second used the product as the aggregator. Thus, it may be assumed that relevance changes the compatibility function in the implication process (Fig. V.6), which can now be represented as

if [(X is A)·δ] and [(Y is B)·δ'] .... then Z is C   (V.6c)

where · is a ⊤ -norm. Relevance is the main justification for additive reasoning.

The introduction of the notion of relevance changes the solution of the Extended Modus Pones, because it changes the fuzzy sets A, B, ..., etc. (Fig. X.6).

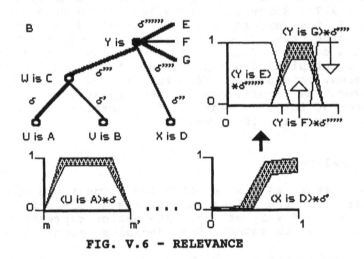

FIG. V.6 - RELEVANCE

V.3 - The "mean" knowledge

The knowledge graphs about the same diagnosis, obtained from different experts, are averaged to obtain a mean graph representing some sort of consensus about the knowledge shared by the experts on this diagnosis.

The following procedure is used to calculate the mean graph (Machado, Rocha and Leäo, 1990):

V.3a - Graph summation

V.7a) A first (any) knowledge graph is taken as the starting graph. A counter with the initial value set one is associated to each arc of this embryonic graph, and a summing device with its initial value set equal to the corresponding relevance δ, is assigned to each of its nodes.

V.7b) The other graphs are aggregated disregarding the labels (jargon) associated with the terminal nodes. Because of this, the aggregation is not of identical graphs but of similar structures. It is done in this way in order to cope with the disagreement that may exist among the volunteers. The probability of identical structures decreases as this disagreement increases. Relaxing the rule and asking for similar structures instead of identical graphs is a key issue for the calculation of the consensus in the case of fuzzy information.

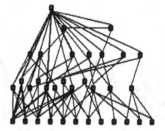

POPULATION GRAPH FOR PDA DIAGNOSIS
EXPERTS

POPULATION GRAPH FOR PDA DIAGNOSIS
NON-EXPERTS

FIG. V.7 - POPULATION GRAPHS FOR PDA OBTAINED FROM DIFFERENT EXPERTS ON CONGENITAL CARDIAC DISEASES AND GENERAL CARDIOLOGIST (NON-EXPERTS)

The width of the arc is related to its frequency in the population graphs.

Whenever a similar structure is found then:

V.7c) the unity is added to the arc counters, and the relevance is summated to the respective node summing device.

Otherwise:

V.7d) the new structure is incorporated to the population graph with arc counters equal to 1 and the node summing devices set equal to the relevance of the corresponding nodes.

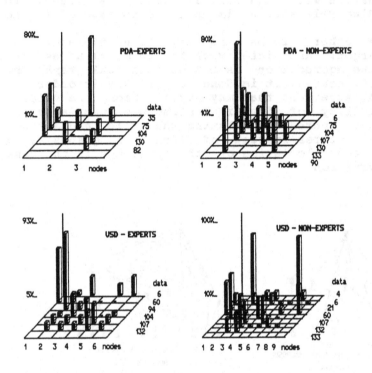

FIG. V.8 - DATA DISTRIBUTION ON TERMINAL NODES OF THE KNOWLEDGE GRAPHS

Fig. V.7 shows the population (aggregated) graphs obtained for the same diagnosis (PDA - Patent Ductus Arterious) from a population of experts and non-experts (Machado, Rocha and Leäo, 1990). The expertise in this case was about Cardiac Congenital Diseases and the non-experts were General Cardiologists.

V.3b - Relevance and labels

The next step in the process of obtaining the mean graph and the consensual knowledge is:

V.8a) to calculate the mean relevance for each node: by dividing the values accumulated at the node summing devices by the sum of the counters of the arcs leaving the node under consideration, and

V.8b) to locate the labels at the terminal nodes of the population graph: the labels (jargon) are attached to the terminal nodes depending on their conditional distribution at these nodes of the population graph (Fig. V.8), so that

V.8c) walking from left to right, the most frequent label is assigned to the node under consideration, depending on whether the label was or was not used at another node yet. If it was already used, then the next most frequent label at the node is picked up as a new candidate.

V.3c - Fuzzy indexes

The final step to obtain the consensual graph demanded the calculation of some indexes (Fig. V.9) measuring the fuzziness of the acquired knowledge (Machado, Rocha and Leäo, 1990):

EXPERTS						NON – EXPERTS					
DIAG.	MEAN EU	MEAN F	MEAN S	DIFFICULTY	AGREEMENT	DIAG.	MEAN EU	MEAN F	MEAN S	DIFFICULTY	AGREEMENT
PDA	.70	1.15	.81	4.1	.46	PDA	.49	1.06	.52	7.8	.21
USD	.61	1.10	.67	6.7	.27	USD	.64	1.05	.64	9.1	.22
ASD	.65	1.13	.75	5.2	.29	ASD	.65	1.09	.71	8.5	.24
AUSD	.61	1.10	.67	6.0	.24	AUSD	.54	1.05	.56	7.9	.22
AOCO	.49	1.08	.52	4.1	.20	AOCO	.53	1.06	.56	7.9	.25
	.62	1.11	.68	6.2	.29		.57	1.06	.60	8.6	.22

FIG. V.9 - AVERAGED VALUES OF THE FUZZY INDEXES

V.9a) Evidence use ratio (EU)

EU = number of terminal nodes / number of evidences used by the population

It assumes value 1 if the individual uses the maximum number of evidences utilized by the entire population.

V.9b) FUZZINESS DEGREE (F)

$$F = \text{number of arcs+1 / number of nodes}$$

It assumes value 1 if the graph is a tree, otherwise it increases as the complexity of the graph is enhanced.

V.9c) SLIMNESS DEGREE (S)

$$S = EU \cdot F$$

The slimness degree is theoretically defined for the whole interval $(0,\infty)$. But since there exists an important disagreement among the volunteers preventing $EU \rightarrow 1$, then S remains bounded in the interval $[0,1]$ (fig. V.9).

V.9d) DIFFICULTY DEGREE (D)

$$D = S_M \cdot \text{maximum number of evidences}$$

S_M stands for the mean slimness calculated for all graphs dealing with the same diagnosis, and the maximum number of evidences is the maximum number of terminal nodes in these graphs. D increases as both the number of terminal nodes and the number of arcs augment.

V.9e) AGREEMENT (A)

$$A = S_M / S_P$$

S_P stands for the slimness degree of the population graph. A will be equal to 1 if S_M is equal to S_P, that is, if the population graph is the mean graph.

V.3d - The averaging

The following algorithm is applied to calculate the mean graph of each diagnosis (Fig. V.10):

V.10) the population graph is pruned until the slimness degree S(P) of the pruned graph P approaches the mean slimness degree Sм(C) calculated for the set C of all KGs about the same diagnosis c.

Fig. V.10 shows the mean graphs for some of the diagnoses studied about Cardiac Congenital Diseases (Leäo and Rocha, 1990; Machado, Rocha and Leäo, 1990). The main results obtained from these studies are:

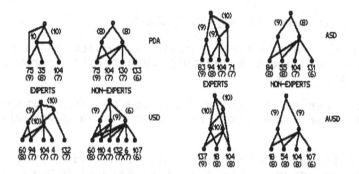

FIG. V.10 – "MEAN" KNOWLEDGE GRAPHS FOR DIFFERENT DIAGNOSES AND DIFFERENT POPULATIONS OF VOLUNTEERS

V.11a) the expert uses less evidences than the non-expert;

V.11b) the expert's graph is less complex than the one obtained from the non-expert;

V.11c) the mean relevance calculated for the experts is in general greater than the one calculated for the non-experts;

V.11d) label (jargon) distribution at the terminal nodes of the mean graph is crisper than the one observed for the non-experts (see Fig. V.8);

V.11e) in general, experts use more alternative pathways than do non-experts to reach the same conclusion;

V.11f) there is a high disagreement among experts, although less than the one observed for non-experts (fig. 9), and

V.11g) the relevance assigned to the terminal nodes decreased from left to right in the same way they were ordered according to their importance to support the hypothesis.

All of this indicates that expertise shrinks the descriptive complexity of the knowledge, while increasing the capacity of good decisions, thus enhancing the functional complexity of the reasoning (Eddy and Clanton, 1982; Kolonder, 1983; Kassirer and Gorry, 1978; Larkin et al, 1980; Milne, 1987). The differences between experts and non-experts is captured by the fuzzy indexes as shown in Fig. V.9.

In conclusion it may said that:

V.12a) experts use a small number of evidences on crisper graphs to speak about a few well characterized frames of a given disease, while

V.12b) non-experts use more evidences on fuzzy graphs to speak about poorly characterized frames of a given disease.

Knowledge on Leprosy acquired with the above methodology was used as the Knowledge Base of a Frame Expert System (Torasso and Console, 1989) to process real data in order to test the performance of the acquired knowledge in classifying patients in one of four possible forms of this disease. Four different experts were interviewed and their knowledge graphs were recoded into non-mutually exclusive frames in order to make the system able to reason independently with the knowledge provided by each expert. The outcomes from the Expert System were compared for cases where biopsy or laboratory tests guaranteed a unique diagnosis. The results showed that each expert had a bias toward one of the four possible forms. Based on a voting paradigm, the Expert Systems agreed with the real data on 90% of cases. SMART KARDS (c) used the population graphs obtained by adding the same individual expert graphs, and a MPNN paradigm to classify the same patients (Chapter IX, section IX.7) with equal success.

JARGON was used (Chapter IX, section IX.11) to extract the standard descriptions of the physical examinations, laboratory tests and histopathology of these patients according to their diagnosis. The results showed that the data base is redundant in respect to each individual jargon lists. However, its contents approached that of the total jargon list because of the disagreement among the experts about the important data supporting the diagnosis. Relevance was only partially related to datum frequency conditioned by the diagnosis, showing that its final value is influenced by other factors (Theoto et al,

1989a). The patterns discovered by JARGON for each Leprosy type are very well correlated with the corresponding population graphs.

All these results validate the present approach of knowledge elicitation and acquisition.

V.4 - Aggregation at the non-terminal nodes

The final step of the knowledge elicitation is to investigate the aggregation process used at the non-terminal nodes to combine the information clustered on it. Most of the time, if not always, the volunteer uses linguistic quantifiers (Zadeh, 1983b) to explain that some but not all of the input data are necessary to guarantee the decision making. In this way, V.6C becomes:

if $Q\{[(X$ is $A)\cdot\delta]$ and $[(Y$ is $B)\cdot\delta']$ ....$\}$ then Z is C

(V.13)

where Q is a linguistic quantifier of the type MOST, AT LEAST N, etc. This kind of quantifier was defined by Zadeh, 1983b, as a proportional quantifier.

Let the proposition be

QRXs are A   (V.14a)

where Q is a proportional quantifier, and R and A are fuzzy sets representing, respectively, the relevance and the prototypical knowledge. The truth σ of the proposition V.14a is calculated into 2 steps (Zadeh, 1983b; Kacprzyk 1986a,b, 1988; Yager, 1990b):

V.14b) to obtain the relative sigma-counting Σ-count(A) of the fuzzy set A given R as

$$s = \Sigma\text{-count}(A \text{ and } R)/\Sigma\text{-count}(R)$$

V.14c) to set the truth of the proposition as

$$\sigma = \mu_Q(s)$$

where $\mu_Q(r)$ measures the compatibility of s with the prototypical knowledge of Q. In general, this membership function is of the type

$$\mu_Q(s) = 0 \text{ for } s \le \alpha_1$$

$$= g(s) \text{ for } \alpha_1 < s < \alpha_2$$

$$= 1 \text{ for } s \ge \alpha_2$$

If δ_i and σ_i represents, respectively, the relevance of and confidence in the ith antecedent of proposition V.13 above, then the relative sigma counting becomes (Kacprzyck, 1988; Yager, 1990c, Zadeh, 1983b):

$$\sigma_a = \sum_{i=1}^{n} \delta_i \cdot \sigma_i \Big/ \sum_{i=1}^{n} \delta_i \quad (V.15a)$$

where n is the number of antecedents in V.13 and $\cdot$ is a $\top$ -norm.

The truth σ_c of (Z is C) in V.13 becomes a function of the compatibility of the antecedents σ_a. Thus σ_c is the compatibility function $\mu_Q(\sigma_a)$ defined in V.14c:

$$\sigma_c = \mu_Q(\sigma_a) \quad\quad (V.15b)$$

$$\mu_Q(\sigma_a) = 0 \text{ for } \sigma_a \leq \alpha_1$$

$$= g(\sigma_a) \text{ for } \alpha_1 < \sigma_a < \alpha_2$$

$$= 1 \text{ for } \sigma_a \geq \alpha_2$$

In this line of reasoning, V.15a represents the aggregation step and equation V.15b the projection step of the EMP resolution in the case of KG.

The adjustment of the thresholds α in V.15b allows the semantic of Q to range from AT LEAST ONE or a true OR, to ALL OF THEM or a true AND, through all necessary values of the expression AT LEAST N. In this way, Q may be viewed as an AND/OR operator, the value of α_1 defining the degree of ANDness (or ORness) of the aggregation. This is in agreement with the proposition of an AND/OR operator proposed by Rocha and colleagues (Rocha et al, 1989; 1990a,b) to describe the aggregation at the non-terminal nodes of KG. This AND/OR supports additive reasoning.

V.5 - Types of non-terminal nodes

Two different types of non-terminal nodes are identified in KG: aggregation and decision nodes (Rocha et al, 1989, 1990a; Theoto et al, 1990).

Aggregation nodes (Fig. V.1) represent fuzzy variables and appear in all KGs, whereas decision nodes (filled nodes in Fig. V.2,5) encode fuzzy linguistic variables and appear only in Procedural KGs.

The basic difference among these types of nodes is that in the case of aggregation node, σ_c is transmitted through all arcs leaving the node, whereas in the case of decision nodes σ_c is recoded in different linguistic labels, each one assigned to one of the arcs leaving the node. Because of this, decision nodes may direct the flow of processing in KG according to the value of σ_a, and they are used to implement propositions of the type:

Given $\quad \sigma_a = Q\{[(X \text{ is } A) \cdot \delta] \text{ and } \dots \}$ $\qquad$ (V.16)

$\qquad$ if $\alpha_{i-1} \leq \sigma_a \leq \alpha_i$ then $(Y \text{ is } l_i)$ is σ_i

where $l_i \in L$ is a term in the set L of linguistic labels associated with the variable Y is C (Zadeh, 1975). The linguistic variable behaves like a filter directing the output of the decision node to specific reasoning pathways. Linguistic variables are used by the expert to implement control knowledge.

V.6 - Declarative knowledge

The Declarative Knowledge is obtained when the list of its terminal nodes are ordered by the importance each node has in supporting the diagnosis. The expert joins these nodes into different clustering describing the distinct frames of the disease he learned to occur among his patients. The declarative KG contains no decision node.

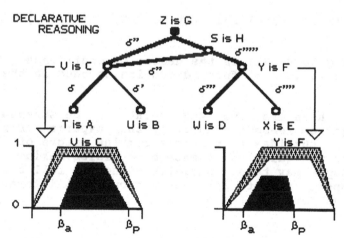

FIG. V.11 - DECLARATIVE REASONING

In many instances, some of these frames are part of bigger clusterings, besides being a valid description of the disease by itself (Fig. V.11). These parallel pathways provide the different frames of the same disease. The expert explains his use of these frames, saying that if he has a high confidence on the data of the initial cluster or germ, he may accept the disease, otherwise he needs to get more information to support the decision making. This additional information is provided by the alternative parallel pathways.

The actual data are matched with the prototypical knowledge defining each fuzzy variable represented at the graph terminal nodes. The calculated degree of matching or confidence is "ANDED" with the relevance of each datum and then aggregated to provide the resulting fuzzy sets at the non-terminal nodes. For example, in the case of the graph in Fig. V.11, the actual data A' and B' associated with the variables T and U are matched with the prototypical knowledge encoded by the fuzzy sets A and B respectively. The calculated degrees of matching are "ANDED" with the relevances δ and δ', assigned to the variables T and U, respectively. These results are aggregated and projected at the non-terminal node V is C (dark areas in Fig. V.11). The result of these operations is the calculation of C'. If the degree of matching between C' and C is very high, then conclusion Z is G is guaranteed, because G' is also very similar to G. Otherwise, information about W and X is taken into consideration in an attempt to increase the confidence that G' is similar to G. The confidence in this conclusion is, therefore, dependent of the degree of this similarity. In this condition, σ_c calculated for

$$\text{if } Q\{[(T \text{ is } A)\cdot\delta] \text{ and } [(V \text{ is } B)\cdot\delta'] \quad ....\} \text{ then } Z \text{ is } C \qquad (V.13)$$

according to V.15b, is the degree of confidence that Z is G is the solution of fuzzy deduction encoded in the knowledge graph.

There are two key points to be stressed in this strategy of reasoning used by the experts. First of all there is the idea that people combine pieces of information in order to "gain" confidence on decision making, and second that people may take a decision without a full confidence in the supporti ng data and even without full information about all these data.

V.6a - Gain of confidence

In general, the relevance of the parent nodes is greater than the relevance assigned to their antecedents in the case of the experts' KG, but not for non-experts' KG (Leäo and Rocha, 1990). Thus, as a general rule

$$\delta'' > \delta \text{ and } \delta' \text{ and } \delta''''' > \delta''' \text{ and } \delta'''' \qquad (V.17)$$

in Fig. V.11. A net gain of confidence is also observed when information spreads from the terminal nodes toward the root (Rocha et al, 1989). For example, confidence on V is C in Fig. V.11 may be greater than the maximum confidence on T is A and U is B.

If Θ in V.5d (or V.15a) is the aggregation function associated with the non-terminal nodes of KG, then the above gain of confidence can be obtained by increasing the slope of the projection function f in V.5e (or g in V.15b) relating σ_c and σ_a. In this line of reasoning, learning involved in expertise is not only related to the datum clustering, but also with the adjustments of the thresholds and the slope of the encoding function. The encoding function is subject to the control of modulators (Chapter III), which may be used to support this learning. The increase of the slope of the projection function has the purpose of reducing the size of the interval of the possible values of σ_a resulting in $\sigma_c \equiv .5$. In other words, the increase of the slope of the projection function reduces the possibility of ambiguity.

V.6b - Support and refutation

The statement made by the expert that

V.18) if I have strong confidence in the supporting data, I make my decision for the disease,

implies a threshold reasoning, in the sense that if σ_c is greater than this threshold α then the decision is accepted. This threshold α will be called Acceptance Threshold, and it correlates with the equality index in the matching process in Eq. V.5 and with the threshold of the axonic encoding function II.14 (Chapter II, section II.4).

A threshold reasoning implies that a decision may be taken with observation of only part of the supporting data. On the one hand, this may be an interesting strategy for cost/benefit optimization of the expert reasoning. On the other hand, this approach may augment the error in a decision process involving competitive hypothesis, unless some other constraints are imposed on the minimal set of observations. These constraints are necessary to guarantee that some other competitive hypothesis does not become a winner if some further pieces of information are obtained. In other words, they are used to guarantee some monotonicity in the decision making reasoning.

Let the following be defined (Machado et al, 1990):

V.19a) Current Acceptance Index β_a as:

$$\beta_a(c) = \sigma_c$$

so that σ_i in V.5d becomes:

σ_i = the actual matching between the observed data and the prototypical knowledge encoded by the corresponding antecedent, and

σ_i = 0 to all unobserved antecedents.

V.19b) Potential Acceptance Index β_p:

$$\beta_p(c) = \sigma_c$$

σ_i = the actual matching between the observed data and the prototypical knowledge encoded by the corresponding antecedent, and

σ_i = 1 in all unobserved antecedents.

V.19c) Maximal Acceptance Index β_M

$$\beta_M(c) = \sigma_c$$

$$\sigma_i = 1 \text{ to all antecedents}$$

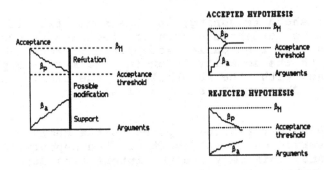

FIG. V.12 - DECISION ON PARTIAL REASONING

It follows from the above (Fig. V.12) that:

V.19d) $\beta_a(c)$ measures the actual support received by the consequent (hypothesis);

V.19e) $\beta_P(c)$ measures the total confidence to be gained if data about all antecedents are obtained;

V.19f) β_M - $\beta_P(c)$ measures the refutation received by the consequent (hypothesis);

V.19g) the ignorance τ about the consequent (hypothesis) is

$$\tau(c) = \beta_P(c) - \beta_a(c)$$

V.19h) if all antecedents are observed then

$$\tau(c) = 0 \quad \text{and} \quad \beta_P(c) = \beta_a(c)$$

In this line of reasoning, the consequent of the fuzzy proposition

V.20a) may be accepted if
$$\beta_a(c) \geq \alpha$$

V.20b) is rejected if
$$\beta_P(c) \leq \alpha$$

V.20c) is accepted if V.20a is fullfilled, and $\beta_P(c')$ associated with any other competing hypothesis is:

$$\beta_P(c') < \beta_a(c)$$

The following definitions also hold

V.21a) exhaustive reasoning implies that decision is made only if $\tau(c)=0$, whereas

V.21b) partial reasoning allows a decision making with $\tau(c)>0$.

This kind of threshold reasoning supports decisions of the type:

V.22a) I may accept (reject) the hypothesis h with a confidence value α ".

Given a knowledge base about a set H of hypothesis, it may be the case that no conclusion about accepting or rejecting any h ϵ H is possible for a given acceptance threshold α. In this situation, α may be continuously reduced (or increased) until a first hypothesis can be accepted (or rejected). This approach supports decisions of the type:

V.22b) My best evaluation is to accept (reject) the hypothesis h with confidence α ".

Whenever $\alpha > 0$ is obtained and a decision is made about a hypothesis h, another $\alpha' < \alpha$ may be investigated to support the following decision:

V.22c) My best evaluation is to accept (reject) the hypothesis h with confidence α, and to accept (reject) the hypothesis h' with confidence α';

In this condition and assuming h' = ˜h as the negation of h, the following holds

V.22d) there exists α and $\alpha˜$ so that

$$\sigma_c(h) + \sigma_c(˜h) = \alpha + \alpha˜$$

While
$$\beta_a(h) \leq \alpha \quad \text{and} \quad \beta_p(h) > \alpha \qquad (V.23a)$$

for any hypothesis in the knowledge base, the expert continues to collect facts to support his decision. The purpose of this inquire is to enhance $\beta_a(h)$ (the support of h) or to decrease $\beta_p(h)$ (the refutation of h), and to reduce $\tau(h)$ (the ignorance about h) as much as possible. Therefore, the inference to be tested first must be that in KG for which the inquiring index $\Gamma(c)$

$$\Gamma(c) = \tau(c)/\beta_p(c) \text{ ---> } 1 \qquad (V.23b)$$

because its solution can guarantee the highest modification of $\beta_a(h)$ if the consequent is proved, or $\beta_p(h)$ if the consequent does not succeed. In this way, the inquiry has the purpose of

V.23c) monotonically increase $\beta_a(h)$, and decrease $\tau(h)$ in order to prove h, or

V.23d) to monotonically decrease $\beta_p(h)$ and $\tau(h)$ in order to reject h.

If there are more than one hypothesis satisfying V.23a, then

V.23e) the hypothesis h exhibiting the highest $\beta_p(h)$ must be investigated first.

The fuzzy sets defined by β_a and β_p are propagated through KG as shown in Fig. V.11 in order to provide the elements for the above decisions.

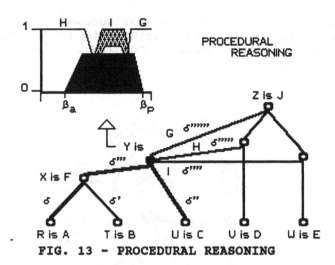

FIG. 13 - PROCEDURAL REASONING

V.7 - Procedural knowledge

The Procedural Knowledge (Rocha et al, 1990a) is obtained if the list of its terminal nodes is temporarilly ordered (Fig. V.13). The expert joins these nodes into different pathways to reach the root representing the diagnosis. These pathways branch from specific vertices called decision nodes (filled node in Fig. V.12). These different branches are associated with the distinct labels of the linguistic variable assigned to the decision node.

In this type of reasoning, if the conclusion at the decision node is:

V.24a) Y is G then the solution Z is J is accepted; if it is:

V.24b) Y is H then information about V is D is asked in an attempt to make a decision; if it is:

V.24c) Y is I then information about W is E is asked with the same purpose.

The presence of the decision node drastically changes the processing in the Procedural Reasoning in respect to the Declarative KG, in order to cope with cost/benefit analysis and with non-monotonic reasoning. Besides this, decision nodes may direct the flux of processing from one KG to another whenever the hypothesis represented in one of them does not succeed in being proved. Because of this, these nodes agglutinate (Fig. V.14) the KGs of a data base into a Knowledge Net (KN).

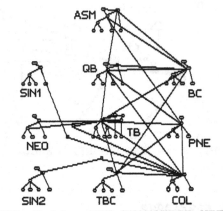

FIG. V.14 - THE KNOWLEDGE NET

V.7a - Cost and benefit

The expert orders the pathways of KG in order to optimize the decision according to their utility, i.e., their cost/benefit relation. The general idea is to (Moskowitz et al, 1988; Rocha et al, 1989, 1990a):

V.25a) choose the pathway p to be explored first according to its cost and the actual confidence on the observed antecedents, so that high risky pathways are chosen only if $\beta_a(p)$ is high.

Let it be given

$$\text{if } Q\{[(X \text{ is } A)\cdot\delta] \text{ and } [(Y \text{ is } B)\cdot\delta'] \ldots\} \text{ then } Z \text{ is } C \quad (V.24b)$$

then the following definitions hold:

V.25c) the benefit $\delta_i(c)$ in solving V.24b is the reduction in the ignorance $\tau(c)$ provided by the observation of the ith antecedent, that is:

$$\delta_i(c) = \tau_{i-1}(c) - \tau_i(c)$$

V.25d) the cost ϕ_i of the ith antecedent is an external value assigned to it in the interval [0,1], to measure at least one of the following: financial cost, time delay, threat to the integrity of the system, etc., in obtaining the information encoded in this antecedent. This value assignment is part of the Procedural Knowledge and it is encoded in the order the terminal nodes are placed in KG: the most costly nodes are placed at the right of KG, that is, associated with high values of i.

V.25e) the utility of a pathway p defined by a consequent c is:

$$\pi_p = \delta_i(c) \ / \ \Phi_i$$

The purpose of the procedural reasoning is to optmize the decision making by giving priority to the pathways exhibiting the highest utility index π_p. This is obtained by ordering the pathways so that those at the left of KG are associated with the highest $\delta_i(c)$ and the smalest Φ_i, and by orientating the inquiring about the antecedents in KG from left to right. Because of the fuzziness of π_p, the threshold reasoning (V.24) represented by the its linguistic encoding at the decision nodes is the best solution for optimizing this process (Fig. V.12). It follows that the procedural knowledge is heavily dependent of the ordering of the propositions in KG., and the decision about the inquiring is determined by V.24 and V.25.

V.8 - Decision making in therapy

Decision making in medical therapy is dependent of different types of data, besides the diagnosis. The following types of information may be identified as influential and the most frequent ones in the decision making about the best therapy:

V.26a) the general state of the patient: factors like hydration, nutrition, temperature, age, discomfort, etc., are taken into consideration. For example, if the hypothesis is infection, the presence of fever as a possible sign of toxemia, may induce the expert to introduce a broad-spectrum antibiotic therapy while waiting for the results of urine culture and antibiotic sensitivity tests (see pathway defined by the nodes CD, IUT and Br in Fig. V.15). The same strategy may be used if discomfort provoked by the illness is great, as may be the case of conjunctivitis. Age may influence the type of surgery chosen or even the choice between clinical treatment and surgery;

V.26b) the state of progression of the disease: in case the disease is progressive, its stage is very influential in the evaluation of the best therapy. For instance, signs of prostatitis (node PR in Fig. V.15) may also influence the decision of a broad-spectrum antibiotic therapy previous to the laboratory results of culture of urine and antibiotic sensitivity tes ts in the case of infection of the urinary tract (IUT);

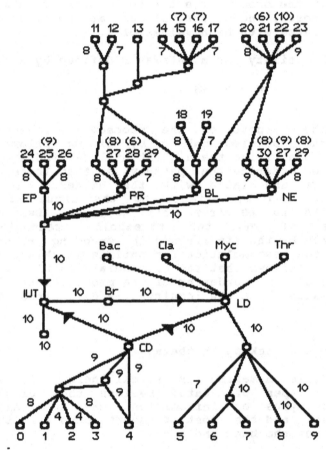

FIG. V.15 - KNOWLEDGE NET ABOUT URETHRITIS

V.26c) associated diseases: if other diseases are present, they may change the decision about the best therapy, or may require stable and acceptable measures of important parameter to support the decision for a specific therapy. For example, stable and acceptable blood-glucose levels are required in the case of associated Diabetes Mellitus;

V.26d) patient's agreement: after explaning to the patient his conditions and treatment/prognosis relations, the expert is influenced by the decision of the patient about the

choice of therapy. In many instances, the outcome of the
patient's decision may be anticipated by the expert
according to his experience with other patients. For
example, in the case of surgery of cataract, the patient's
decision is well correlated with the degree of visual
deficiency and professional activity. In some instances, the
family's agreement is taken into consideration because of
some mental incapacity of the patient.

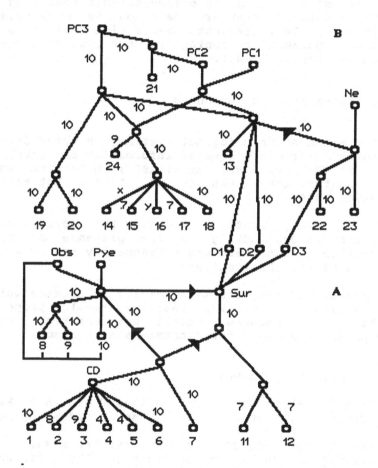

FIG. V.16 - KNOWLEDGE NET ABOUT LITHIASIS

The expert uses this kind of data for a reasoning of
the type:

V.27a) Decision is X unless the Exception is Confirmed
then Decision is Y.

For example, after the diagnosis of Lithiasis is established, the rule assigned to the node labelled P in Fig. V.16 is

Ask Pyelography (node Pye) unless Anuria (node 7) is Present
   If Anuria is Present then Surgery (node Sur) is Proposed
                                                         (V.27b)

This type of reasoning is called default reasoning (Reiter, 1980) because a standard decision is made unless some exception or less frequent observation justifies another solution. Default reasoning is one type of non-monotonic reasoning (Bobrow, 1980).

V.9 - Non-monotonic reasoning

        Non-monotonic logical systems are those in which the introduction of new pieces of information can invalidate old decisions (Bobrow, 1980; Davis, 1980; McDermott and Doyle, 1980; Reiter, 1980; Wah et al, 1989). But according to Winograd, 1980:

"they are very useful in modeling the beliefs of active processes which, acting in the presence of incomplete information, must make and subsequently revise assumptions in light of new observations".

As a matter of fact, non-monotonicity is pointed out (Wah et al, 1989) as one of the most remarkable properties of intelligence, because intelligent entities are non-deterministic rather than deterministic systems.

        Let the rule be:

V.27c) Wait Laboratory Result unless There Are Fever ....
   if Fever is Present ... then Use Broad-spectrum Therapy

assigned to the node CD in Fig. V.15. The result about the sensitivity of the actual bacteria to the different tested antibiotics (nodes 6 and 7 in Fig. V.15) may confirm the actual drug in use as adequate to the actual patient, or it may require the change of therapy.

        Because of this necessity of truth maintenance, it is wise to maintain these knowledge systems as minimal as possible, or in the words of Davis, 1980:

V.27d) only those objects should be assumed to exist which are minimally required by the context.

But this is one of the basic properties of expertise (see section V.11) since the knowledge graphs and nets obtained from the experts are always smaller and simpler than those provided by the non-experts (Figs. V.6,9).

Besides directing the processing flux inside KG according to the utility of the pathways, decision nodes are also in charge of diverting this flux toward another KG whenever $\beta_P(h)$ in the first one falls below the acceptance threshold α for h.

Whenever no solution or any contradiction is encountered in the actual KN, the thresholds in this net may be changed to increase or to reduce the size of the minimal set of inferences in use, and to permit a revision of the actual beliefs. The axonic thresholds are under control in the brain, and many special chemicals are available for such a control (Rocha, 1990d). This is in line with the definition of truth maintenance systems (McDermott and Doyle, 1980).

V.9a - Default reasoning

A default has three parts (Bobrow, 1980): a pre-requisite or a Tester; a consequent; and a set of assumptions or Restrictor. Essentially, the Tester is a set of propositions used to determine if it is appropriate to enforce the application of the Restrictor which, if satisfied, implies the alternative solution of the problem (Yager, 1990a). For example, given

$$\text{if (X is A) and [(Y is B) is possible] then} \quad \text{(V.28a)}$$
$$\text{Z is C else W is D}$$

the Tester is the space of all interpretation for which (X is A) tends to be true, and the Restrictor is the set of all interpretations which exclude (Y is B). If the Restrictor (Y is B) tends to be true then the decision is (Z is C), otherwise the conclusion is (W is D). The example in V.27 is a rule of this type.

Let the following general formula of default reasoning to solve V.28a be:

$$\text{if } [(X \text{ is } A) \cdot \delta] \text{ and } [(Y \text{ is } B) \cdot \delta'] \text{ then R is L} \quad \text{(V.28b)}$$

$$L = \{ \text{ apply, not apply } \}, \text{ and}$$

$$\text{if R is (apply) then Z is C else W is D}$$

In this context, the solution of V.28a becomes:

Given $[(X \text{ is } A) \cdot \delta]$ and $[(Y \text{ is } B) \cdot \delta']$,

if $\sigma_a \geq \alpha$ then Z is C else W is D (V.28c)

The following types of default are defined depending on the behavior of the aggregation used to calculate σ_a:

V.28d) Default type 1 - used for deciding about different outputs (therapies in the above examples):

$$\sigma_a \geq \alpha \text{ if } (Y \text{ is } B) \text{ holds}$$
$$\text{otherwise: } \sigma_a < \alpha$$

V.28e) Default type 2 - used to decide about competitive hypothesis

$$\sigma_a \geq \alpha \text{ if } (Y \text{ is } B) \text{ holds or absent}$$
$$\text{otherwise: } \sigma_a < \alpha$$

This kind of reasoning is non-monotonic, because, e.g. in the absence of any information about $(Y \text{ is } B)$, $(Z \text{ is } C)$ is inferred. If information about $(Y \text{ is } B)$ is provided in the future, then the previous conclusion may be invalidated if $(Y \text{ is } B)$ is proved to be true. Let the case bethe one of default type 1.

Since the Tester has to be observed in order that the default can be applied:

$$0 < \beta_a(c) < \alpha < \beta_p(c) \qquad (V.28f)$$

Where α is the acceptance threshold for Z is C.

If no information is reported about $(Y \text{ is } B)$

$\sigma_W = \beta_a(c)$ so that $(W \text{ is } D)$ is accepted   (V.28g)

because $\sigma_a <$ than α. Also:

$\tau(c) > 0$ because $\beta_p(c) > \beta_a(c)$ (V.28h)

In the case information is provided and $(Y \text{ is } B)$ does not hold:

V.28g is satisfied

and

$\beta_p(c) \longrightarrow \beta_a(c)$, so that $\tau(c) \longrightarrow 0$ (V.28i)

In this condition, $(W \text{ is } D)$ is accepted with confidence smaller than α.

Finally, if (Y is B) is proved to be true

$$\beta_a(c) ---> \beta_p(c) > \alpha$$
and (Z is C) is accepted
with $\tau(c) --> 0$ (V.28j)

Thus, the conclusion (W is D) can be modified by the observation of (Y is B), but in this case $\tau(c)$ decreases. Default types 1 and 2 are monotonically decreasing concerning $\tau(c)$.

There is another type of default used by the expert

V.28k) Default type 3 - (W is D) is taken as try another observation of (Y is B), so that:

$$\sigma_a \geq \alpha \text{ if (Y is B) holds or is absent}$$
otherwise and after some trials: $\alpha ---> \sigma_a$
and (Z is C) is accepted (or rejected)

This type of default is used if σ_a is above .5 and the observation of (Y is B) is not an error-free observation. In this situation, negation of (Y is B) is assumed to be an error which could not invalidate the decision. Because of this, the expert is initialy unable to disregard the hypothesis, and he prefers to assume the possibility of the error. If after some trials he is not able to prove the error, then he accepts (rejects) the hypothesis in presence of some ignorance. If (Y is B) is proved to be true the confidence increases and the ignorance is reduced. This type of default reasoning is called also non-monotonic reasoning induced by resource limitations (Winograd, 1980).

Non-monotonic reasoning imposes special conditions to deffuzzification of the Extended Modus Ponens, because of the notion of typicality implicit in this kind of thinking. Typicality will favor some points in C as the most important for any comparison with C'. Also, the choice of the singleton $c \in C'$ to represent the final result of the implication is influenced by the default reasoning. Methods like those proposed by Katai et al, 1990a, favoring one or the other limit of the constraint-intervals composing C' as the eligible singleton, are more in line with the way the experts defuzzify the implication in the case of non-monotonic reasoning.

V.10 - The uncertainty state space

Uncertainty is a multidimensional concept (Klir, 1989; Shaffer, 1976; Smithson, 1987; Zadeh, 1987) each of its dimensions being privileged by certain mathematical theories e.g. probability, possibility, etc. (Klir, 1989).

The experimental data on expert reasoning shows that man uses different kinds of knowledge representation to work different dimensions of uncertainty. On the one hand, Declarative Knowledge handles two dimensions: relevance or possibility and confidence or similarity. On the other hand, Procedural Knowledge takes profit from ordering the input data to introduce a third axis in the uncertainty space in order to handle uncertainty of utility.

The AND/OR operator defined in V.15 may handle decision in the bidimensional uncertainty space of the Declarative Knowledge. This is done according to the procedures described in V.20.

The processing of uncertainty in the case of the Procedural Knowledge is not supported by the aggregation of the 3 axes of uncertainty of its confidence state space. Instead of this, a threshold reasoning (Pauker and Kassirer, 1980; Moskowitz et al, 1988) orders the relevance/confidence surfaces for accepting, inquiring, alternating and refusing hypothesis (Fig. V.17) in this confidence space, and decision becomes dependent of the capacity of climbing or jumping from one to another of these surfaces.

The expert navigates the knowledge graph (Fig. V.2 and 12) from left to right in order to gain confidence either in establishing a diagnosis or asking high costly data to support his decision (Kassirer et al, 1987; Rocha et al, 1990). At the first steps, when confidence is not great, it is possible to refuse any hypothesis of disease, and consider the individual as healthy (Eddy and Clanton, 1982; Milne, 1987). However, as more data is gathered, less free is the decision. As confidence increases, the only possible outcome is a diagnosis (default reasoning type 3, V.28k). This is attained maintaining the actual hypothesis or changing to another one in the knowledge net.

In general, when the decision made is to change hypothesis, the jump is from an "Alternative" surface in the first hypothesis space toward an "Inquiring" surface in the other space (pathway J in Fig. V.17) with some gain of confidence. This process explains why the expert may avoid the use of negative information on reasoning with competing hypotheses. This is a strong strategy employed by all, except one of the experts interviewed in this research. The jargon lists provided by these exp erts are composed only by positive information like Fever, Pain,, etc. and never by negative findings as Absence of Fever, Absence of Pain, etc.

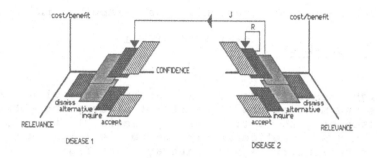

FIG. V.17 - NAVIGATION IN THE KNOWLEDGE NET

If the confidence in positive information is low, the reasoning may jump to an "alternative" surface in the confidence space of another hypothesis. This is equivalent to the use of negative information to support the alternative hypothesis, without the risk of being inconsistent if additive reasoning is being used, as is the case of the additive reasoning (section V.6a).

Let positive d and negative information ˜d be used. Now, if $\sigma(d)$ and $\sigma(\tilde{d})$ ---> .5 and additive learning is used then

$$\sigma(h) \text{ and } \sigma(\tilde{h}) > .5 \qquad (IV.29a)$$

for the hypothesis h and ˜h supported by d and ˜d, respectively. This is a consequence of the slope of f in V.5e being greater than 1. But from V.22d

$$\sigma(h) + \sigma(\tilde{h}) = \alpha + \alpha^{\tilde{}} \qquad (IV.29b)$$

where $\alpha(h)$ and $\alpha^{\tilde{}}(h)$ are the acceptance thresholds for h and ˜h, respectively. Thus

$$\alpha^{\tilde{}}(h) \text{ --> } \alpha(H) \qquad (IV.29c)$$

as

$$\sigma(h) \text{ and } \alpha(\tilde{h}) \text{ ---> } 1 \qquad (IV.29d)$$

which invalidates VI.22c.

This kind of inconsistency is avoided if either

V.30a) negative information is not used in additive reasoning, or

V.30b) the following constraint is imposed in IV.22d

$$\alpha(d) + \alpha^{\sim}(d) < 1$$

in order to prohibit

$$\sigma(d), \sigma(^{\sim}d) \text{ ---> } .5$$

High costly inquiry is associated with high confidence and most of the time, if not always, the obtained datum is used only to confirm the hypothesis. If this is not possible, because confidence in datum is low, then the decision is to inquiry again (pathway R in Fig. 17 and the recursive pathway from node Obs in Fig. V.15). This is a consequence from the fact that real data is not error-free and additive reasoning tends to sequester confidence in the highest quarter of the Confidence space as the number of used evidences increases, even if they are less trustful. This corresponds to the default type 3 reasoning.

V.11 - MPNN supports expertise

Each KG provides the structure for a fuzzy deduction because the processing supported by it maximizes the confidence in the decision making. The knowledge net KN formed by the KGs associated with the different diagnoses is the fuzzy knowledge base associated with expertise. It was shown in Chapter II that MPNN supports fuzzy reasoning. In this way, the expert knowledge may be implemented into a modular MPNN, each MPNN module associated with one of these KGs.

The KG may be read as a MPNN because:

V.31a) the update neuron is a fuzzy device being able to process the extended modus ponens required by the solution of a fuzzy implication (Chapter II, section II.6.3);

V.31b) the aggregation supported by the update neuron (II.14) is the same kind of quantified aggregation (V.15) used by the expert to reason with the collected data;

V.31c) the axonic encoding (II.18) is related to linguistic variables (II.27) in the same way the expert uses the filtering properties of these variables (V.16) to navigate KN;

V.31d) the axonic thresholds can be put under control (see II.16, Chapter II, section II.4) to support the threshold reasoning discussed for both monotonic (V.22,21,22) and non-monotonic reasonings (V.28);

V.31e) MPNN can process a multi-dimensional uncertainty space, since:

V.31f) uncertainty of matching is related to the axonic activation σ_c;

V.31g) uncertainty of frequency is related to the synaptic strength s_i;

V.32h) the ordering of the input neurons may be used to encode the uncertainty of utility in the same way the position of the terminal node reflects the utility of the information it represents. Ordering of neurons is a well established property of brain circuits and is easily produced by the embryogenic processing. Because MPNN can be programmed by a similar embryogenic process (Chapter III, section III.9) ordering may be easily enforced in MPNN to account for the encoding of utility uncertainty.

Besides all of this, the emergent properties of MPNN discussed in Chapter IV are the same basic properties of the expert reasoning disclosed in this research.

V.12 - Properties of the expert reasoning

The following properties of expertise follow from the contents of the previous sections:

V.32a) the expert knowledge graph is smaller and crisper than the graph provided by the general cardiologist of the same congenital cardiac diseases;

V.32b) the complexity of the expert knowledge graph was directly correlated with the difficulty of the diagnosis;

V.32c) the expert provided precise knowledge about rare diseases even when he had never treated a patient with them. Non-experts refused to provide knowledge even for complex but not rare diagnosis;

V.32d) although great among experts, disagreement is smaller than among non-experts;

V.32e) knowledge elicited from different experts and used as the knowledge base for a second generation expert system, discloses bias to different diagnoses among the experts;

V.32f) relevance values associated to the arcs of the expert knowledge graph are higher than the values provided by the non-expert;

V.32g) relevance is weakly related to the conditional frequency of data given the diagnosis calculated from the expert's data base;

V.32h) relevance values increase from terminal nodes toward the root with a net gain, so that values assigned to the incoming arcs of the root are higher than the maximun for the other arcs. This is very true in the case of the experts;

V.32i) the experts gathered information to have a net gain of confidence for decision making, because they used quantified agglutination;

V.32j) Contrary to the non-expert, the expert organizes his knowledge in multiple alternative pathways to reason with partial data;

V.32k) the ordering of the terminal nodes in the expert procedural KG perfectly encodes the utility of the information represented by the node;

V.32l) the thresholds at the decision nodes are dependent of this utility ordering;

Similar properties were also associated with the expert reasoning by other authors (Eddy and Clanton, 1982; Kassirer and Gorry, 1978; Miller and Masarie, 1990; Pauker and Kassirer, 1980; Sanchez, 1989).

Properties V.32 are derived from the following properties of MPNN discussed in Chapters III and IV:

V.33a) Well Learned σ-Models or MPNNs (WLMs) are produced in a non-homogeneous environment similar to that experienced by the expert;

V.33b) Encoding by WLM tends to be context free, whereas it is context dependent in not well learned models (LM);

V.33c) WLM tend to be trees with small number of nodes, whereas larger graphs tend to represent LM;

V.33d) Time required for processing in WLM is less than that required by LM;

V.33e) WLM have strong autonomy or self-reproduction, being easily recreated in their original brain. Recall is easy for WLM;

V.33f) WLM are strong germs for the development of new models about the changing world or for enhancing the comprehension of the surrounding environment. WLM are evolutive;

V.33b) WLM may be sociogenetically inherited;

V.33c) Deductive learning changes the structure provided by the inductive learning. The strength of the linkages in WLM as well as their encoding functions are dependent of both types of learning;

V.33i) WLM exhibit strong mobility or exogenous-repro-
duction in environments of low entropy, such as specialized
environments. However, each newborn model will lightly
differ from its ancestor, reflecting the individuality of
the new fostering brain, and

V.33j) WLM have small descriptive complexity and high
functional entropy. The entropies of WLM are related to the
entropy of the knowledge being modeled.

As a consequence of V.33, it can be said that MPNN
implements the expert reasoning. Examples of two different
MPNN systems implementating the expert reasoning will be
discussed in Chapter VII and Chapter IX.

ACKNOWLEDGMENT

Fernando Giorno was who first proposed me to adapt
as a general knowledge acquisition tool, the methodology I
and M. Theoto were using for language understanding
analaysis (Chapter VIII). Beatriz Leäo took in charge the
responsability of showing the efficacy of this adapted tool
in eliciting the expert knowledge. The discussions with
Ricardo J. Machado were very important in establishing the
general rules of the Knowledge Graph navigation.

I am in debt with the students: A.T. Sato, C.C.
Gravina, J.L. Roque, J.V.L. Cardoso Jr., R.J. Czerwinski and
Y. Irokawa of the course of Artificial Intelligence Applied
to Petroleum, Faculty of Mechanical Engineering, UNICAMP,
because they applied the knowledge acquisition tool
discussed in this chapter, to acquire knowledge in the field
of Offshore Well Maintenance; used this knowledge in the
Knowledge Base Fuzzy Expert System, and successfuly tested
the model with real data provided by Petrobás (Rocha et al,
in preparation).

CHAPTER VI

MODULAR NETS

VI.1 - Modularity of knowledge

Modularity of knowledge is one of the properties disclosed by the experimental investigation of expertise (Chapter V). The expert reasoning models are aggregated into complex nets according to the classifications, diagnoses or procedures used to organize the activities within the field of expertise. Decision nodes link this reasoning models in an Expert Knowledge Net. Modular reasoning is also proposed to solve many other problems. Chapter IV discussed the idea of using germs and haloes as building blocks for the complex reasoning involved in such activities like scheduling, research, etc.

Modular programming is a very popular approach in computer sciences, but it is also proposed to organize some biological activities. For example, elementar motor actions supported by spinal circuits are proposed to furnish a set of basic models or building blocks in the programming and organization of the motor control (Handelman and Stengel, 1987). In this approach, complex movements of the repertoire of a given animal are obtained as adequate combinations of the building blocks after a training period (Chapter III, section 11). Movement pathways are obtained as time sequences of these basic movememts. To control walking, swimming, etc. becomes a matter of planning these sequences.

Modularity is also a solution discovered by nature concerning the evolution of the neural systems. It begins with the construction of the ganglionic systems in primitive animals, but reaches its efficiency in structures like the cerebellum and the cortex.

The cerebellum is one of the most important neural systems involved in the learning and control of motor activities. Its structure is organized around the Purkinje cell in an essentially modular fashion. These modules are sequentially linked by a fibre system running parallel to the cerebellar surface. This structure allows the temporal ordering of the movement to be encoded in the spatial sequence of these modules.

The cortex is the most recent acquisition of the nervous systems of the most developed animals in the earth. It is considered the place of many of the most important intellectual activities of man. The cortex is a very modular structure.

VI.2 - Modularity of the cortex

The human cortex has an area of about 2500 cm2 with at least 10 thousand million neurons. About 5% of this area is special for receiving sensory inputs from eyes, ears, skin and projecting motor outputs. The dominant component (over 90%) is believed to be specially related to mental events (Eccles, 1981), associating and integrating all kinds of information.

The cortex operates in a modular fashion, its neurons being organized in columns of about 400 μm diameter and 4000 elements, about 2000 of which are pyramidal cells (Eccles, 1981; Valverde, 1986; McConnell, 1988). The elements of each of these modules are disposed in six layers and a vertical arrangement prevails in their connections (Fig. VI.1).

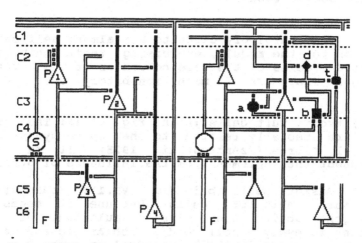

FIG. V.1 - THE STRUCTURE OF THE CORTICAL COLUMNS

Information from the thalamus and other cortical areas (F in Fig. VI.1) arrives at the Stellate cell S in the cortical layer C4. The S cells form one of the main inputs to the Pyramidal cells (P1 to 2 in Fig. VI.1) in layers 2/3. The other component of inputs to these cells are the parallel fibres in the cortical layer 1. The P cells (P1 to P4 in Fig. VI.1) are the main output elements of the columns, although they exchange synapsis between themselves and some of them in a recurrent mode. The output is directed to other cortical areas and to subcortical systems. Some P cells (P4 in Fig. 4) have recurrent axon collaterals which are the main source of the parallel fibres in layer 1. These recurrent axons make synaptic contacts at close and long distances (Eccles, 1981; Valverde, 1988). It is estimated that each column can establish important connections with around 60 other modules in the same and in the contralateral hemispheres (Eccles, 1981 and Valverde, 1988). All these contacts are made by means of excitatory synapsis.

It is proposed (Mountcastle, 1978; Szentagothai, 1978) that each module can be functionally subdivided into many minicolumns because of the extremely narrow spread (50 μm) of the vertically-directed axons. Therefore, the functionally processing circuits must have low structural complexity, despite the huge number of elements in each column. The possibility is that well closed circuits are packed inside the module, each one of them representing slight different associations of almost the same set of input. This process could be effective for memory representation of sets of observations about the same problem, such as storage of individual cases of the same classification task.

The inhibitory circuits are almost self-contained in relation to the module, exercising their inhibitory action over the P cells of the same column or at most over neighbor modules. There are 4 main inhibitory circuits (Eccles, 1981):

VI.1a) Axonic Tuft cells (t in Fig. VI.1): These inhibitory cells are specially related to the spine synapses made by parallel fibres (Szentagothai, 1975). It may be used to disconnect the columns from the influence of other modules.

VI.1b) Basket Cells (b in Fig. VI.1): These are the most important inhibitory cells because it seems that by convergent action they give a multitude of inhibitory synapses to every P cell body. Both Axonic and Basket cells exercize a graded inhibitory control over the P cells, which may be important for a competitive behavior among sub-modules in the same column.

VI.1c) Axon-axonic cells (a in Fig. VI.1): They act upon the initial portion of the axons of the P cells in the layers 2 and 3. These cells are distributed over several hundred P cell axons and in each axon there is convergence from about five (2-14) cells (Szentagothai, 1978). The control exercised by these cells is different from the action of Basket cells which could be a graded effect dependent of the relative intensity of the excitatory and inhibitory synaptic action. The axonic inhibition is more likely to be total or not at all. The fact that this circuit is restricted to layers 2 and 3 may point to some special features of this gating control to represent control knowledge (see section IX).

VI.1d) Cellule à Double Bouquet (d in Fig. VI.1): They are

also restricted to layers 2 and 3 and are inhibitory to all other inhibitory cells. This process of disinhibition of the P cells was proposed to exert special effects on the organization of the minicolumns of each module, but it remains up to the moment mostly unclear (Eccles, 1981).

The study of the distribution of sensory information in the primary receiving areas (somatic, auditory and visual areas in Fig. VI.2) discloses much of the functional aspects of the columnar arrangement of the cortex (e.g. Mountcastle et al, 1964; Poggio and Mountcastle, 1963; Rocha, 1985). Each column is related to some functional aspect of the analysis of the sensory input, such as receptive field, stimulus orientation, contrast processing, etc. Besides, different processing spaces may be defined for each primary sensory information (e.g. Ballard, 1986).

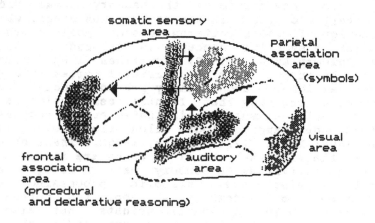

FIG. VI.2 - INTEGRATIVE FUNCTIONS OF THE CORTEX

The analysis of the cortical somatic sensory representation shows that the columns are ordered in some esthetic area mimicking the body's topology, with information from the feet represented in the upper part and from the face and from the head in the lower part of the hemisphere contralateral to the stimulated side. The body is represented many times, each representation enjoying a different neighborhood. This means that the columnar ordering is a partial order, e.g. most of the columns in the hand area process information arriving from the receptors in the hand, But some other modules are related to other different parts of th e body (e.g. mouth, arms, etc.) which maintain functional relations to the hand (Rocha, 1980, 1985).

Partial ordering of columns and the pattern of multiple representation of the same information in different neighborhoods is also observed for the other sensory areas, and at least for the parietal associating region (Hyvarinen, 1982; Sakata et al, 1980). Columnar partial ordering may be considered a general plan for cortical processing.

The main input for the parietal association area (Fig. V.2) is provided by primary sensory areas (e.g. Hyvarinen, 1982, Luria, 1974). The result is that parietal columns are related to the processing of complex sensory information involved in specific patterns of stimulation (e.g. Sakata et al, 1980). Also, it seems possible to establish links on this level, between the incoming information and its semantics, as in the case of word recognition in the speech areas (e.g. Luria, 1974).

It is possible to consider that pattern recognition is the basic rule of the sensory areas and of the preliminary processing in the association areas, whereas the association of these patterns to some specific meanings to establish a symbolic concept is at hand in the parietal areas (Goldman-Rackic, 1988; Hyvarinen, 1982; Luria, 1974). Most of the output from the parietal cortex is directed to the frontal lobe. Neuropathology provided enough information to support the view that frontal association areas are involved with the most sophisticated human reasoning, such as text understanding, logic calculations, etc. (Greco and Rocha, 1988; Luria, 1974; Rocha, 1990c). These observations support the view that neural networks are involved from the most basic activities on sensory pattern recognition to the highest sophisticated symbolic processing of this information. To perform this job, the cortex encodes knowledge not only on synaptic weights, but also on the types of neuron composing the columns and on the chemicals handling their synaptic transactions.

VI.3 - Modular MPNN

In current applications of neural nets, one generally starts with some sort of a black box where all or most of all kinds of connections among neurons are allowed. The net is then subjected to some kind of learning, which changes the connecti ons inside the box, so that after some training, some knowlegdge is modeled inside the net. This approach is time consuming because the complexity of the learning is proportional to the difference between the entropy of the net and of the system to be modeled (Chapter IV, section IV.4). The net's entropy enhances as the number of connections increases and the initial weights are equalized. Thus the black box approach tends to augment the length of the training period by increasing the difference

of entropy between the net and the model. Modular MPNN
(Rocha, 1990b and Rocha et al, 1990e) is the alternative to
this hard approach because it allows some initial knowledge
to be crafted into the modules. This initial knowledge is
genetically or sociogenetically inherited by MPNN (Chapter
IV, section IV.8). The notion of modularity is strongly
linked with that of MPNN. Because of this, the term MPNN
will be used from now on as Modular Multi-Purpose Neural
Nets.

A MPNN is a neural network composed of specific
sub-nets or modules, each module executing a defined
processing, and the entire behavior of MPNN depending on how
these modules are combined. Thus, modules are defined and
combined to build the entire net, in the same way cortical
columns are associated to support the human reasoning.
Because the modules are small nets, their programming by
means of an adequate language like that introduced in
Chapter III, becomes an easy task. The combination of
modules into a large net becomes similar to the building of
a program with any kind of Object Oriented Language. SMART
KARDS(c), described in Chapter IX, is an intelligent system
which takes advantage of this kind of approach to program
neural nets.

Besides being crafted, MPNN may be trained in the
same way classic neural nets are built. Combining these
different techniques, one may craft the modules and then use
inductive learning to adjust the weights of connections
inside and between modules. Selection of these modules
provides the germs for new deductive learnings. In this way,
new modules and/or new nets may be created by a random
process, or associative learning or by formal models of
learning (Chapter IV). The structure of the modules can also
be specifically modified according to instructions provided
by the user, in the same way humans learn by being told at
home, school, etc. and/or by parents, friends, teachers,
etc. By the same process, germs and haloes may be "told" to
generate new modules or new nets. Jargon described in
Chapter VIII takes advantage of all these learning
strategies to discover and to acquire knowledge about the
contents of language data bases.

The crafting of the required modules of a MPNN
involves:

VI.2a) the selection of the types of neuron to be used: part
of the initial knowledge and of the specificity of the
module is dependent of the types of neurons composing it.
The library £ of neurons used to craft the MPNN can be
specified by its genetic G and may be programmed with the
language L(G) supported by G (Chapter III);

VI.2b) the distribution and ordering of these neurons inside
the module: MPNN can encode different types of uncertainty,

some of which are dependent of the ordering of their input neurons in the module. Layer distribution and ordering are specific issues of the embryogenic process used to build the MPNN (Chapter III), and

VI.2c) the wiring of the connections between the neurons or the definition of the allowed synapses in the module: the initial wiring of the module may be specified by the genetic G of the neurons and the substrate capacity of its layers (Chapter III, section III.9). The L(G) defined by this genetics not only encodes the module's initial knowledge but also provides the processing language to carry out the computations assigned to the module.

VI.4 - The library £ of neurons

This section presents and discusses some types of neurons which are important devices for crafting MPNNs. The different types introduced here are not intended to be an exhaustive listing neither of the processing components of MPNN nor of their properties.

The neurons N composing the library £ of a given MPNN have the following general structure:

$$N = \{ \{ W_p \}, W_o, T, R, C, \Theta, \{ \alpha, g \}, \{ f \}, L(G) \}$$
(VI.3a)

combining two different types of processings:

VI.3b) electrical processing: the different electrical activities v_i elicited by the distinct n pre-synaptic cells N_i upon the post-synaptic N_j are aggregated into a total activity:

$$v_j = \overset{n}{\underset{i=1}{\Theta}} v_i$$

where the semantic meaning of Θ is dependent of the chemical recoding on the pre-synaptic level and on the axonic encoding function g_j:

$$\text{if } v_j \begin{cases} \leq \alpha_1 \text{ then } w_j = w_i \\ \geq \alpha_2 \text{ then } w_j = w_s \\ \text{otherwise } w = g_j(v_j) \end{cases}$$

g_j is either a $\top$-norm or a $\top$-conorm. The different branches of the same axon exhibits different filtering properties, so that

$$\text{if } \alpha_k < a_j < \alpha_{k+1} \text{ then } w \in W_k$$

VI.3c) chemical processing: supported by the language L(G) defined by the genetics G, constructed from the simple encoding alphabet D, the substrate dictionary S and the grammar ⚡. The semantic of L(G) is defined by the set A of actions, whose purposes are to modify the behavior of the synapsis and/or the processing capacity of the neuron. The activation of the post-synaptic cell N_j by the transmitter released by the pre-synaptic cell N_i activates these control molecules c_j and triggers these actions:

$$t_i \; \Gamma \; r_j \; \gg \; c_j \; \text{---}> \; a \; \epsilon \; A$$

The amount m_i of t_i released by w_i at the terminal branch of N_i is:

$$m_i = f \; (w_i, M(t_i))$$

where f is a $\top$ -norm or $\bot$ -conorm and $M(t_i)$ is the total amount of t_i in N_i. The effect induced at the post-synaptic cell is:

$$v = m \; \hat{} \; M(r) \; * \; \mu(t,r) \; \cdot \; v_0$$

where $M(r)$ is the amount of post-synaptic receptor to bind the pre-synaptic transmitter; $\mu(t,r)$ is the binding affinity between these two chemicals, and v_0 is the post-synaptic activity triggered by one quantum of transmitter.

The following is an extended library from Rocha and Yager, 1992, of neurons supported bu VI.3:

VI.4a) Crisp Neuron (CN):
$$\text{if } 1/s = 0$$

where s is the slope of the encoding function g in VI.3b.

VI.4b) Fuzzy Neuron (FN):
$$\text{if } 1/s > 0$$

The output of the FN is an α-cut if g is the identity function, otherwise it is an α-level set (Negoita and Ralescu, 1975).

VI.4c) Inverse Fuzzy Neuron (IFN) - if g is a monotonic decreasing function.

IFN exhibits spontaneous activity different from zero since the output for $v_j = 0$ must be different from 0, and decreases if $v_j > 0$. IFN are usefull devices for calculating negation.

VI.4d) Full Range (FFN) - Full Range FN also exhibits spontaneous activity for $v_j = \alpha$; increases this activity if $v_j > \alpha$; decreases it if $v_j < \alpha$, and leaves it unchanged if $v_j = \alpha$.

VI.4e) Recurrent Neuron (RN) - one of its pre-synaptic input is its own output. In general, this synapsis is a modulator (MRN) or a gate synapsis (GRN). Recurrent neurons are useful to implement max-min operations in MPNN as well as to act as resetting devices.

VI.4f) Fuzzy Decision Neuron (FDN) - A FDN is a fuzzy neuron that spreads its activation differently throughout its axonic terminal branches, depending on their filtering properties. The filtering characteristics are defined by specific values of α and g associated to these axonic terminals. In this line of reasoning, a Decision Fuzzy Neuron is defined by one type of aggregation Θ and a family of thresholds α and encoding functions g. Thus:

$$FDN = \{ \Theta, \{\alpha\}, \{g\}, T, R, C \}$$

and

$$card(\{\alpha\}), card(\{g\}) > 1$$

where card stands for the cardinality of the corresponding family.

FDN are important devices for implementing ordering other than spatial ones in MPNN. They are also important for calculating some types of negation and for implementing controled inferences of the type IF THEN ... ELSE.

VI.4g) Aggregation neuron (AN): The cardinality of $\{\alpha\}$ and $\{g\}$ of AN, contrary to FDN, is equal to 1.

VI.4h) Gating neuron (GN): The slope of g in GN is very high so that it tends to fire as a Yes-No device. Besides, it tends to contact the post-synaptic neuron near the axon, so that most of its action is to quickly approach (or recede) the post-synaptic neuron to (from) its threshold. This kind of neuron is very important for the physiology of the brain (Allen and Tsukahara, 1974; Eccles, 1981).

VI.4i) Matching or receptor neuron (MN): MN is a neuron which has only one source S of input. This source is not another neuron, but it is a source of energy in the outside word. So, the primary calculation performed by MN is of the type

$$S \text{ is } A$$

Its structure is reduced to

$$MN = \{ \ S, \ W_0, \ \{\alpha\}, \ \{g\}, \ T, \ \{ \ f \ \} \ \}$$

VI.4j) Effector neuron (EN): EN is a neuron whose output set W_0 is used to control an effector device outside the MPNN.

VI.4k) Quantifier Neurons (QN): The threshold encoding in IV.3b plays an important role in defining the properties of the neurons of MPNN. In this way:

If α_1 --> 0 then QN tends to function as an OR device,

elseif α_1 --> $\max(v_j)$ then QN tends to function as an AND

otherwise QN functions as an AND/OR device

the degree of ORness (ANDness) depending on α_1.

True AND and OR devices are obtained if the slope of g is also increased, making QN crisper as α_1 --> $\max(v_j)$ or 0, respectively. So, the degree of ORness (ANDness) is also dependent of the slope of g.

As a matter of fact, the value of α_i and the slope of g are crucial factors defining the caracteristics of the linguistic quantifier Q (see Chapter II, section II.6) associated to each neuron in MPNN. Controlling these parameters, QN may be programmed according to the semantics of AT LEAST N, MOST OF, MANY, FEW, etc.

The chemical processing on the synaptic level may be used to program some other properties of the neurons in £:

VI.4l) Competition - controllers activated by one neuron can reduce the synaptic transmission from other neurons in MPNN;

VI.4m) Cooperation - controllers activated by one neuron may enhance the synaptic transmission from other neurons in MPNN;

VI.4n) Competition and cooperation among controllers triggered at the same post-synaptic neuron by different pre-synaptic cells may be also programmed, enhancing the computational power of the neurons in £.

Taking advantage of all this properties, the user may program the MPNN modules for both numeric processing as in the case of classic neural nets, and/or for algorithmic processing as discussed in Chapter III. Also, properties of the different types of neurons described above can be combined into a single neuron increasing its computational capability.

VI.5 - Basic circuits

The above types of neurons may be used to implement some basic circuits inside the modules of MPNN to perform some specific function. In this way, the MPNN module can be considered modular structure, too, in the same way the cortical column can be subdivided into minicolumns.

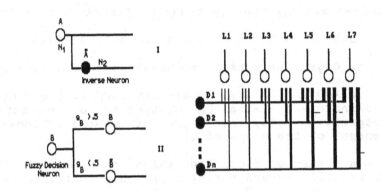

FIG. VI.3 - NEGATION AND ORDERING
Modified from Rocha and Yager, 1992

The circuits in Fig. VI.3 implement complementation and ordering in MPNN. In the circuit VI.3AI, the spontaneous discharge of N_2 is required to be equal to the maximum activation of N_1 and the encoding function g has to be the identity function, if complementation of N_1 is to be calculated in circuit VI.3-I as:

$$N_i = 1 - \overline{N}_i \qquad (VI.5a)$$

Otherwise, a non-conventional complementation will be calculated as:

$$N_i + \overline{N}_i <> 1 \qquad (VI.5b)$$

In the Circuit VI.3A-II, complementary concepts are defined for the same measure (B) according to the degree of activation of B. In this case:

$$\overline{B} + B <= 1 \qquad (VI.5c)$$

The circuit in Fig. VI.3b orders information in MPNN. The input is provided by the decision neurons D_1 to D_n making contact with the output (OR) neurons L_1 to L_7 by means of axonic branches having different filtering properties, so that L_7 is activated only if activation at the decision neurons attains high level, while L_1 is activated only if the activation attains low level in the decision neurons. In this context L_1 to L_7 process a fuzzy ordering from, e.g. LOW to HIGH, the membership function of each linguistic term being determined by the g functions associated to each axonic branch and to each L neuron. This circuit can implement the OWA operator proposed by Yager, 1988b, if all L_i converge to a output neuron N_0 with synaptic weights equal to the values of the averaging vector assigned to the OWA operator. In this condition the output of N_0 is the expected ordered weighted averaging.

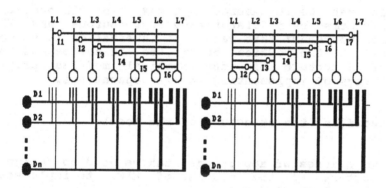

FIG. VI.4 - A MIN AND MAX CIRCUIT
Modified from Rocha and Yager, 1992

Max and Min operations are very important in Fuzzy Logic (e.g. Zadeh, 1965, 1975, 1985). The implementation of these operations in MPNN requires either special neural circuits or special properties for neurons and synapsis (see Chapter II, section II.6). Following, two circuits derived from that in Fig. VI.3b will be discussed to implement Min and Max operations in MPNN. They are not intended to exhaust the subject since there are other solutions.

The circuit in Fig. VI.4a calculates the minimum activation observed in the decision neurons D_1 to D_7.

The difference between this circuit and that of Fig. VI.3b is that here the neuron L_j fires according to the value of the input provided by the decision neurons, only if none of the inverse neurons I_i, $i < j$, is activated, otherwise its output is 0. Each inverse neuron I_i is also a gating neuron, because its activity controls the threshold of L_j, $j > i$. Thus, L_i fires only if no other L_j, $j < i$ is activated. This L_i represents the minimum output provided by the decision neurons D_j.

The circuit in Fig. VI.4B calculates the maximum activation observed in the decision neurons D_1 to D_7. The circuit is similar to that of Fig. VI.4a. The difference being that a neuron L_j fires only if none of the inverse neurons I_i, $j > i$, is activated. The output of this L_i represents the maximum output provided by the decision neurons D_j. The system of parallel fibres in the cortical layer C_1 provides the structure required by the circuits in Fig. VI.4.

It must be remembered that max and min operations may be performed by modulator recurrent neurons MRN. It is interesting to remark that recurrent innervation is a common occurrence in the cortical column,too. It may be concluded that the structure of the cortical modules provides different ways of implementing max-min operations.

VI.6 - Specifying the structure of the module

Two classes of approaches can be used to design the basic structure of the modules of MPNN to implement a specific initial knowledge K_0:

VI.6a) theoretical: in this case, some formal knowledge about the process to be modeled exists and it may be used to craft the internal structure of the modules and of the net.

This may be the case in developing MPNN for process control purposes. In many instances, engineering may provide precious informations about the process to be controlled which can be crafted into the net or its modules. Also, causal knowledge in medicine and other areas can be used for the same purpose. The algorithmic procedures derived for this type of knowledge may be encoded into the MPNN module taking advantage of the properties of the neurons in £.

VI.6b) experimental: in this case it is possible to obtain the structure of K_0 from experts or from the previous analysis of the specialized data base.

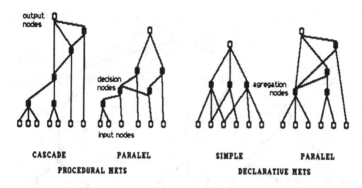

FIG. VI.5 - BASIC STRUCTURES OF EXPERT REASONING

The analysis of the structure of the knowledge graphs KG obtained from the experts (see Chapter V) pointed to (Fig. VI.5):

VI.7a) the existence of 4 basic types of nodes or neurons: input, aggregation, decision and output nodes, and

VI.7b) 3 basic ways of organizing nodes or neurons in the knowledge net: simple, cascade and parallel, which can be used as

VI.7c) modules for building declarative and procedural knowledge nets.

Part of the nodes in the KG are used to index the expert knowledge, since the information stored in the first terminal nodes acts as triggers to bring the hypothesis represented in the KG to the focus of attention (Rocha et al, 1988; Eddy and Clanton, 1982; Kassirer and Gorry, 1978). At the beginning of the consultation, the physician listens to the patient and pick some pieces of information as key words to think about specific diseases. The moment he has a few possible hypotheses, the expert starts an active inquiry of the other pieces of information required for a decision making about these diseases (Rocha et al, 1988).

These properties of the expert knowledge may be encoded in MPNN modules with the following structure (Fig. VI.6):

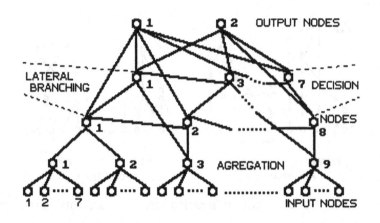

FIG. VI.6 - THE MPNN-MODULE

VI.8a) 63 input nodes aggregated into 9 clusters by means neurons of the types VI.4i and VI.4k;

VI.8b) two layers of 8 and 7 decision nodes of the types VI.4f and VI.4K, respectively, and

VI.8c) 2 or more output nodes of type VI.4e to represent slight different frames of the same hypothesis.

The connections between these layers are:

VI.8d) the two first input clusters converge to the first node of the first decision layer, and are assumed to be the indexing or triggering cluster. The other clusters of the aggregation layer diverge to the other neurons in both decision layers;

VI.8e) each decision node in the first layer branches to the next node in the same layer; to the correspondent node in the second decision layer and to the first output node;

VI.8f) each decision node in the second layer branches to the next node in the same layer and to the other output nodes;

VI.8g) all decision nodes have also a lateral branching directed to other modules of the net;

VI.8h) all output nodes have a recurrent branching to control its encoding function, because they represent the objective function to be optimized according to the constraints represented by the aggregated clusters. The output neurons are of the type MRN defined in VI.4e;

VI.8I) The characteristics of the QN neurons in the aggregation and decision layers are set initially for an AND/OR type of processing (see VI.4K), but it may be controlled by modulator neurons in order to implement the threshold reasoning discussed in Chapter V.

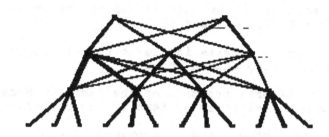

FIG. VI.7 - THE PROGRAMMING OF THE MPNN-MODULE

This type of net can be programmed with the same simple L(G) used in Chapter III to encode some type of MPNNs (Fig. IV.7).

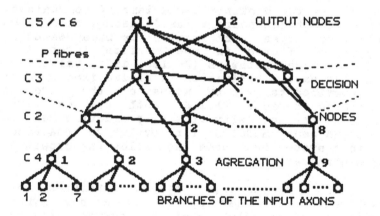

FIG. VI.8 - CORTICAL COLUMNS AND MPNN-MODULES

This kind of structure allows the implementation of simple, cascade and parallel knowledge representation, besides having a close correspondence with the basic structure of the cortical columns (Fig. VI.8). In this line of view:

VI.9a) the input nodes corresponds to the branches of the input axons arriving from the thalamus and other cortical columns;

VI.9b) the agregation nodes are considered to be the Stellate cells of the cortical layer 4;

VI.9c) decision nodes correspond to the pyramidal cells in the cortical layers 2 and 3;

VI.9d) pyramidal cells in the cortical layers 5 and 6 are associated to the output nodes;

VI.9e) the lateral branches of the decision nodes correlates with the parallel fibre systems in the cortical layer 1, and

VI.9f) the output nodes provides (mainly contralateral) cortico-cortical pathways and cortico-subcortical fibres.

VI.7 - The computational structure of MPNN

The flow of information inside the cortex is governed by two distinct mechanisms if communication among columns or between neurons inside each module is considered. In the first case, information is broadcasted by means of the parallel fibres in the cortical layer 1 or by the cortico-cortical pathways (Eccles, 1981, Valverde, 1986). In the second case, activation is mailed from specific neurons to determined targets by means of the predefined wiring illustrated at Figs. VI.1,6. Both systems were developed because of the necessity of easily creating new connections inside and among modules for evolutive learning (Chapter IV), in a system that does not allow the growth of axons at long distances.

The broadcasting system furnishes the adequate structure for the development of a new synapsis among columns placed at distant sites, because whenever necessary the parallel axon can branch over a cell of the desired colunm. This kind of system maintains the potentiality of creating new synapses whenever necessary, without the cost of their maintainance, if the associations they represent are not useful. The control of the intermodular connections may be programmed with L(G). On the one hand, whenever a new synapsis is required to be established, control molecules can specify the type of receptor to be produced from the pre-synaptic precursor molecule and/or to modify the affinity between this transmitter and the post-synaptic receptor. On the other hand, opposite effects may be exerted by the controllers if the purpose is to eliminate an undesirable synapsis.

The operation of the broadcasting system requires some sort of address matching for the message running in the parallel system to be effective on finding the adequate target. This address matching system is also supported by L(G). Any axon can produce precursors for different types of transmitters. The choice of one of them is dependent of controllers provided by the post-synaptic cells as well as of the filtering characteristics of the pre-synaptic axon. Controllers can also change the affinity of the post-synaptic receptor with the released transmitter. In this way, the spike train travelling the parallel axon may be considered a message that will be used by those cells having the specified receptor (address) for the possible transmitters released by this axon.

The broadcasting system may be implemented at low cost in artificial MPNN (Fig. VI.9) by using the concept of bus, where the output messages from each module are channeled, and from where each MPNN-Column reads the data addressed to it (Data Bus in Fig. VI.9).

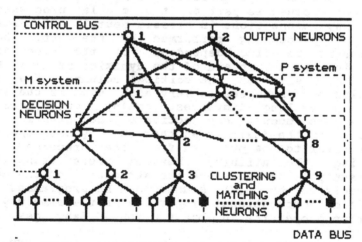

FIG. VI-9 - THE ARTIFICAL CORTICAL COLUMN

The output messages in the data bus must have the following basic structure:

VI.11a) the code identifying the neuron, that is, the type of transmitters $t_i \in T$ released into the bus, and

VI.11b) the degree of activation of the neuron, that is, the amount of the released transmitters.

The matching neurons are the elements responsible for getting data from the Data Bus to be processed by the module. They must have the description of the code to be accepted in the same way the pos-synaptic cell contains receptors ($r \in R$) for the transmitters ($t \in T$) used by the pre-synaptic cell. This can be specified by a pre-programming of the MPNN-net or learned, in the same line the post-synaptic receptor can be genetically encoded or their production triggered by the use of a pathway (Byrne, 1987; Goodman et al, 1984; McConnell, 1988).

In this way, data running in the Data Bus are messages to be accepted by specified types of modules to perform some type of processing over these messages and to return the results to the very same Data Bus. This kind of processing is the same utilized in Object Oriented Languages (OOL). The MPNN modules can be considered objects in the language L(G), to process specified messages according to defined specifications represented by their wiring and their composition of neurons.

The control bus in Fig. VI.9 is used to broadcast the modulator for learning control purposes.

The mail system distributing information inside each cortical column is part of the modular processing in the same way connections in classic neural nets determine the type of processing performed by the net. The modular processing is also dependent of the type of neurons composing the column and of the dynamics of L(G). Because of the small size of the cortical columns, the freedom of building new synapses is very high even if the axon cannot grow up to long distances. Again, controllers released inside the module can activate the creation of a new synapsis or the death of another one, by both specifying the transmitter to be produced from the precursor and/or by changing the affinity between these transmitters and post-synaptic cells. Since the structure of the columns can be specified by L(G), modules can be programmed to inherit specific computational properties in the same way an object inherits properties in Object Oriented Languages.

MPNN modules are aggregated into MPNN nets in a parallel structure for knowledge processing. Messages provided by the user (a human being or another net) run in the Data Bus and are processed by specified modules in the net, which may exchange messages among themselves through the Data Bus, too. In the same way, the result of the processing in the net is driven to the user through the same Data Bus. The distinction between these messages in the Data Bus is guaranteed by the address matching. In this way, the interface between the MPNN and the user is defined as another object. In this framework, the computational structure of MPNN is very similar to that of Object Oriented Languages. SMART KARDS(c) (Chapter IX) takes advantage of this similarity to associate the MPNN technology with a Data Base Language, in order to provide an intelligent environment to manage the expert world.

Other MPNN topologies may be developed for purposes other than the present one of simulating the expert reasoning. An example is the topology proposed in Chapter II to implement fuzzy control.

VI.8 - MPNN hierarchy

Different MPNNs may be hierachically organized (Fig. VI.10) into high order neural systems in the same way different neural nuclei are serially organized to support many hierarchical types of processing in the brain (Fig. VI.2). These MPNN systems or brains are composed of at least two basic types of MPNN nets (Rocha and Yager, 1992): afferent and efferent nets.

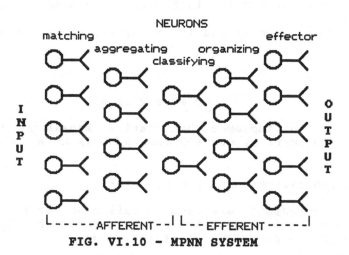

FIG. VI.10 - MPNN SYSTEM

VI.8a - Afferent Nets

The Afferent MPNN (Fig. VI.10) is composed of two layers with the following types of neurons (Rocha and Yager, 1992):

VI.10a) Matching or Sensory (Input) Layer: matching neurons encode some prototypical knowledge into their encoding functions (see Chapter I) about some label assigned to it. This label is provided by the formal language L(N) used to label the MPNN system (Chapter IV, section IV.6). The neurons at this layer represent propositions of the type

X is A

where X is the label and A is the fuzzy set defined by the encoding function VI.3b.

A spatial ordering may be assigned to this layer, with the purpose of encoding a third type of uncertainty besides confidence and relevance, which are represented, respectively, by the degree of activation of the neuron and by the weight of its synapses.

VI.10b) Aggregation (Hidden) Layers: composed by 1 or more layers of ANs and/or DFNs. AN neurons support the following type of calculus:

if Q{ (RX is A) and (RY is B) .... } then Z is C

whereas FNs process:

if Q{ (X is A) and (RY is B) .... } is K then Z is C

elseif Q{ (X is A) and (RY is B) .... } is L then W if D

.
.
.

or else V is E

AN is used to implement Declarative Knowledge and DFN is used to implement Procedural Knowledge in MPNN.

VI.10c) Classification (Output) Layer

MRM neurons are used in this layer if the knowledge representation is punctual. In this case, each output neuron represents a classification of the type:

Z is C

and a Max operation selects the output neuron of highest activation to represent the result of the inference

If X is A then Z if F
or
If Y is B then W is G
or

.
.
.

encoded in the MPNN:

QN neurons are used in this layer if the knowledge representation in the MPNN is distributed. In this case, each output neuron calculates one of the membership values of the collection of classes representing the decision. The reasoning in this net is:

```
IF
        Q { If X is A then Z
                AND
            If Y is B then W is G
                AND
                        .
                        .
                        .
                                        }
                                THEN V is H
```

It is interesting to remark that the same MAPI structure supporting the Extended Modus Ponens processing is preserved in the structure of the Afferent MPNN: the

VI.10d) matching is performed by the input layer;

VI.10e) aggregation is calculated in the Hidden Layers;

VI.10f) projection may be considered the main task of the the output layer, and

IV.10g) inverse-matching is performed by the Efferent Net associated with it.

The afferent MPNN whose input layer is composed by sensory neurons is called Sensory MPNN, and its role is to fuzzy classify the afferent data provided by the external world. They are related to the Sensory Systems in the brain, processing information collected by the receptors and performing a Low Level Fuzzy Pattern Recognition.

The output layer of a Sensory MPNN can provide the input to another MPNN. In this way, complex MPNN systems can be constructed to perform High Level Symbolic Processing and they can be used for High Level Fuzzy Pattern Recognition. Similar to the real brain (Fig. VI.2), the processing in MPNN systems can progress from an Analog/ Discrete Conversion on the low level MPNN toward a Discrete/ Label or Symbolic Conversion on the high level of the circuit.

VI.8b - Efferent nets

The Efferent MPNN (Fig. VI.10) is composed of two layers with the following of neurons (Rocha and Yager, 1992):

VI.10g) Organizing (Planning or Coordinating) Layer: the FDNs of this layer recruit neurons to perform a task in response to the fuzzy classification provided by the afferent net. The complexity of this processing will be reflected in the number of Organizing Layers. Some of them are related to the Planning of the Actions (e.g, Associative Areas in the Brain), some others to the Coordination of the chosen Actions (e.g. Cerebellum), etc.

VI.10h) Effector Layer: composed by fuzzy EN controlling the effector devices performing the Label or Symbolic to Analog Conversion of the final output of the system. These effector devices furnish a family of output functions which are used as building blocks of the repertoire of Actions of the MPNN system.

The neurons EN in this layer and the controlled effector devices generate a basis set A of elementary feedfoward or feedback actions supported by specific control circuits (e.g. Chapter III, section III.11). In this way, the ENs take charge of controlling separated components of the entire output of the system. This basis set A provides the building blocks used by the associating layers to plan the desired control according to the actual classification. Planning is a symbolic task processed by MPNNs taking advantage of the properties of their formal language L(G). The complexity of the efferent net may increase if specific MPNNs are programmed to account for different organizing activities such as planning, supervising, etc., in the same way different cerebral nuclei are chained in the motor control system of the animals.

A variety of structures may be used as effector devices of a MPNN system. Among them:

VI.10i) other MPNN circuits: this is a much used strategy in the evolution of the animals to increase the computational capabitily of their brains. In this way, it is possible to specialize MPNN circuits for important functions such as Learning Control, Language, etc. For example, the use of a Language MPNN system as an output device (see Chapter 4, section IV.6) of other MPNN systems greatly increases the possibility to observe these systems which is a key issue for Deductive Learning and for communication among animals.

VI.10j) other computational hardware and software devices: EN may be programmed to control other computational devices like disks, printers, etc., or even other pieces of software if necessary. This greatly increases the capability of the MPNN circuits and allows the user to combine different computational technologies to solve his problems.

VI.10k) general effector devices: any machine which may be controlled through an D/A conveter may be used as an efferent device of MPNNs.

VI.9 - The learning control

Learning in MPNN can be accomplished by means of different and complex strategies combining both Inductive, Deductive and Inheritance procedures (Chapter IV). Evolutive learning takes advantage of all knowledge in MPNN in order to develop the best models to guarantee the success of the MPNN system. The complexity of the evolutive learning control pressed nature to develop special neural circuits to control learning and memory. These circuits compose the Limbic System.

Initially, this system is in charge of controlling the degree of satisfaction of the basic goals required for the animal's survival: e.g. food, water, sheltering, sex, etc. The degree of satisfaction of these goals is measured by two different circuits: reward and punishment circuits, which evaluate the goal satisfaction in pleasant and unpleasant sensations. Because of this there is some dispute if reward and punishment are the best terms with which to name these circuits. The limbic system is also involved in satisfying these goals by controlling the animal's behavior toobtain food, water, sheltering, sex, etc. As a consequence, this system is involved in determining the degree of motivation, emotion and arousal of the animal.

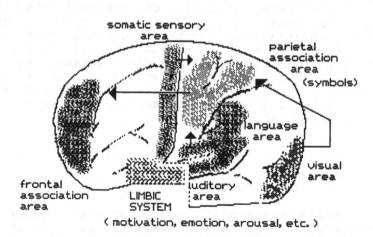

FIG. VI.11 - THE LIMBIC SYSTEM

The actions of the Limbic System over the other neural circuits and the body are in general broad actions required to maintain the homeostasis of the system. Because of this, most of its neurons were selected by nature to produce hormones or neuropeptides and to release these chemicals into the blood stream (hormones) or through axons (neuropeptides) over wide areas of the brain.

The cerebral evolution resulted in the development of the cortex, a structure which greatly increases the computational capacity of the brain. Most of the activities processed in this new structure are not directly correlated with the basic survival goals of the animal. Despite this, the cortex establishes strong and important connections with the Limbic System. As a matter of fact, part of the cortex integrate the limbic system, being called the Limbic Cortex. In this way, the same old system continues in charge of the measure of the goal satisfaction even if not anymore correlated with the basic necessities of the animal.

In virtue of the complexity of the control of the Evolutive Learning, special nets are proposed to process the learning in MPNN systems, in the same way that the Limbic System is correlated with this task in the real brain. These special nets influence the other circuits in the MPNN system by means of a Control Bus (Fig. VI.9), where they deliver their messages.

The Control Bus may have some dedicated lines reaching specific MPNN nets in the systems in order to mimick the behavior of the neuropeptides, but it can also have general lines simulating the hormonal distribution of information. The first process is called a partial or restricted broadcasting and the second one is called general broadcasting system. The partial broadcasting reduces the complexity of the t^r encoding because part of the addressing is encoded in the specialization of the lines, and part of it relies on the t^r matching discussed before (Chapter III). The general broadcasting is totaly dependent of this latter matching.

VI.10 - Conclusion

The computational power of MPNN is greatly increased in comparison with classic neural nets, because besides their parallel processing capacity, the MPNNs may be hierachically organized into MPNN systems. This puts partially in sequence the processing in these systems, in the same way it renders reason ing a temporal task in the real brain.

The compartimentalization of the processing inside modules of the MPNN system greatly favors the programming of these systems in the same way Object Oriented Languages reduce the complexity of the programming and maintainance of the software.

The specialization of some of the nets in the MPNN system to control learning is another advantage of the present approach. Different learning strategies may be programmed taking advantage of the power of the L(G) in specifying the structure of the MPNN nets. Because of the symbolic capacity of this language, no restriction is imposed upon the learning strategies to be used. Both inductive and deductive learnings are easily programmed.

The next chapters will present some results obtained with MPNN systems using some of these strategies.

ACKNOWLEDGEMENT

The discussions with Ronald R. Yager were very important for some of the notions of modularity presented in this chapter. Also, ideas exchanged with Takeshi Yamakawa, Torao Yanaru and Eiji Uchino, at Iizuka, during the month of February, 1990 were the germs for developing the notions about how to implement MPNNs as a software language, and how to start to think a hardware to support this language.

CHAPTER VII

NEXTOOL:
A MPNN CLASSIFYING SYSTEM

VII.1 - Some initial words about classification

To solve a classification task is to assign a physical object or an event under analysis to one or more of several pre-specified categories. This is done by computing a degree of similarity (or membership) of the actual data with the prototypical descriptions or patterns of these classes. The prototypical knowledge is encoded in a set of relations between characteristic pieces of information (signs, symptomns, etc.), and it is easily explained by means of the fuzzy knowledge nets discussed in Chapter V. The computation of the similarity between the actual object or event and these prototypical descriptions is the result of the navigation in these nets. Each prototypical pattern is associated to a module in the knowledge net, so that the degree of similarity or the confidence in the assumption that the pattern does exist in the actual data, is a function of the degree of activation attained at the output nodes of these knowledge graphs. This degree of activation is obtained in two steps:

VII.1a) first, each piece of information is matched against some expected datum definitions encoded in the input neurons of these knowledge graphs. These standard definitions specify either acceptable ranges or qualitative properties of the fuzzy variables supporting the classification task. The initial matching at the input layer provides a degree of confidence that the actual data are these variables, so that

VII.1b) in the sequence, the degree of activation of the input nodes is spread in the graph being powered by the strength of the arcs connecting the input to the output nodes. At each intermediate node, different entries are associated according to some learned standard data clusterings, and a new degree of activation is calculated taking into account both the relevance of each piece of information and the fuzzy quantifier aggregating these informations. The degree of activation calculated at the output nodes are, therefore, dependent of the pathways used to spread the input matchings.

The rules the expert provides for navigating the knowledge net are easily implemented in the MPNN because the MPNN neuron is a fuzzy logic device (Chapter II, section II.6.3). In this way, the expert knowledge may be used to program the MPNN nets for classification tasks. But MPNNs are also learning devices which can use both inductive and

deductive learning techniques to acquire information from
data bases.

The present chapter introduces and discusses NEXTOOL
(Machado et al, 1989, 1990a,b, 1991a,b,c), a MPNN
classifying system which can be programmed with the
knowledge provided by one or many experts, as well as to
classify the objects of a data base. Alternatively, NEXTOOL
adapts the expert knowledge to the contents of the given
data base. The system uses the notion of punishment and
reward in Hebbian Paradigm to implement inductive learning
by means of modification of the synaptic weights of the MPNN
modules; and uses many of the ideas proposed by the Genetic
Algorithm Approach (Holland, 1975; Booker et al., 1989) to
modify the structure of its initial knowledge for a
deductive learning.

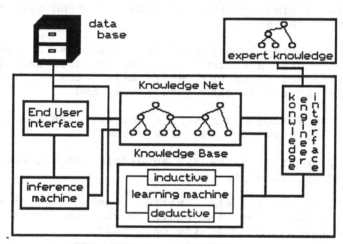

FIG. VII.1 - NEXTOOL

VII.2 - The general structure of NEXTOOL

NEXTOOL (Fig. VII.1) is a shell for building expert
reasoning systems for classification tasks providing the
following facilities (Machado et al, 1990, 1991):

VII.2a) Connectionist knowledge base: the NEXTOOL knowledge
base is a connectionist representation of the knowledge
necessary to reasoning, to build and to dialogue with both
the user and the knowledge engineer. MPNN modules are used
to store the knowledge acquired from the expert or learned
from the data base. The net provided by these modules form
the Expert Knowledge Net (EKN). Another MPNN net plays the
role of a Semantic Net (SN), used to encode the genetic and

the embryogenic rules used to build the EKN. The SN supports the user's interface and is the main knowledge engineering tool to implement the expert knowledge in the system;

VII.2b) Learning machine: is an evolutive learning machine used to acquire knowledge from the data base or to adapt the knowledge provided by the expert to the realities of the this data base or observed word. The inductive learning is supported by a punishment-reward process changing the strength of the synapsis of the EKN. The deductive learning uses many ideas imported from Genetic Algorithm (Holland, 1975; Booker et al., 1989) to change the structure of the modules of the EKN, adding neurons and pathways to them or using parts of the existing modules to create new subnets in the EKN;

VII.2c) Inference machine: is the actual interpreter built in APLII to simulate the behavior of MPNN nets in a serial IBM 3090 machine;

VII.2d) User's interface: is used as a communication interface between NEXTOOL and both the user and the observed word represented by the data base. The communication is supported by the SN discussed above, and

VII.3e) Knowledge engineer's interface: the SN provides the tool to the knowledge engineer either

VII.3ea) to program the EKN according to the knowledge net provided by the expert, or

VII.3eb) to design the initial topology of the EKN which will be modified by the learning machine according to the information provided by the data base.

VI.3 - The expert knowledge net

The EKN is implemented as a MPNN composed by multiple modules of 3 or more layers, where:

VII.4a) the input layer is composed by matching or sensory neurons (see Chapter VI, section VI.4): This layer is in charge of encoding the prototypical knowledge about the variables (signs, symptoms, laboratory test results, etc.) which support the decison making. This prototypical knowledge corresponds to the definition of the fuzzy sets associated with the input variables. The axonic encoding function (see Chapter II, eq. II.14) is implemented as the fuzzy restrictions (see Chapter X, section X.4) defining these fuzzy sets. In this way, the actual value of the incoming information is matched against these restrictions, and the axonic activation encodes the membership of the actual datum with the fuzzy set associated with the input variable. This degree of membership expresses the confidence that the actual datum is the expected pattern to support the decision making;

VII.4b) the intermediate layers are composed of aggregating or decision nodes, depending on the type of knowledge to be encoded: because many types of $\top$ -norms may be used to support fuzzy aggregation, NEXTOOL is provided with a library £ of different neurons, which are chosen by the user to implement a specific kind of fuzzy reasoning. The user can easily implement a new type of neuron, since each cell in the library is the APLII program running the desired calculation. In the examples to be discussed in this chapter, the neurons of the hidden layers are min neurons, and implement the classic fuzzy reasoning proposed by Zadeh, 1975;

VII.4c) the output layer is composed by OR neurons: although many types of $\top$ -conorms could be used to implement this fuzzy operator, the max-rule is used because any fuzzy deduction is considered, here, as proposed by Zadeh, 1975, as the solution of a mathematical program (see Chapters II, section II.6.4 and Chaper X, section X.7) to maximize the confidence in the decision making;

VII.4d) the confidence assigned by the input neurons is propagated in EKN taking into consideration the relevance of each variable in supporting the decision making: this means that the actual axonic activation transmitted from the pre-synaptic to the post-synaptic neuron is powered by the strength of the synapsis, whose value is set according to the relevance of the information represented by the pre-synaptic neuron to support the decision making at the post-synaptic cell, and

VII.4e) two different values of confidence are propagated in EKN: each time information about confidence is transmitted from the input to the output layer, the values of the actual and the potential activation, calculated according to eqs. V.19 (Chapter V, section V.6). To implement this type of calculation, the output of the input neurons are set initially equal to 1, the value required to calculate the value of the potential activation, and modified if any piece of information is matched by the neuron. In this case, the axonic activation is set according to this degree of matching, which is the value required to calculate the actual activation.

The initial topology of the EKN modules are either:

VII.5a) the knowledge graphs provided by the expert (Fig. VII.2a), or

VII.5b) that necessary to provide all clusters of the input

neurons from size 1 to size m, if NEXTOOL is supposed to
have acquired the expert knowledge from the data base (Fig.
VI.2b): this type of topology was called by Machado and
Rocha, 1989, the Combinatorial Topology of Order m. The
Model of Order M contains as many modules as clusters of
size 1, 2, ... m of the variables represented at the input
node. This means that the number p of pathways in EKN
implementing the Model of Order m is

$$p = \sum_{i=1}^{m} C(n,i) \qquad (VII.5c)$$

where C(n,i) represents the combination of size i of n
elements. The actual value of m is chosen according to the
entropy of the observed word to be modelled (see section
VII.7a), however, it must be restrained to low values
because of the exponential growth of the number of required
elements at the associative (intermediate) layers. This is
not a serious constraint to implement the expert reasoning,
since one of the major properties of this type of thinking
is to have a low descriptive complexity (see property V.31,
Chapter V, section V.12). This means that a small number of
evidences n is used to support the decision making, which
are combined in cluster of small size.

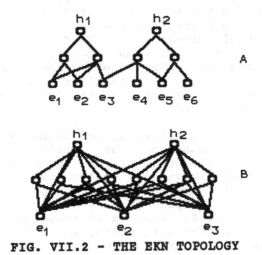

FIG. VII.2 - THE EKN TOPOLOGY

VI.4 - The semantic net

The semantic network (SN) is used to represent the
concepts of the problem domain (object and evidences) and
their relations. These pieces of information are used to
encode the genetic and the embryogenic rules to construct

the EKN modules and to connect them to a knowledge net. The SN is used to specify the EKN. The problem domain is described i SN at two levels (Levesque and Mylopoulos, 1979 and Mylopoulos et al, 1983):

VII.6a) the Intensional Semantic Network (ISN): involving only the classes of objects and a set of primitive relations (Fig. VII.3a), and

VII.6b) the Extensional Semantic Network (ESN): where the object classes and their relations are instantiated (Fig. VII.3b).

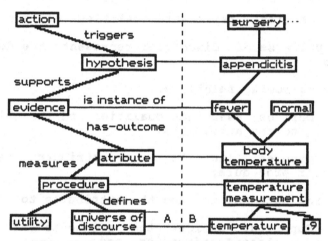

FIG. VII.3 - THE SEMANTIC NET

This organization of the semantic network allows a clear differentiation between expressions at the conceptual level and the statements on the extensional level (Machado and Rocha, 1992). The ISN provides a framework for the representation of abstract semantic relations between concepts in order to formulate the semantics of particular subject areas. In this way, the ISN encodes a metaknowledge about the domain problem. The ESN is intended to represent the semantics of statements about concrete objects of respective subject areas. In this way, the ESN describes the domain problem.

The SN is implemented as a MPNN system. But before discussing this point, let some points in the Semantic Network theory be introduced.

The example of Fig. VII.3 illustrates the use of SN. The metaknowledge about the domain problem is encoded in the ISN by means of the following concepts (Fig. VII.3a):

VI.7a) Action: represents the result of the decision making;

VI.7b) Hypothesis: represents the categories of a classification problem used to the decision making;

VI.7c) Evidences: represents the arguments for the decision making;

VI.7d) Attributes: represents the features which provide information about the evidences;

IV.7e) Procedure: represents the tasks executed to obtain the values of the attributes, and

IV.7f) Utility: represents qualities of the attributes;

IV.7g) Universe of discourse: represents the domain of the evidence;

and the following relations:

IV.8a) Defines: associate qualities and the universe of the discourse to the attributes;

IV.8b) Measures: indicates the attributes measured by a particular procedure;

IV.8c) Has outcome: relates the attributes to the evidences;

IV.8d) Supports: assigns the arguments for the decision making of a classificafication, and

IV.8e) Triggers: shows the action to be executed according to the decision making.

There is a general relation in the Semantic Network Theory called Is_instance_of, which is used to map the ISN into the ESN (Fig. VII.3). For example, this relation can declare temperature and .9 as instances of the universe of discourse and utility, respectively; Temperature Measurement as a instance of Procedure; Fever and Normal Temperature as instances of Attribute, etc.

The concepts in VII.7 and the relations in VII.8 allow the user to describe the domain problem, that is to build the corresponding ESN. In the example of Fig. VII.3, the ISN (Fig. VII.3a) is used to encode the following description of a medical reasoning encoded in the ESN (Fig. VII.3b):

VII.9a) the universe of discourse and the utility of the procedure Measure of the Body Temperature are temperature and .9, respectively. The value of the body temperature is used to evaluate the evidence Fever, which is one of the arguments for decision making about Appendicitis which in turn implies the action Surgery.

The description of the domain problem is provided by:

VII.9b) the knowledge net obtained interviewing the expert (Fig. VII.4): in this case, NEXTOOL may refine this initial knowledge according to the contents of the data base; or

VII.9c) a theoretical knowledge obtained from text books: in this case, NEXTOOL will learn how to classify the contents of the data base.

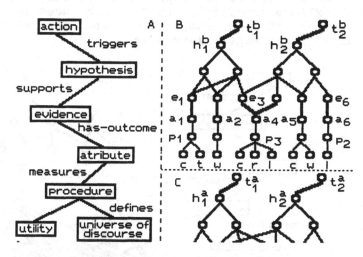

FIG. VII.4 - THE EXTENSIONAL SEMANTIC NETS AND THE KNOWLEDGE GRAPHS

If the domain description is provided by a knowledge net, then as many evidences (nodes e_i in Fig. VII.4b), hypotheses (nodes h_i), actions (nodes t_i), etc. are defined in the ESN as the corresponding input and output nodes of the knowledge graphs KG (Fig. VII.4b). The evidences are the fuzzy variables assigned to the input nodes of the knowledge graph and whose fuzzy sets are defined by the restrictions the expert associated to the procedures to obtain their values. The attributes (nodes a_i) the expert associated with these evidences and the procedures (nodes p_i) he uses to measure these attributes are specified in the ESN by means of the adequate nodes and arcs. The different clusters (intermediate nodes of the KG) of evidences supporting the same hypothesis are represented in ESN by means of imaginary unlabelled concepts (unlabelled nodes in Figs. VII.4b and c).

In the case knowledge is obtained from a population of experts, their different graphs are introduced in NEXTOOL, which may use each knowledge net as a different view of the same problem (Fig. VII.4B and C), or may calculate and use the consensus knowledge obtained from the individual graphs by using different techiniques (Leäo and Rocha, 1990; Machado, Rocha and Leäo, 1990 and Chapter V, section V.3).

VII.5 - Writing the ESN into the MPNNs of the EKN

The metaknowledge represented in the ISN and the domain knowledge encoded in the ESN are used to create the MPNN modules of the EKN, which are the computational structures of the expert knowledge.

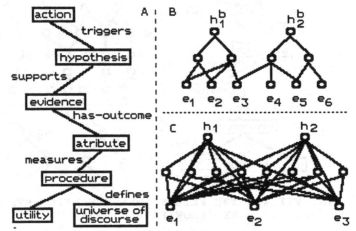

FIG. VII.5 - TRANSLATING THE ESN INTO EKN

The mapping of the ESN into the EKN is processed in two steps:

VII.10a) choice of neurons: the different types of neurons to compose the EKN are chosen among those of the library £ having either the properties required by the procedures attached to the input nodes, or the aggregation properties of the other non-terminal nodes of the ESN. These aggregation properties are specified by the different semantic relations linking the correspondent nodes in the ESN according to the metaknowledge encoded in the ISN. For example, the relation supports assigns an AND/OR aggregating neuron to the clustering neurons of the intermediate layers of the MPNN modules and an OR neuron to their output neurons. In this condition, each neuron in the EKN represents an object in the ESN; and

VII.10b) design of the topology: this design is dependent of the origin of the domain knowledge. In the case of VII.9b, the knowledge graph is used as the template to define the wiring of the MPNN modules (e.g. FIG. VII.5B), whereas in the case of VII.9c the topology of these modules is that provided by the Combinatorial Topology of Order M (Fig. VII.5C), the value of m being dependent of the variability of the domain to be modeled.

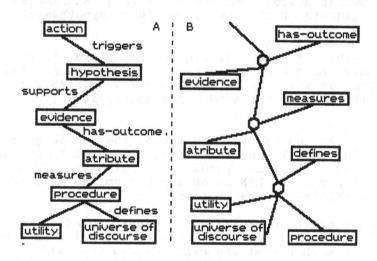

FIG. VII.6 - MAPPING THE SN INTO THE MPNN

VII.6 - Using MPNNs to encode SNs

The MPNNs can be used to implement both the ISN and the ESN. The following procedure is used to encode the ISN (Fig. VII.6a) into a MPNN (Fig. VII.6)b:

VII.11a) each node of the semantic net (e.g. utility, procedure, etc.) is associated to an input node of the MPNN;

VII.11b) each arc or relation of the semantic net (e.g. is_part_of, measures, etc.) is associated to an output neuron of the MPNN, and

VII.11c) the neurons of the intermediate MPNN layer are used to relate the concepts represented at the input MPNN neurons to the relations represented in the output MPNN neurons.

The following procedure is used to encode the ESN into a MPNN:

VII.12a) a n level hierarchic MPNN system of modules is used to implement the following concepts: attributes (level 1), hypotheses (level 2), actions (level 3), ... concept n (level n), respectively, so that

VII.12b) at each level, the nodes of the ESN map directly into neurons of the MPNN preserving the topology of the semantic net on this level; and

VII.12c) the required associations are established between the output neurons of one level (e.g. attributes) with the input neurons of the next level (e.g. hypotheses).

The result of the application of this procedure is to guarantee that the topology of the MPNN net is the same of the ESN, so that the net shown in Fig. VII.4B can be read as both the ESN and the MPNN implementing it.

The SN is used to encode the knowledge required to build the EKN. The SN is the key tool used to construct both the User and the Knowledge Engineer's Interface. The embryogenesis of EKN is obtained from the knowledge encoded in the ISN and ESN as follows:

VII.13a) start to navigate the MPNN representing the ESN at the input nodes of the level hypothesis. In other words, activate each one of the evidence neurons, and

VII.13b) transfer this activation to the input nodes of the MPNN representing the atribute level, which are the nodes attributes, and spread this activation to the procedure and the utility neurons associated with this attribute;

VII.13c) the activation of these neurons are transferred to the correspondent nodes of the MPNN representing the ISN;

VII.13d) the activation of the corresponding ISN neurons result in the activation of the relation neurons defines and measures (see Fig. VII.6b) which determine the choice of the adequate neuron from the library £ to represent the activated evidence in the EKN. This choice is dependent of the information stored in the nodes of the corresponding attribute MPNN;

VII.13e) now spread the activation from the evidence neuron toward the hypothesis neuron in the MPNN representing the hypothesis level, and

VII.13f) transfer the resultant activation to the corresponding ISN neurons, in order to choose the adequate neuron in the library £ to reset the neurons of the intermediate layer of EKN;

VII.13g) once the hypothesis level of the ESN is read, then move the activation of the hypothesis nodes to the input nodes of the action level, and

VII.13h) navigate this new net, spreading its activity to the MPNN representing this level in the ESN, etc., etc.

VII.7 - The inductive learning rules of NEXTOOL

Inductive learning is used both to refine the knowledge expert according to the contents of a given data base, and to acquire the knowledge stored in this very same data base. The learning paradigm is that of supervised learning, since the cases stored in the data base must have a clear classification, which is used to decide for a Punishment or Reward of the synapses of the MPNN models activated by the contents of each case. The entire process of inductive learning is implemented in NEXTOOL by means of the following algorithms: Punishment/Reward Algorithm and Pruning and Normalization Algorithm.

VII.7a) The Punishment/Reward Algorithm

Let the evidential flow v in a synapsis be defined as the product of the

VII.14a) activation a_i of the pre-synaptic excitatory neuron by the weight s_i of the synapsis this cell makes with the post-synaptic nerve cell:

$$v = a_i . s_i, \text{ or}$$

VII.14b) one minus the activation a_i of the pre-synaptic inhibitory neuron by the weight s_i of the synapsis this cell makes with the post-synaptic nerve cell:

$$v = (1 - a_i) . s_i$$

The algorithm is the following:

Set to each synapsis of the network
-
    the accumulators for rewards and punishment
-
    set the initial values of these accumulators equal to 1
End

Set to network arcs
-
    the weights equal to 1
End

For each training example in the data base
-
    propagates the confidence in the evidences from the
    input toward the output nodes
-
    marks the patways reaching the output nodes with
    evidential flow greater than 0
-
    For each synapsis of these pathways
    -
        if the output node is the correct classification
        -
            increases the reward accumulator by a value equal
            to
$$h = v . a_o$$

            where a_o is the activation of the output neuron
        or else
        -
            increases the punishment accumulator by the value
            h above
        end
    end
end

        Only one iteration on the training examples is
necessary to the calculation of the learning, since the
topology of the Combinatorial Module Order M is a feedfoward
topology.

VII.7b - Pruning and Normalization Algorithm

        The following is processed

For each synapsis in the network
-
    compute the net accumulator value
    -
        NETACC = rewards - punishments accumulator values
    -
        if NETACC $\leq$ 0
        -
            remove the synapsis from the network
        or else
        -
            if punishments accumulator value > 0
            -
                compute the arc weight as

$$s_i = \text{NETACC} / \text{MAXNET}$$

$$\text{MAXNET} = \max_{\text{MPNN}} (\text{NETACC})$$

or else

‒

compute the arc weight as

$$s_i = \sqrt{T_{ACC}} + (1 - \sqrt{T_{ACC}}) / \text{MAXNET}$$

    end

  end

‒

if arc weight < the pruning threshold T_{ACC}

‒

delete arc

  end

end

    The goal of the pruning process is to remove all the weak and negative synapses (NETAAC < 0), and that of the normalization is to set the value of the synaptic weight in the interval [0,1].

VII.7c - Complexity and order

    If no initial knowledge is provided to NEXTOOL by means of an expert knowledge net, it has to learn from scratch. This means that the Combinatorial Model M to be created must have a variability greater than that of the external world W to be modeled (see propositions V.10, 11 and 12, Chapter IV, section IV.3). However, the difficulty of learning increases if the variability of the Combinatorial Model becomes much bigger than that of W. Thus, the actual value of m must be chosen according to the variability of the external world W, in order to avoid the combinatorial explosion which can rapidly increase the difficulty of learning.

    Let the structural variability h(MPNN) of the MPNN implementing the Combinatorial Model M be measured as the Shannon's entropy

$$h(\text{MPNN}) = - \sum_{i=1}^{n} p(r_i) \, \log p(r_i) \qquad (VII.15a)$$

where n is the number of pathways of modules r_i generated according to the number of evidences in the domain problem, and the order of the chosen combinatorial model. The probability $p(r_i)$ of the module r_i to be activated in the MPNN is dependent of the strength of its connections. h(MPNN) is maximal if $p(r_i)$ is equal to 1/n for all modules r_i in MPNN:

$$\max(h(MPNN)) = \log n \qquad (VII.15b)$$

or from VII.5c

$$\max(h(MPNN)) = \log \sum_{i=1}^{n} C(q,i) \qquad (VII.15c)$$

where q is the number of evidences in W.

Now, let h(W) be the entropy of the world W. For sake of simplicity, let it be considered the case of W being composed by discrete events w_i associated with a probability of occurrence $p(w_i)$. In this context

$$h(W) = - \sum_{i=1}^{q} p(w_i) \log p(w_i) \qquad (VII.16)$$

Each pathway r_i of the MPNN may be an attempt to classify the patterns of occurrence of events in W. Induction changes the connectivity of the MPNN, increasing the probability of some modules being activated and reducing the frequency of some other pathways be activated. This requires the entropy $h(MPNN_0)$ of the initial model to be (Chapter IV, section IV.3):

$$h(MPNN_0) > h(W) \qquad (VII.17a)$$

or from VII.5

$$\log \sum_{i=1}^{n} C(q,i) > h(W) \qquad (VII.17b)$$

Eq. VII.17b is used to determine the order m of the Combinatorial Model to be used to learn W. However, the computational cost (memory and time) of the combinatorial model is an exponential function of m. Thus, the modeling of high entropic worlds would require

$$\log \sum_{i=1}^{n} C(q,i) < h(W) \qquad (VII.17c)$$

to avoid the exponential explosion of costs. In this case, the learning must increase the entropy of the MPNN, by creating new synapsis from a low order Combinatorial Model. This learning approach is the Deductive Learning discussed in Chapter IV, section IV.8. NEXTOOL's deductive learning paradigm borrows some ideas from Genetics Algorithm (Holland, 1975).

VII.8 - Deductive learning

Genetic algorithms are adaptive search algorithms inspired in models of heredity and evolution in the field of population genetics. In NEXTOOL, they are used to generate conjecture about W, to be tested by the induction learning tool. The modules of MPNN are seen as an evolving population of pathways (individuals) presenting different degrees of fitness. This fitness is the possible maximal evidential flow in the module. Each training example from W provides to the modules it activates an opportunity to reproduce if they have a degree of fitness greater than the average of the MPNN. The reproduction probability is proportional to the evidential flow produced by the training example.

The reproduction is performed by the application of genetic operators. The following genetic operators are available:

VII.18a) Crossover: exchanges parts of the input elements between two parent modules;

VII.18b) Addition: adds an evidence (input) node to the module;

VII.18c) Elimination: deletes an evidence (input) node in the module, and

VII.18d) Substitution: replaces evidences (input) nodes in the module.

The genetic operations are randomly selected and operate circumscribed within the limits of the training examples, in this way reducing, the search space.

The maximum size of each module population is limited to avoid uncontrolled growth. During an iteration an additional buffer space is open to receive the generated offspring, which are added to the population if they do not match any other module in the population. The size of this buffer is the limit of the population growth. Inductive learning modifies the fitness of the entire population (parents and offspring). At the end of the training, all the weakest modules are eliminated to reset the population sizes to their original values. In this way, the deductive learning fills the gap between the initial knowledge of MPNN and the world to be modeled (eq. VII.17c).

VII.9 - The evolutive learning engine

Inductive and deductive strategies are combined in NEXTOOL to support its Evolutive Learning Engine. This engine operates according to the following algorithm:

VII.19a) Preparation:

For each category c of the domain problem
-
compute a reproduction threshold $\alpha(c)$

as the average of the fitness of its MPNN modules

VII.19b) Reproduction (deductive learning)

For each training example

- identify the correct category c of the example

- propagate the example in the MPNN

- if activation a_c of the output neuron c > $\alpha(c)$

    - for each module of the modules reaching c

        - select r_i for reprodution if $a_i \geq \alpha(c)$
        and give it a probability of reproduction
        proportional to its current evidential flow

        - randomly select a genetic operator

        - generate the offspring from r_i

        - add the offspring to the buffer space if it
        is original and there is available
        space still
    or else
        for each module r_i reaching c

            - reconstruct r_i holding its input
            elements belonging to the example

            - replace the others by additional elements
            from the example
        end
    end
end
end

VII.19c) Fitness Evaluation (inductive learning)

Make all weights of the resultant MPNN equal to 1

- Apply the Punishment and Reward algorithm

- Apply the Pruning and Normalization algorithm

- Compute the fitness of each module in the MPNN

VII.19d) Population Reduction

Eliminate the weaker pathways until the buffer space is
empty

VII.19e) MPNN testing

Use a set of test examples to evaluate the performance
of MPNN in classifying these examples
-
if performance is above a acceptable threshold
-
    stop
or else
-
    go to VII.19a
end

VII.10 - Deciding about inductive and deductive learning

The decision about the choice of Inductive and
Deductive learning supported by eqs. VII.17 was investigated
(Machado and Rocha, 1992) in the case of three toy problem
domains (John and Mary, XOR and TC) exhibiting different
entropies (Fig. VII.7a).

A

TOY PROBLEM	USED ORDER	WORLD ENTROPY	ENTROPY OF $MPNN_0$		
			order 1	order 2	order 2
JOHN & MARY	1	1.25	3.0	5.3	6.5
XOR	2	2.00	2.0	3.3	3.8
T-C	3	8.85	4.1	7.4	9.9

B

NUMBER OF PATHWAYS	MPNN ENTROPY	I L ERROR RATE	EL ERROR RATE
291	8.18	.33	.33
391	8.61	.08	.00
535	9.06	.00	.00

FIG. VII.7 - LEARNING OF DIFFERENT WORLDS

John & Mary (Machado and Rocha, 1989) form a tiny
medical differential diagnosis problem involving the
diseases d_1 and d_2 and the symptoms e_i, e_2, e_3 and
e_4. The used training data base is

        John(d_1,e_1,e_2,e_3)
        Diana(d_1,e_1,e_2,e_4)
        Mary(d_2,e_1,e_3,e_4)
        Peter(d_2,e_2,e_3,e_4)

The XOR is the classical difficult problem for neural nets

by Minsky and Pappert, 1969. T-C is a discrimination problem
between the letters T and C, written in a 3x3 matrix,
subject to shift rotation. Both positive and negative
evidences were represented in the input layer in order to
avoid the use of inhibitory synapsis. The results shown in
Fig. VII.7 show the order of the combinatorial model
required to implement a MPNN able to learn to solve this
problem using the inductive learning approach. The results
support the condition imposed by IV.17b.

The following approach was used in order to study
the choice between deductive and inductive learning
according to eqs. VII.17:

VII.20a) distinct MPNNs exhibiting different initial
entropies were created by a partial deletion of some modules
from a MPNN of order 3 (Fig. VII.7b);

VII.20b) these MPNNs were used as the initial topology of a
EKN to learn the T-C problem, and

VII.20c) two different learnings were performed using the
inductive approach described in section VI, and the
evolutive learning proposed in section IX.

The results of these simulations are presented in
Fig. VII.7b. They clearly show that the error rate in
classifying test examples after learning, decreases as
H(MPNNo) approaches H(W) in the case of the two learning
strategies. Besides, a good learning was obtained with the
evolutive approach using an initial topology of entropy
lower than the variability of the world W.

The combination of the inductive and deductive tools
employed in the evolutive engine of NEXTOOL provides a
powerful and flexible learning mechanism to construct
intelligent systems. The use of the Genetic Algorithm in
NEXTOOL substantially differs from other efforts in the
literature (Montana and Davis, 1989; Muhlenbein, 1990;
Whitley et. al., 1990) to combine this technology with
neural net theory. Also, the evolutive paradigm used here
differs from the proposals of Hall and Romaniuk, 1990. Also,
JARGON (Chapter VIII) makes a different use of the evolutive
learning to acquire knowledge from language data bases.

VII.11 - The inference machine

The inference machine IM works on two levels: the SN
and EKN levels. On the semantic level, the IM analyzes the
user's goals during a consultation and determines the best
sequence of application of reasoning models to reach them,
and activates the corresponding EKN modules or nets to
process the adequate information (Machado et al, 1991a). For

instance, if the user wants to know how to repair a defective machine, the IM determines that firstly a diagnostic reasoning model, and secondly a repair selection model must be used. This navigation of the SN automatically selects the corresponding EKNs to be used as the knowledge processors. In other words, the IM selects the adequate MPNN to process the required reasoning.

The goals of the IM at the EKN level are (Machado et al, 1991a):

VII.21a) to compute the degree of possibility of each hypothesis and to present those having a possibility greater than the acceptance threshold as being the classification solution. This implements the threshold reasoning discussed in Chapter V and is processed by propagating the available input evidences forward in the corresponding MPNN modules. Note that IM can express its indecision that one object is similar to several hypotheses, as humans frequently do in ambiguous situations;

VII.21b) to use the decision made on the hypothesis level to guide the choice of the action, etc. This is done by the same kind of processing discussed in VII.21a. Whenever it is necessary to obtain more information to support any decision on the EKN level, the IM is able

IV.21c) to determine the next optimal question to be asked to the user, that is, to control the flux of inquiry. This decision takes into consideration both the Potential Acceptance Index (see V.19b, Chapter V, section V.6) and the Utility Index (see V.26e, Chapter V, section V.7) of each pathway reaching the current EKN nodel, and

VI.21d) to explain to the user the reasoning employed to reach a problem solution. Since each EKN neuron processess a fuzzy rule, it is quite simple to provide an explanation by means of the chain of rules corresponding to the activated EKN neurons.

VII.12 - The interfaces with the external world

NEXTOOL uses two different interfaces in order to communicate with the user and the knowledge engineer. Both of them provide a set of windows and menus associated with defined funtions necessary to the definition, management and maintenance of a given problem domain (Fig. VII.8). The following is a brief description of these functions:

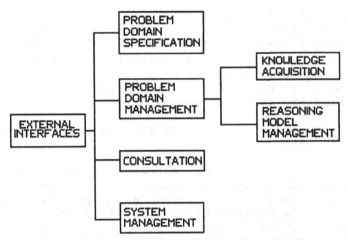

FIG. VII.8 - INTERFACE FUNCTIONS

VII.22a) Problem domain specification: a set of windows and menus to construct the required ISN and ESN related to a problem domain. These semantic nets encode the expert or theoretical knowledge necessary to build the EKN to be used as the reasoning processor in the specified area of human specialization. The semantic nets are used as friendly interfaces allowing the knowlegde engineer to introduce all specifications of a given application into the system, no matter if the initial knowledge is detailed by the expert or it is just a general description of the problem obtained from text books;

VII.22b) Problem domain view management: different experts handle distinct cases of a given area of expertise, and consequently can develop different heuristic knowledge about a given problem domain. The variability of the knowledge graphs provided by distinct experts is high (see Chapter V, section V.3). These different views of the same problem can be used by NEXTOOL either as alternative strategies to solve the case or to build a consensus knowledge. The user may choose o ne of these alternative EKN as the reasoning engine of his application;

VII.22c) Knowledge acquisition: both theoretical and expert knowledge can be used to construct an application. In the first case, some general information is encoded in the semantic nets to guide the genesis of a Combinatorial EKN of order m, the value of m being determined according to the entropy of the world to be modeled. In the second case, the interface is used to translate the expert knowledge nets into both the ESN and EKN;

VII.22d) Reasoning models management: the output levels of the EKN to be displayed and used for decision making are chosen by the user to define the different reasoning models defined in the ISN. The simplest reasoning model is classification, by means of which a category is assigned to an event in the worl d W to be modeled. The decision making about the classification can be used to support a choice of therapy, prognosis, etc. This other decisions define the other reasoning models associated with a given application;

VII.22e) Consultation: performed in two steps, called active and passive phases. During the passive phase, the user enters a set of initial data whose purpose is to activate a set of initial most probable hypotheses to solve the problem. For instance, in the case of the medical consultation, this set of initial data corresponds to the information provided by the patient about his disease. This set of initial hypotheses compose the Consultation Focus. The passive phase may end either because a maximum number of hypoteses is brought to the consultation focus or because the user stops to supply new evidences. During the active phase, NEXTOOL tries to prove or refute each one of these initial hypotheses by means of an inquiry guided by the rule described in VII.21c, and

VII.22f) System management: provides a set of maintenance activities such as: defining the library of neurons to be used by EKN; defining genetic operators; setting standard parameters of both the Inductive and Deductive Learning strategies, etc.

VII.13 - Learning from a medical data base

A data base of 378 real cases of Uremia, Nephritis, Lithiasis and Hypertension was used to test the learning capabilities of NEXTOOL. Most of the contents of this data base is written and verbal information, whereas some other data are numerical values of laboratory tests.

SMART KARDS(c) (see Chapter IX) was used to store all this information in different cards of the patient's folders: history of the disease containing the information about the patient's complaints; history of familiar diseases; morbid antecedents; physical examination; and cards for laboratory results of different tests. Four different cabinets: Uremia, Nephritis, Lithiasis and Hypertension, were used to store the information about the patients assigned one of the above diagnoses.

JARGON (see Chapter VIII) was used to discover the specialized language used to describe verbal information and to define the evidences to be used by NEXTOOL. JARGON was also used to standardize the verbal information of these evidences in the data base. The other set of evidences used by NEXTOOL was provided by laboratory test results and the name of these tests. Smart Kards ran in a PC environment and NEXTOOL was implemented in a 3090 IBM machine. A comunication protocol was developed to allow NEXTOOL to read the data from the cabinets.

A total of 18 procedures, 121 attributes, 255 evidences and 4 hypotheses were represented in the ESN. This knowledge was provided by the analysis JARGON did of the verbal information in the data base and by the expert information about the fuzzy variable associated with the laboratory tests. 250 cases were randomly selected as the training set and the other 128 cases were used as testing cases. Inductive learni ng was used to train a Combinatorial EKN of order 2. To avoid the combinatorial explosion of applying this model to 255 evidences, the learning was tested with data provided by the history of the disease; the history of familiar diseases; morbid antecedents, and physical examinations. This means that a total of 58 evidences were used. The decision making was implemented as winner takes it all.

The results of this simulation are presented in Fig. VII.9. The misclassification error rate was equal to 0.26%. Most of these errors (63%) were Nephritis cases classified as Uremia cases. These errors are justified since, according to the experts, the discrimination between these two diagnoses requires laboratory test data. Although 58 evidences were used, the results of the study showed that only 27 evidences were actually necessary to support the decision making, because the remaining evidence nodes were disconnected from the hypothesis nodes.

CLASS COMPUTED BY NEXTOOL

		A	B	C	D	N
CLASS ASSIGNED BY THE EXPERT	A	36	0	3	1	3
	B	18	6	2	0	6
	C	0	0	32	0	0
	D	0	0	0	21	0

A - UREMIA   B - NEPHRITIS   C - LITHIASIS

D - HIPERTENSION   N - NO CLASSIFICATION

FIG. VII.9 - THE PERFORMANCE OF NEXTOOL

The inspection of the MPNN modules being used by NEXTOOL to support the decision making about these diagnoses, revealed that some of the decisions were made using only one datum. This is an uncommon characteristc of the expert reasoning. As a matter of fact, none of the knowledge graphs analyzed in Chapter V shows this feature. The worst is that some of the decisions were based in informations provided by the history of familiar diseases or morbid antecedents. These types of information are never used in isolation by the experts to support any decision in medicine. The discrepancy between the NEXTOOL and the expert learning may be justified by the fact that the students are taught at the medical school not to rely on this kind of information (history of familiar diseases and morbid antecedents) to make decisons, but to use them as auxiliary data. This metaknowledge can be implemented in the ISN and used by the Punishment/Reward Algorithm to modify the synaptic weight depending on the relevance of the input data according to this metaknowledge. This strategy is now being implemented in NEXTOOL.

CLASS COMPUTED BY
NEXTOOL

		A	B	C	D	N
	A	42	0	4	1	0
CLASS ASSIGNED BY THE EXPERT	B	3	20	0	3	0
	C	0	0	38	0	0
	D	0	0	0	22	0

A - UREMIA   B - NEPHRITIS   C - LITHIASIS

D - HIPERTENSION   N - NO CLASSIFICATION

FIG. VII-10 - THE IMPROVEMENT OF THE PERFORMANCE

The quantitaive performance of NEXTOOL may be improved by:

VII.23a) allowing the occurrence of an inhibitory synapsis;

VII.23b) allowing the occurrence of pathognomonic pathways, that is, direct linkages between some very specific evidences with defined output classes, and by

VII.23c) introducing a default type of non-monotonic reasoning: if no information exists about a given evidence then consider its negation.

The introduction of these properties in both the learning and the reasoning machines improved the performance of NEXTOOL in handling the above renal diseases data base to 93% of correct classifications (Fig. VII.10).

VII.14 - Conclusion

NEXTOOL combines a hybrid scheme for knowledge representation that seems to be a powerful and flexible tool for developing heuristic classification systems (Machado et al, 1991b). It combines the expressiveness of semantic networks, the naturalness of fuzzy logic and the learning power of both inductive and deductive learning strategies.

The semantic networks give the system the ability to represent symbolic concepts, to structure and organize the problem domain knowledge, and to provide high level inference mechanisms such as the choice of the best reasoning models to solve a particular task.

The learning capability provided by the inductive and the deductive strategies supply a very potent tool to make of artificial intelligent systems structures very adaptive to a changing environment. Some interesting capabilities of such systems are: learning from scratch; automatic conversion of external (expert graphs) knowledge into EKN; continuous knowledge refinement, etc.

The inference and inquiry processes are low cost processes in NEXTOOL because they are supported by local decision and acyclic networks. Also, the description of NEXTOOL decisions is a very natural set of fuzzy rules of the type

if X is A and Y is B .... then Z is C

because its reasoning is supported by fuzzy logics.

ACKNOWLEDGEMENT

NEXTOOL was created by Ricardo J. Machado following the guidelines we discussed as part of a joint research program between Scientific Center Rio - IBM-Brazil, Escola Paulista de Medicina and UNICAMP, in the period 1988-1990. Ricardo has been an unvaluable partner and the source of many interesting ideas we have worked togheter.

CHAPTER VIII

JARGON
A NEURAL ENVIRONMENT FOR
LANGUAGE PROCESSING

VIII.1 - Jargon: a specialized subset of natural language

Human languages play 3 different functions:

VIII.1a) communication: they are used to move models from one MPNN to another;

VIII.1b) cognition: they provide a set of operators used to craft a new model from a previous knowledge. Thus, they play an important role in deductive learning, and

VIII.1c) archive: they store a set of reasoning models used by the culture serviced by them.

Human languages provide us with a core of common meanings centered on the basic schemes, scripts or frames related to our survival both in the physical world and within society (Washabaugh, 1980). In this sense, human utterances form a closed system of self-referred meanings. However, humans also modify the meanings of the words to speak of their individualities (Olson, 1980). In this use, human utterances become an open system of meanings referred to each individual context. Because of this, human languages should be treated as partially closed systems, where beliefs are always evaluated with respect to both the language itself and the context of the speaker (Greco and Rocha, 1988; Rocha and Rocha, 1985; Rocha, 1990a,b).

Language is also a cognitive tool. On the one hand, it provides the terms used to label at least the input and output neurons of the natural MPNNs (Chapter IV, section IV.6). In this way, it provides the basic symbols to speak of our reasoning. On the other hand, it provides some basic operators for deductive learning (see Chapter IV, section IV.8), which permit mutations by addition, deletion or knowledge association by means of MPNN crossing over. The conjunction BUT is an example of such operator. In general, this conjunction indicates that something contrary to the common knowledge must be combined to the piece of information antecedant to it. For instance, in I had lunch BUT .... this conjunction prompts this piece of information to be associated with the negation of the most usual consequences of having lunch, such as being full, being satisfied, etc., in order to describe an infrequent situation like I had lunch BUT I am still hungry. As a cognitive tool, language has also to be a partially opened system, since it has to adapt itself to speak of the new learned models at the same time it has to preserve the meaning in the already learned MPNNs.

In essence, human languages are fuzzy systems. A fuzzy set is a partially closed set, whose degree of closure can assume any possible logical value from completely open to completely closed, according to the system it is modeling. The degree of semantic closure of any language varies among other things with the degree of learning and the context of use.

As the strength of knowledge increases and defines a human specialization, the degree of restriction of the semantic of the language used to describe these learned models increases. The closure of the semantic associated with this specialization defines a jargon J(L) or specialized language as a subset of the entire language L (Rocha and Rocha, 1985; Sager, 1987). In this way, different contexts of use of L created within each one of the semantic of L assumes specific values depending on the models used in these specialized contexts. Inquiry is another process for the context closure of the use of L. This is a consequence of the fact that inquiry is used to obtain defined pieces of information to a specific learning. In the case of inductive learning, the closure of the context is a necessity to increase the possibility of repeated observation of the same fact. In the case of deductive learning, attention is focused upon pieces of information which are compatible with the model being explored.

A jargon J(L) is composed by a restricted dictionary D of terms concerning all words W of the language L, and a small subset P of all possible productions of this language. The jargon J uses these terms and productions to speak of a reduced subset M of meanings of the entire semantic S of L (Rocha and Theoto, 1991c). In this context:

$$J(L) = \{ D, R, M \} \qquad \text{(VIII.2a)}$$

$$R : D \times W \longrightarrow [0,1] \qquad \text{(VIII.2b)}$$

$$P: D^n \longrightarrow [0,1] \qquad \text{(VIII.2d)}$$

$$R : P \times M \times S \longrightarrow [0,1] \qquad \text{(VIII.2d)}$$

where the restriction R selects the dictionary D from W and specicifies the meanings M of P. The restriction R is either the expert knowledge in a field of specialization of human activity (e.g. medicine, law); the subject of the inquiring in the interview, or the scope of the data base (Rocha and Rocha, 1985; Rocha and Theoto, 1991c), etc.

VIII.2 - Theme and rheme

Any text or dialogue has a theme or subject and a (set of) rheme(s) or what is said about the theme (Sgall et al, 1973; Rocha and Rocha, 1985). For instance, the previous section may have "JARGON" as the theme, and its definition, development, formalization, etc., as possible rhemes of this theme. In the same way, Theme and Rheme are the central subject of this section. Here, the author's intended rhemes for this theme are definition, and use of these concepts in speech understanding. The declaration of the theme is a procedure to increase the closure of the text or dia logue. In other words, the theme refines R in (VIII.2).

People associate a personal degree of confidence with each piece of information as they pick it up from speech according to their previous knowlege. However, they have to wait until they have at least a grasp of the theme of the communication in order to assess the relevance of these pieces of information to support the chosen theme (Rocha, 1990a). If the Theme is related to some knowledge of the listener/reader it is quickly recognized, otherwise people must attempt to construct some initial scheme of the speech, assembling all the pieces of information into a network of meanings, guided by their confidence in each received piece of information and using deductive operators, either logical or linguistic. Once the theme is identified, the listener/reader proceeds with the identification of the possible rhemes. The consistency of the speech is then assessed by the relevance each piece of information has to support the models related to the theme and rhemes. Again, if a previous knowledge exists supporting the theme and/or the rheme, the speech can be considered consistent even if from the linguistic point of view it is badly constructed. Otherwise, the consistency is assessed according to the relevance of the incoming information to support some general knowledge stored in models of the language itself. In this condition, the speech has to be carefully constructed according to the rules of the used language.

The understanding of a given speech is closely related to its capacity to address some germs (see Chapter IV, section IV.8) either in some specialized knowledge of the speaker/listener or in the common knowledge they share in their culture and language. The closure of the speech is high in the first case and low in the latter. The complexity of the semantic analysis required by the language processing is low in the case of any jargon and increases as the specialization of the context decreases. The declaration or recognition of the theme of the speech helps to reduce the complexity of this analysis. Because the choice of the theme and rheme is knowledge dependent, the understanding of the same speech is flavoured by each decoder according to their own past experiences. Rocha and colleagues (Greco et. al., 1984; Greco and Rocha, 1987; Theoto et al., 1987; Theoto, 1990) developed a method to study the comprehension of a text by a given population of listeners. The results found by these authors are discussed in the next section.

VIII.3 - Investigating speech understanding

A text was tape-recorded and played to volunteers, who were asked to recall it and to perform some activities related to the construction of a graph representing his text decoding (Fig. VIII.1a). The text was selected taking into consideration the interest it could arise in the target population, because this was crucial to guarantee the attention of the volunteer focused upon the experimental task.

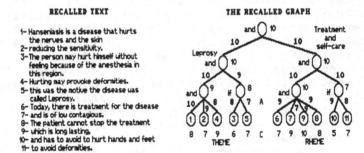

FIG. VIII.1 - RECALLED TEXT AND GRAPH

After the listening session, the individuals were requested to recall the text in a written form (Fig. VIII.1b) and to point out its theme and its rheme(s). The volunteer was allowed to repeat the listening session if he judged this would improve his understanding of the text.

The phrases of the recalled text were sequentially numbered, and the individuals were asked to split the recalled phrases into two subsets, one of them containing the phrases related to the theme and the other containing the phrases judged as supporting the rheme(s). These ordered subsets were taken as the terminal nodes of two subgraphs used to represent the understanding of the theme and of the rheme, respectively (Fig. VIII.1b). The subgraphs were built by asking the volunteers to join these terminal nodes into non-terminal nodes in the same way they assumed the recalled phrases had to be combined to support the theme and rheme, respectively. The same procedure was applied to the secondary, tertiary, etc., nodes, until the root representing the theme or rheme was reached.

After the recalled graphs were obtained, the volunteers were requested to assign a value of relevance for each arc of the graphs (Fig. VIII.1b) according to the importance the information represented at the leaving node was judged to have to support the theme or rheme. In the sequence, the individuals associated with each node of the graphs the degree of confidence triggered by information represented at the node. Finally, they were asked to associated the logic connective they used to join the information at each non-terminal nodes (Fig. VIII.1b).

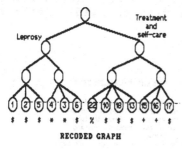

RECODED GRAPH

RECODED TEXT

1- Hanseniasis typically affects the skin and nerves
2- provoking spot and anesthesia
5- Because these injuries deform people
4- and may hurt himself unknowingly
3- Because of this anesthesia, the patient loose the sensation to light touch, pin prick or temperature.
6- Hanseniasis is also called Leprosy.
22- Todays, there exists treatment for the disease
10- Today, its is considered to be a poorly communicable disease
18- The treatment is long
13- and has to be submitted to a continuous treatment
15- in order to avoid injuries
16- if he has anesthetic hands and feet
17- This avoid body deformation.

FIG. VIII.2 - RECODED GRAPH AND TEXT

The labels (recalled phrases) assigned to the terminal nodes were recodified according to their correspondence to the phrases of the original text (Fig. VIII.2). This was done by one of the researches after the end of the interviewing session. The purpose of this recodification was to have an uniform description of the terminal nodes necessary to the calculation of the conditional distribution of the labels over the terminal nodes of the text graphs. The recodification was done in a copy of the recalled graph because extra nodes were introduced in the recodified graph whenever a recalled phrase corresponded to a merge of original phrases (nodes marked * and + in Fig. VIII.2b). Any recalled phrase unrelated to any original phrase received the same label. In the example of Fig. VIII.2, these extra phrases were labeled as phrase 22, once the original text has 21 phrases; the other numbers at the recodified graphs correspond to the sequential ordering of the phrases in the original text.

The averaging of the recodified graphs (Fig. VIII.3) was obtained by calculating the mean number of levels for both the theme and rheme subgraphs and the mean number of the nodes at each level. These mean values were assumed to be, respectively, the number of levels and nodes of the mean graph for a given population. The arcs of this graph were obtained from the analysis of the frequency of the corresponding arcs in the recodified graphs. The most frequent arcs in this population were maintained in the mean graph. This was done because the inspection of the recalled graphs showed a great variability of their connectivity, so that any graph summation even if based on similarity as in the case of the knowlege graphs in Chapter V, section V.3, would be useless. A threshold of 40% for including the arc in the mean graph was sufficient to maintain consistency between the node and level averagings and the arc counting, avoiding disconnected graphs as the final result.

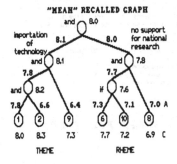

"MEAN" RECALLED TEXT

1- Brazil spends about 3 billion dollars each year in technology importation,
2- the equivalent to 8 times the total value of investments in applied and basic research in the country.
9- Despite this, we have some good examples of technology creation, such as ___
6- Scientific and technology research do not receive the necessary backing,
10- because Brazil faces today the problems of a narrow-minded technocracy at CNPq,
8- Even Lubrax-4 from Petrobras has a foreign formula

FIG. VIII.3 - MEAN RECODED GRAPH AND TEXT

The phrases were attached to the terminal nodes of the mean graph according to their conditional distribution over these nodes so that

"The most frequent phrase at a given terminal node was considered its label if it was not a label for any previous node, otherwise the next most frequent phrase at the node was picked up as the next candidate."

Fig. VIII.3 shows one of the mean graphs and texts obtained in this way for a group of students (Greco and Rocha, 1987).

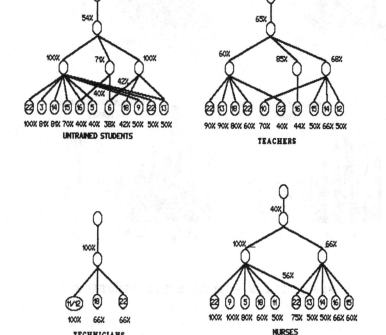

FIG. VIII.4 - MEAN GRAPHS FOR DIFFERENT POPULATIONS

Experiments on 3 different texts and involving around 300 people disclose some interesting properties of the text decoding (Greco and Rocha, 1987,1988; Theoto et al., 1987; Theoto, 1990):

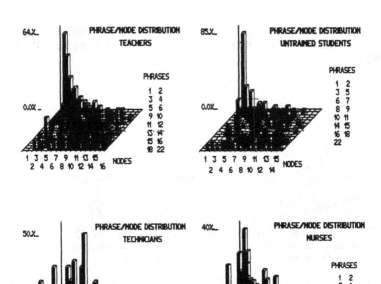

FIG. VIII.5 - PHRASE DISTRIBUTION

VIII.3a) In general, half of the original phrases were recalled by each volunteer. As a matter of fact, the probability of the phrase being recalled was a linear function of the mean confidence it shared in the population (Greco and Rocha, 1987,1988). In general, people recalled just those phrases eliciting confidence > .5. Only in very rare cases, they assigned confidence values < .5 to nodes of the recalled graph;

VIII.3b) The degree of fuzziness of the mean graph was not high, although it depended (Theoto, 1990) on the type of the text and on the skillness of the target population (Figs. VIII.4 and VIII.5). Less skilled people produced graphs smaller than well educated populations (Fig. VIII.4). In general, the mean graph may be viewed almost as a tree, for which just few (not all) nodes can have around two ancestors;

A TEXT ABOUT LEPROSY

Hanseniasis is a disease (1) which hurts the nerves and the skin (2), provoking spots and anesthesia (3). Because of this anesthesia, the patient does not feel pain and temperature in the affected regions (4), and he may hurt himself (5). These hurtings may deform people (6). Because of this, Hanseniasis was called Leprosy in the past (7). Also, there was a poor knowledge about the disease (8) and its treatment in the past (9). The patients were discriminated (10). Today, it is known that Hanseniasis is pourly contaminative (11), and it is curable (12). To be cured (13), the patient must not interrupt the treatment (14), which is prolonged (15). The treatment yelds good results (16) and avoids trasmission (17). The patient must take care (18) to avoid to hurt himself because of the anesthesia (19). In this way he may avoid body deformation (20).

NURSES

Hanseniasis is a disease (1) which hurts the nerves and the skin (2), provoking anesthesia (3) and the patient does not feel pain and temperature in the affected regions (4). Hanseniasis was called Leprosy in the past (7). Hanseniasis is a contagious disease and patients were discriminated (10), because of the deformation. The treatment is prolonged (15). It is curable (12). The patient must not interrupt the treatment (14). The patient must take care (18) to avoid to hurt himself because of the anesthesia (19).

UNTRAINED STUDENTS

Hanseniasis is a disease (1) which hurts the nerves and the skin (2), provoking spots and anesthesia (3) and deformities. Hanseniasis was called Leprosy in the past (7). The patient does not feel pain and temperature in the affected regions (4), he may hurt himself (5). The patient must take care (18) to avoid to hurt himself because of the anesthesia (19). Today, it is known that Hanseniasis is pourly contaminative (11), and it is curable. (12). The patient must not interrupt the treatment (14), which is prolonged (15).

CLASSROOM STUDENTS

Hanseniasis is a disease (1) which hurts the nerves and the skin (2), provoking anesthesia (3). Today, it is known that Hanseniasis is pourly contaminative (11), and it is curable (12). Because of this anesthesia, the patient does not feel pain and temperature in the affected regions (4), and he patient must take care (18) to avoid to hurt himself because of the anesthesia (19). These hurtings may deform people (6). Because of this, Hanseniasis was called Leprosy in the past (7). The treatment is prolonged (15). The patients were discriminated (10) the illness is a stigme. The patient must not interrupt the treatment (14).

TRAINED STUDENTS

Hanseniasis is a disease (1) which hurts the nerves and the skin (2), provoking spots and anesthesia (3) and the patient does not feel pain and temperature in the affected regions (4), and he may hurt himself (5). These hurtings may deform people (6). Today, it is known that Hanseniasis is pourly contaminative (11). There was no treatment in the past, today it is curable (12). The patient must not interrupt the treatment (14), which is prolonged (15). The patient must take care (18) to avoid to hurt himself (19).

TEACHERS

Hanseniasis is a disease (1) which hurts the nerves and the skin (2), provoking anesthesia (3) and the patient does not feel pain and temperature in the affected regions (4), and he may hurt himself (5). These hurtings may deform people (6). The illness is curable (12). The patient must not interrupt the treatment (14), which is prolonged (15). The illness was considered to be highly contaminative in the past. Today, it is known that Hanseniasis is pourly contaminative (11). The patient must be teached to avoid to hurt himself because of the anesthesia (19). The patient must take care (18) and must not interrupt the treatment (14).

FIG. VIII.6 - ORIGINAL AND MEAN TEXTS ON LEPROSY

VIII.3c) The distribution of the recoded phrases at the terminal nodes depended on the degree of education and expertise of the listener (Fig. VIII.5). Non-homogeneity on phrase distribution increases with education and expertise as one can observe in Fig. VIII.5 if Technicians and Nursing Untrained Students are compared with Nursing Teachers and Nurses. The text studied in this case speaks of stigma, treatment and self-care on Leprosy (Theoto, 1990) and it was planned as a text for Health Education Program on Leprosy. The interviewed technicians and nurses were involved with this kind of program;

VIII.3d) Although fuzzy, the terminal node phrase distribution was not flat, so that the proposed algorithm for terminal label assignment operated well;

VIII.3e) The contents of the mean graph provides a mean text which describes the mean comprehension of this text by the studied population;

VIII.3f) This mean text was always meaningful, requiring only minor adjustments for syntactic correctness (Fig. VIII.3 and FIG. VIII.6);

VIII.3g) The contents of the mean texts correlated with the background knowledge of the studied populations. For example, in the case of the text about Leprosy studied by Theoto, 1990, speaking about signals and symptoms of the disease; stigma and discrimination; treatment, and self-care (Fig. VIII.6):

VIII.3ga) all groups spoken about signals and symptoms; contagion and treatment;

VIII.3gb) all groups introduced at least one extra phrase of their own in the texts (not numbered phrases in Fig. VIII.6); the meaning of this phrase varying from group to group;

VIII.3gc) classroon students which attended theoretical classes about Leprosy, favored information about the stigma and its correlation with deformation;

VIII.3gd) the trained students who also received practical training spoke of deformation, but did not correlate it with stigma, and stressed the inexistence of treatment in the past;

VIII.3ge) among all the students,the untrained ones produced the shortest texts;

VIII.3gf) the teachers produced the longest texts and reproduced the phrases almost in the same order they appeared in the original text, while the nurses were short and precise in reproducing the text;

VIII.3gg) the technicians produced very bad recallings because of their very low degree of education.

The main conclusion from these experimental studies on language comprehension is that language decoding is dependent of the individual knowledge about the theme and rhemes of speech, but despite this variability, a consensus exists in different populations about the contents of the same text. This consensus provides a good description of the restriction of the semantics induced either by the specialization or the inquiry.

VIII.4 - The theoretical backgroung supporting JARGON

As a partially opened system, the human language provides a very important and adequate cognitive tool for any intelligent system, since this entity must be non-deterministic (Wah et al, 1989). Because of this, the processing of the language can be a very hard task for artificial intelligent systems. However, the human language is also a partially closed system, the degree of closure being directly related to the specialization of the context in which this language is used. If the processing of language is restrained to be performed in specialized contexts, the complexity of the semantic analysis will decrease considerably (Rocha et al, 1992; Sager, 1987). Since the speech understanding is closely dependent of the user's knowledge, then it may be proposed that the competence of the human being on language comprehension is achieved step by step, in each of his many specialized contexts of relations familiar, social, emotional, professional, etc.

JARGON is a MPNN system intended to acquire knowledge from natural language data bases about specialized contexts, taking advantage of the fact that the complexity of the required language analysis in this condition can be low. In this way, its competence is restricted by the semantics of these specialized environments. As a matter of fact, the power of JARGON as a knowledge extractor is closely related to the closure of the investigated environment. Since JARGON may use both inductive and deductive learning strategies, it is hoped that in the future it becomes capable to increase its linguistic competence by moving from one to another context, whenever required for the understanding of a complex speech. In its actual infancy, however, its competence remains constrained by specialization.

VIII.4a - The cerebral organization for language processing

Cerebral language processing is supported by a hierarchical assembly of MPNNs, each one of them specialized in analyzing some specific aspects of speech. Let us have a brief view of this hierarchy in the case of the neural system processing of the verbal utterances.

The incoming sound is decomposed into the set of its basic frequencies at the cochlea, because the ciliary receptors located at different places of the cochlea are distinctly activated according to the different frequency components of the incoming sound. The information provided by these receptors is aggregated in the first cochlear nuclei, so that some neurons in these nuclei specialize to fire in the presence of defined frequency patterns. These patterns are characterized not only by a group of specific frequencies but also by a specific timing between these frequencies. These neurons are sensitive to the formants and the voice onset time (VOT) of the human phonemes, which are recognized at these early stages of the verbal sound processing.

The output of these phoneme neurons is aggregated at thalamic and cortical areas, where some nerve cells become representatives of words (e.g. Brown et al., 1976; Kutas and Hillyard, 1980; Luria, 1974; Neville, Kutas and Schmidt, 1982; Rocha, 1990a,b,c). The activity of these ensemble neurons can be recorded and analyzed in the electro-encephalogram (e.g. Glasser and Ruchkin, 1976; Greco and Rocha, 1987,1988). The analysis of the event-related activity (ERA) in the electroencephalogram (EEG) associated with the word recognition revealed some interesting properties of the language processing.

All sound components of a word are not necessarily required for its recognition. Analysis of the ERA components disclosed that the earliest signal related to the word analysis appear around 100ms after the beginning of the voicing of this word. Late components may appear up to 700-800ms. However, most of the word recognition tasks is correlated with some positive waves peaking around 300ms. These and many other results favor the idea of a semantic analysis of the words founded on a expectancy controlled process (McCallun et al, 1983; Rocha, 1990a). According to this hypothesis, as soon as the received information activates some hypotheses about the incoming sound, this knowledge may be used to guide the linguistic analysis. For example, the recognition of a verb can prompt the system for its complements; the acceptance of a thematic phrase can preclude the cerebral processing of a competing information, etc. This guided analysis may attain important conclusions and decisions even before the word end, dismissing its full analysis. If the expectancy is confirmed the analysis may continue with the next piece of information, and the rest of the sound is redundant. However, if the expectancy is broken, the late waves are correlated with the task of discovering the mistake. This explains why the competence of people in recognizing verbal information uttered at high speed and in noisy environments increases with their mastering of the language being learned.

The output of the word neurons is directed to the next processing station. Here, the language syntax is effective in aggregating these words into phrases, but practice engraves idiomatic phrases in specialized neural circuits (Luria, 1974) and specialization creates jargon phrases. Greco and Rocha, 1988 and Rocha, 1990a showed that the EEG activity is closely related to both the confidence in the information provided by the phrases of a text and with the acceptance of these phrases to compose the theme or rheme of the text understanding. High ERA is associated with high confidence and high probability in accepting the phrase. But both measures are also dependent of the previous knowledge of the listener.

Finally, the output of phrase neurons at the parietal cortex is projected to the frontal areas where they are combined to create new or to recall known themes and rhemes (Luria, 1974). Different kinds of knowledge can be used to glue the recognized phrases into these complex ideas. Specific knowledge may privilege some key information to confirm or reject activated hypothesis. For instance, the expert privileges the data he needs to decision making about diagnoses being considered. General knowledge may be used to combine the same phrases according to different keys. For example, case gramatics (Fillmore, 1968) may be used to aggregate phrases according to the inquiry about who is the agent; who is the patient; when, where, why, the action took place, etc. But the incoming pieces of information may be used also to derive some new knowledge by means of deductive learning.

VIII.4b - The computational structure

JARGON is composed by 3 MPNNs hierarchically organized (Fig. VIII.7), whose purpose is to discover the commom contents, if any, of a group of texts in a natural language data base (NLDB). The first net is the Word Net (WN) whose job is to scan the texts and to learn the most frequent and meaningfull words in these texts. WN provides the input for the Phrase Net (PN), who is in charge of discovering the most frequent word associations (phrases) in the NLDB. These phrases are used as the input in the Text Net (TN) which is responsible for finding the possible text patterns in the data base. JARGON processes the contents of the NLDB in 3 steps: first it learns the most frequent words; in the sequence it discovers the most reliable word strings and finally it processes the possible summaries of the NLDB.

Each net is composed of a variable number of subnets or modules. The general structure of the modules is specified by their genetics G. The general rules encoded in G specifies, for instance, the number of layers in the net, the minimum and maximum size of the module, the compatibilities among the different types of neurons in the net, etc., and are the guidelines used by JARGON to create as many modules as necessary to accommodate the different words, phrases and texts it finds in NLDB. Evolutive learning (see Chapter IV) guides the genesis of the JARGON's modules.

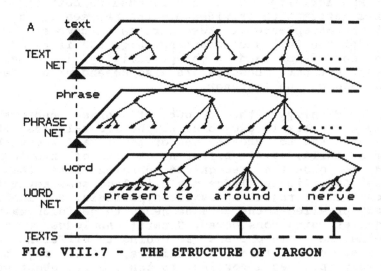

FIG. VIII.7 - THE STRUCTURE OF JARGON

The adjustment of the entropy of the net according to the variability of the environment it tries to represent is the central idea of the Evolutive Learning. The main steps of this process are:

VIII.4a) Genesis: a module is created whenever necessary to represent a new instance founded in NLDB during the training period. The structure of each new module is dependent of the pattern of the instance to be represented, and it is constrained by its genetics G. This process is discussed in great details in the next sections.

VIII.4b) Adaptation: the strength of the synapsis inside each module is dependent of the frequency in the data base of the instance it represents. Frequent well formed patterns are associated with strong modules having strong synapses, because repetition of the same instance during the training increases the strength of the conectivity inside the module, in the same way the fitness of the population is modulated by its ability to cope with the environment in the Genetic Algorithm theory. This is also the idea in Neural Darwinism, since those modules generated at the embryogenic period and which are most successful in representing the environment are rewarded by having their synapses strenghtened.

VIII.4c) Selection: only meaningfull modules are allowed to remain in the net after the training period. Selection is the key instrument in any evolutive theory since it was first proposed by Darwin. Selection is accomplished here by two different mechanisms: automatic and selective pruning.

Automatic pruning eliminates all modules whose strength is below a defined threshold. The value of this threshold is set dependent of the structural variability of both WN and PN whereas it is an ad hoc definition in TN used to set the degree of confidence the user may assign to the output of the net. In the case of WN and PN, the idea is to maximize the entropy of the dictionaries of words and phrases represented by the modules of the corresponding nets. In this way, the pruning threshold is obtained as the solution of a fuzzy linear mathematical program maximizing this entropy (Rocha and Theoto, 1991c). This algorithm is discussed below.

Selective pruning is the key tool the user has to complement the knowledge JARGON acquired about the restricted semantic in NLDB. After the automatic pruning in WN or PN, JARGON asks the user about the meaningfulness of the remaining modules, allowing him to eliminate those words or strings of words he judges not useful for the comprehension of the semantics used in NLDB. The user is allowed to eliminate also any word or word associations he wants for any other reason. He is asked to refine, if necessary, the semantics of the proposed phrases (word strings) by selecting or modifying one of the phrases used during the training to build the corresponding module, or even by providing any meaning by himself.

VIII.4c - Automatic pruning

The process of adaptation of the modules is common to all the JARGON's nets. Basically, the strength of the synapsis of the module is increased whenever the activation of this module by the training example is greater than .5. The augmentation of the synaptic strength is obtained by increasing the amount of transmitters at the pre-synaptic neurons n_i and of the receptors and controllers of the post-synaptic neuron n_k, by an amount that is equal to the degree of activation a_k of the post-synaptic neuron. Thus, if $t_i(m)$, $r_j(m)$ and $c_k(m)$ are the amounts of the tramsmitters, receptors and controllers at the mth iteraction of JARGON with NLDB, then:

$$t_i(m) = t_i(m-1) + a_k \quad \text{(VIII.5a)}$$

$$r_j(m) = r_j(m-1) + a_k \quad \text{(VIII.5b)}$$

$$c_k(m) = c_k(m-1) + a_k \quad \text{(VIII.5c)}$$

In this way, the synaptic weights of the modules correlate with the frequency in the NLDB of the instances they represent, and the structural entropy of the net reflects that of the data base.

The structural entropy h(MPNN) of nets of JARGON is calculated by normalizing the synaptic weight of the modules and using these values as the arguments of the Shannon's function:

$$h(MPNN) = \sum_{i=1}^{n} w_i \log w_i \quad \text{(VIII.6a)}$$

where w_i is the weight of the strongest synapsis of the germ of the module i.

The first step in characterizing the jargon J(L) is to obtain its dictionary D and the set of P of productions composed by terms of D. Both D and P must describe the most significant meanings in the restricted semantic M. The amount of information provided by D and P is dependent, among other things, of the entropy of their elements, because both very frequent and rare words or phrases are meaningless. Thus, to remain in WN and PN after the training period, the dictionaries D and P may be obtained by maximizing the structural entropy of these networks by means of fuzzy mathematical programming (Rocha and Theoto, 1991c).

Let the entropy h(S) of the jargon dictionary D or the production set P be calculated as

$$h(S) = - \sum_{i=1}^{n} p_i . \log p_i \quad \text{(VIII.7a)}$$

where p_i is the probability of the word i in D or a production in P, and it correlates with w_i in VIII.6a. The maximization of h(S) is obtained as the solution of:

$$\max h(S) = - \sum_{i=1}^{n} p_i . \log p_i \quad \text{(VIII.7b)}$$

$$h(s) > \alpha \quad \text{(VIII.7c)}$$

$$\begin{array}{c} n \\ \sum_{i=1} p_i = 1 \qquad \text{(VIII.7d)} \end{array}$$

The restriction in (VIII.c) is intended to maintain the mean entropy of the words or phrases above the fuzzy threshold α. The value of this threshold is one of the parameters defined by the user and one of the measures of the closure of the semantics of $J(L)$.

VIII.5 - The word net WN

The word net WN is a modular MPNN, whose modules are 3 layer nets. The input to the WN's modules is the ASCII code of the words in the NLDB. The genetics of WN allows the creation of modules having a maximum number of input neurons. This maximum is set according to the language used in the NLDB. The WN modules are constrained to have a minimum size, too. This is because in general very short words are less likely to carry any important information in a specialized language. Most of the small words are prepositions, articles, etc. which may be considered as meaningless for the purpose of recovering the jargon $J(L)$. However, JARGON is allowed to incorporate small words as exceptions whenever necessary. The number of the neurons in the clustering and output layers are dependent of the structure of the word the module represents.

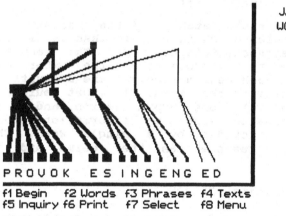

JARGON
WORD NET

PROVOK ES INGENG ED

f1 Begin   f2 Words   f3 Phrases f4 Texts
f5 Inquiry f6 Print   f7 Select   f8 Menu

JR/06/91
13:48:48

FIG. VIII-9 - AN EXAMPLE OF A WN MODULE

The genetics of WN allows as many receptors to be created in the input layer as the number of symbols in the ASCII code. This is because each letter is assumed to represent a different transmitter and each word is assumed to be a string of transmitters. So each letter of a given

word in NLDB may be associated with a specific input neuron
of a WN module by assigning the adequate receptor to this
neuron (Fig. VIII.9). In this way, the words in the NLDB may
activate different WN modules according to the matching
between its characters and the receptors of the module's
input neurons.

Whenever the total matching of the incoming word
(considered as a string of transmitters) with one or more WN
modules is greater than .5, the corresponding weights of
their synapses are increased proportionally to the degree of
this matching, and if necessary new neurons are added to the
module to represent new letters found in the word. By this
process, either typing errors are discovered or sufixes
(halloes) are learned to be combined with a germ (prefix) to
form different words (Fig. VIII.9). This is specially useful
in languages like Portuguese, where the verb conjugation is
very rich. In this way, the verb and its conjugation may be
learned by a few WN modules, as a set of germs and a family
of haloes. The germ is composed by all initial characters up
to the first transmitter/receptor mismatching. The haloes
are composed by all the remaining substrings necessary to
form all the words activating the module. Germs and
suffixes are clustered by different neurons in the
associative layer. Different words are produced by combining
germs and haloes at different output neurons. Germs serve as
indexes for the words in the NLDB. The minimum and maximum
sizes of a germ are restricted by the minimum and maximum
module sizes, respectively.

Whenever the matching of the incoming word with all
existing modules of WN is smaller than .5, a new module is
created to represent this new word. This module is created
with as many input neurons as are the number of letters in
the word, and each neuron receives the adequate receptor to
match one letter of the word. All input neurons are assumed
to be linked with one output neuron, and no intermediate
neuron is created, since no information about germs and
suffixes are provided by a single word. These direct
linkages are weakened as the germs and haloes are
strengthened.

Summarizing, JARGON

VIII.8a) creates one WN module to represent each new word it
finds in the NLDB;

VIII.8b) a new word is discovered whenever the matching of
its character string with all the already existing WN
modules is smaller than .5;

VIII.8c) whenever the matching of the incoming string with
any already existing WN module is greater than .5, the word
it represents is considered equal to the incoming word if
the matching is equal to 1, otherwise these two words are

assumed to be similar strings and the structure of the modules is changed to describe the differences between them;

VIII.8d) similar strings share a germ and are differentiated by their haloes;

VIII.8e) the germ is composed by the initial characters of the similar strings up the first difference detected;

VIII.8f) the haloes are composed by complementar substrings of the words strings concerning the germ string;

VIII.8g) germs and haloes are identified by different neurons in the intermediate layer of the WN module, and

VIII.8h) the different words are represented by distinct output neurons of the WN module

```
which/which\&£! 78                                                    A
disease_illnes/disease,illnes,ilsness\&£! 53
treatment_medica/treatment,medication,medicadion\£! 52
no_dont_doesnt_without/no,dont,doesnt,without\£! 49
deform_hurting_mutilate/deform,deforation,mutilate\£! 49
patient/patient\£! 43
hands_feet_region/hands,feet,region\£! 40
anesthe_analgesic/anesthesia,anesthetized,analgesic\£! 36
hanseniasis/hanseniasis\£! 32
provok_caus/provokes,provoking,causes,causing\£! 32
contagi_contaminati/contagious,contagiaus,contaminative\£! 29
pain_temperature/pain,temperature\£! 29
one/one\£! 25
care_avoid_protect/care,careful_avoid,protect\£! 24
people_individual/people,individual\£! 23
known_called/known,called\£! 23
skin/skin\£!22
hurt_affect//hurts,hurting,affects,affecting,affected\£! 22
hurt_damage/hurts,hurting,damages,damaged\£! 19
despite/despite\£! 29
_may/may\£! 19


which/which\coj&£c! 78                                                B
disease_illnes/disease,illnes,illness\suj& cop&£c! 53
treatment_medica/treatment,medication,medicadion\ADT&ADU&ADJ$COJ$£v! 52
no_dont_doesnt_without/no,dont,doesnt,without\adv&£c! 49
deform_hurting_mutilate/deform,deforation,mutilate\UTD&AUX&ADU&£v! 49
patient/patient\suj&£c! 43
hands_feet_region/hands,feet,region\ADJvtd& ADJvti&£c! 40
anesthe_analgesic/anesthesia,anesthetized,analgesic\adj&£c! 36
hanseniasis/hanseniasis\COP&COJ&£v! 32
provok_caus/provokes,provoking,causes,causing\SUJ&COJ&UTD&UTD&AUX&£v! 32
contagi_contaminati/contagious,contagiaus,contaminative\SUJ&ADV&COJ&AUX&£v! 29
pain_temperature/pain,temperature\UTDvtd&£c! 29
one/one\adj&£c! 25
care_avoid_protect/care,careful,avoid,protect\SUJ$AUX$COJ&UTV&£v! 24
people_individual/people,individual\suj&£c! 23
known_called/known,called\ADU&SUJ&AUX&COJ&£v! 23
skin/skin\UTDvtd&£c! 22
hurt_affect/hurts,hurting,affects,affecting,affected\UTD&UTD&SUJ&UIT&£v! 22
hurt_damage/hurts,hurting,damages,damaged\UTD&UTD&SUJ& vtv&£vc! 19
despite/despite\coj&£c! 29
_may/may\aux&£s! 19
```

FIG. VIII.10 - AN EXAMPLE OF DICTIONARY

The survival of a module in WN is dependent of the frequency with which it is activated by the words of the

NLDB. Whenever the computer working memory (CWM) of JARGON approaches a given limit during the training phase, the old weak modules in WN are killed. Newly weak modules are saved from this killing. The size of CWM defines the type (short, medium, etc.) of memory in WN. This size is initially specified by the user, and it is adjusted by JARGON to avoid frequent killings. Because of this, the final value of CWM is another measure of the closure of the $J(L)$ to be discovered in the NLDB. A small CWM means a very restricted $J(L)$. The killing eliminates infrequent words in the NLDB from WN. The frequency of the killing defines the semantics of this fuzzy quantifier infrequent.

After the training period, WN is submitted to a final automatic pruning, whose purpose is to maximize the informative capacity of the dictionary D (Fig. VIII.10a) produced by WN. This pruning is supported by the mathematical programming in eq.VIII.7. The WN modules having the structural entropy smaller than the threshold α in eq. VIII.7c are eliminated from WN in order to maxime h(WN). The threshold α is specified by the user. High values of α mean producing restrictive jargons $J(L)$. The user may experience different values of α to obtain the best dictionary D for his purpose, since JARGON always saves the WN produced at the end of the training period to restart the analysis.

The maximization of h(N), however, is not enough to characterize the entire restricted semantics being used in the NLDB. The second step of module selection in WN requires the participation of the user.

It is very common children asking their parents about the meaning of words they have encountered, instead of trying to discover this meaning by experiencing with the word. This is also a common strategy used by students, who ask the teacher or the expert about the meaning of special words of the jargon in use. This is to learn by being told. After scanning the data base and learning the initial dictionary, JARGON asks the user to refine his knowledge about the restricted semantics of the data base. JARGON shows the words it knows and the user may accept or eliminate them, or he may teach the system synonym relations. By this way, JARGON refines its dictionary in WN by being told about the semantics by the user. Fig. VIII.10a shows part of the dictionary created by JARGON from the data base of texts about Leprosy (LDB) used in Chaper III, section VIII.3 after being refined by the user. The string before the delimiter / contains the germs of the words showed between the delimiters / and \. Synonyms declared by the user are separed in the germ string by the underscore character _.

VIII.6 - The phrase net PN

The utterances in the NLDB are of two types:

VIII.9a) descriptive: they contains the definition or description of a symbol, for example: Hansen's disease is infectious; or

VIII.9b) declarative or procedural: they describe an action involving the elements of the phrase, e. g.: The disease hurts the skin and the nerves.

The key word in the case of the descriptive phrases is a symbol, about which some characteristics are stated. In the case of the declarative or procedural phrases, the central word is the verb describing the action. The simplest syntax used by JARGON takes this into consideration. JARGON requests the user to point out the verbs and symbols in the dictionary provided by WN (words assigned as fv in Fig. VIII.10b), as well as to mark the other words as complements (words described as fc in Fig. VIII.10b). A word may be verb in some utterances and complement in other phrases. In this case, it is labelled fvc. The output neurons of the word modules in WN are allowed to produce their transmitters according to this syntax. They will produce different transmitters for verbs and complements.

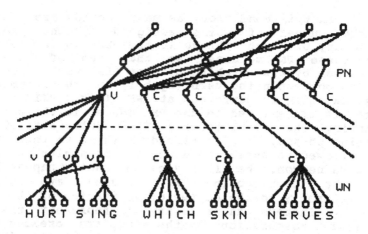

FIG. VIII-11 - THE STRUCTURE OF THE PN MODULE

If no other information is provided about the syntax of the language L used in the NLDB, JARGON will try to learn the phrases in this data base as associations between one verb or symbol and as many complements specified by the user. The maximum size of the modules created in PN is this specified maximum number of complements plus 1. Each PN module is accepted to receive input from two or more verb

(symbols) modules in WN only if they are declared synonyms.
Only the first input neuron in each PN module is allowed to
produce the required receptor to bind the verb transmitter
released by the output neurons in WN (Fig. VIII.11). If a
synapsis is established between a verb WN module and a PN
module, the other PN neurons in this module are allowed to
produce the receptor required to bind the transmitter
released by complementary WN modules.

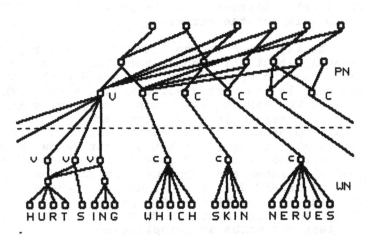

FIG. VIII-12 - AN EXAMPLE OF THE PN MODULE

        In this condition, as many modules are created in PN
as the number of verbs assigned by the user in the
dictionary D. Each phrase in the NLDB is initially matched
with these PN modules, so that one of them will be
activated by the matching between its verb (symbol) neuron
and the corresponding verb (symbol) in the incoming phrase.
This matching is performed at WN (Fig. VIII.11) and its
degree is transmitted to the PN module via the corresponding
synapsis. The complementary WN modules activated by the
incoming phrase, if not already linked to this PN module,
are allowed to establish a new synapsis with the first free
input PN neuron. The already established synapses between PN
and the corresponing WN modules are strengthened
proportionally to the degree of activation of the PN module.
As many clusters as necessary to represent the verb and
complement associations in the NLDB, are created by linking
the input PN neurons to the neurons in the associative layer
of the PN module (Fig. VIII.12). The linkages between the
associative and output layers are used to represent the
different phrases in the NLDB sharing the same verb or
symbol. Each of these phrases is represented by one output
PN neuron.

        Summarizing, the user

VIII.10a) defines the verbs and complements in D, and

VIII.10b) this information is used to define the transmitters produced by the WN modules.

Using this information, JARGON

VIII.11a) creates as many PN modules as the number of verbs in D, so that

VIII.11b) the first input neuron in PN is the verb (symbol) neuron which is linked to a verb (symbol) WN module. This verb neuron defines the PN module;

VIII.11c) the other input neurons are called non-verb neurons and they are linked to the WN modules representing the complements of the verb defining the PN module;

VIII.11d) the neurons in the intermediate layer of the PN module represent the distinct clusters of the verb complements as founded in the NLDB, and

VIII.11e) the output neurons of the PN module represent all phrases of the NLDB sharing the same verb.

Because of the low combinatorial restriction imposed by this simple verb/complement syntax, many badly formed phrases are learned by JARGON in this condition. This number increases as the phrase segmentation in NDLB becomes poorer. The ASCII character 46 (.) is used to break the NLDB strings into phrases. A poor punctuation increases the size of the "utterances" partioned by JARGON in the NLDB and favors the creation of meaningless phrases at PN. Poor punctuation was the most frequent finding in the real NLDBs JARGON analyzed up to now.

At the end of the training period, the same automatic pruning described by eq. VIII.7 is used to eliminate the less entropic significative modules in PN. Besides this, JARGON shows to the user the phrases it learned, in order he selects the most promising word combinations as representatives of the phrases in the NLDB. The user is also requested to provide a semantic meaning to the phrases represented in the remaining PN modules (phrases after the delimiter $ in Fig. VIII.14). The mechanism of this meaning assigment procedure will be discussed in section VIII.8.

VIII.7 - Implementing the syntax

The Formal Genetic Code (FGC) described in Chapter

III, section II.5 may be used by the user to encode a syntax of the language L stronger than the above verb/complement syntax. For example, it may be used to encode a syntax founded over the concepts or classes: Subject (suj), Adverbs (adv), Adjectives (adj), Conjunctions (coj), Direct Transitive Complement (vtd), Indirect Transitive Complement (vti), Auxiliary verbs (aux), etc.

The FGC may be programmed to produce transmitters, receptors and controllers specifying these classes, so that the t^r affinity is used to encode the syntatic rules specifying the class concatenations in this syntax. In this way (Fig. VIII.13):

VIII.12a) transmitters such as suj, adj, adv, vtd,, etc., are produced to encode the classes: Subject, Adjective, Adverb, etc., respectively. These transmitters are assigned to the complement WN modules;

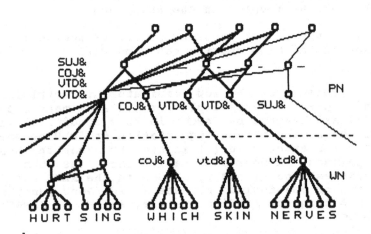

FIG. VIII.13 - THE STRUCTURE OF THE PN MODULE
UNDER A SYNTAX OF L

VIII.12b) receptors such as SUJ, ADJ, ADV, VTD,, etc., are associated with the same classes as above. These receptors are assigned to the non-verb input neurons of the PN modules. Whenever the PN module is activated, these neurons can concatenate with their complementary WN modules;

VIII.12c) controllers of the type vtdADJ, sujADJ, vtdADV,, etc. may be associated to the non-verb PN input neurons in order to implement syntactic conditional rules. In this context, whenever the adequate vtd, suj,, etc., input neuro is activated, it can release its controller ADJ, or ADV,, etc.:

SUJ ˆ sujADJ » ADJ, VTD ˆ vtdADV » ADV, etc.

which will act as a receptor in a neighboring neuron. This neuron may now concatenate with the corresponding adjective, adverb, etc., WN modules.

Whenever a FGC is used to encode the syntax of L (Fig. VIII.10b):

VIII.12d) the user defines as the complements and verbs in D as before;

VIII.12e) the user assigns the different categories to the complement terms of D, using the corresponding strings (suj, adj, adv, etc.) of the FGC. Since the same term of D may belong to different categories, each category is ended by the delimiter &;

VIII.12f) the user describes the verb syntax, by declaring the categories it may accept. These categories are specified as receptors, thus they are encoded by the corresponding strings (SUJ, ADJ, ADV, etc.) of the FGC composed by capital letters, each category being ended by the delimiter &, and

VIII.12g) the complementary terms of D are allowed to ask for specific complements of their own (e.g. sujADJ, vtdADV, etc.). In this case, the corresponding receptor (ADJ, ADV, etc.) is added to the transmitter string (suj, vtd, etc.), as the controller to be used to label an input neuron of PN. This procedure allows to condition the acceptance of some syntactic categories by PN to the presence of some specific combination of categories. For example:

hands_feet_region/hands,feet,region\ADJvtd&   ADJvti&£c

in the dictionary of Fig. VIII.10b is allowed to incorporate the word

anesthes_analgesic/anesthesia,anesthetized,analgesic\adj&£c

after concatenating with a verb like

provok_cause/ .... \SUJ&COJVTDAUX£v

to represent the NLDB phrase

provoking anesthetized hands and feet.

Using this information, JARGON (Fig. VIII.13):

VIII.13a) creates as many PN modules as the number of verbs in D, so that

VIII.13b) the first input neuron in PN is the verb (symbol) neuron, which is linked to a verb (symbol) WN module. This verb neuron defines the PN module;

VIII.13c) the output neurons of the non-verb WN modules are informed by the verb neuron about the type(s) of class(es) the word assigned to it belongs to. This information is used to activate the adequate gene of its genetics G to produce the adequate transmitter. In the example of Fig. VIII.10b, the Subject output WN neurons (Fig. VIII.14) are instructed to produce the transmitter suj"; the Adjective neurons are instantiated to release the transmitter adj", etc. In this way

VIII.13d) non-verb neurons are assigned to produce the corresponding receptors for the different syntatic categories accepted by the verb and their complements. In this way, as many input neurons are created in the PN module as the number of syntactic classes required by its verb, each one of them receiving one adequate receptor (e.g. SUJ, ADJ, etc.). By this process, each PN input neuron becomes specialized in recognizing one of the syntactic classes accepted by the verb defining its PN module (Fig. VIII.13). These neurons may produce controllers to be used as receptors by other neurons in the PN modules. This implements the conditional processing described in VIII.12g,.

Now,

VIII.14a) each phrase in the NLDB activates the PN module corresponding to its verb (symbol), according to the degree of the matching between this verb (symbol) and its correspodent WN module;

VIII.14b) the other words of the NLDB phrase activates their corresponding complement WN modules;

VIII.14c) only those non-verb WN modules assigned with the adequate categorical transmitter are allowed to bind the non-verb PN neurons;

VIII.14d) the strength of the activated synapses is increased proportionally to the degree of activation of the PN module;

VIII.14e) as many different neurons are created in the intermediate layer of the PN module as necessary to represent the distinct syntactic clustering of the words of the incoming phrases, and

VIII.14d) as many different output neurons are created in the PN module as required to represent the different phrases in the NLDB sharing the verb defining the PN module.

The use of FGC to encode a syntax of the language L greatly reduces the combinatorial explosion induced by a poor NLDB phrase segmentation, because it restricts the genesis of the synapses according to the t^r affinity. This affinity encodes rules of the chosen syntax. The more restrictive is this syntax, greater is the effect in containing the combinatorial explosion. The best effect of the FGC encoding of the L syntax is, however, the improvement of the quality of the phrases produced by PN. The syntax used in the example of Fig. VIII.14 reduced the number of phrases produced by PN in 60% in the case of the Leprosy Data Base (LDB). The user discarded only 6 of the phrases proposed by JARGON as nonsense. Besides this, most of the word strings proposed by JARGON were very similar to the phrase chosen by the user to define its meaning (see Fig. VIII.14).

```
hanseniase/disease_illnes/05CDE002$hanseniasis is a disease¿% 20
hurt_affect/which/skin/nerves/05CDE007$which affects the skin and nerves¿% 16
hurt_affect/hands_feet/05CDE007$affecting hands and feet¿% 16
cause_provok/deform_hurtings_mutilate/05CDE021$provokes deforation¿% 25
cause_provok/spots/05CDE022$provoking spots¿% 10
feel/doesnt/pain_temperature/05CDE027$doesnt feel pain and temperature¿% 8
hurt_damage/patient/may/05CDE029$the patient may hurt himself¿% 8
know_called/leprosy/05CDE032$it was called leprosy¿% 10
know_called/no_without/05CDE033$not well known¿% ?
contagi_contaminati/poorly/05CDE035$poorly contaminative¿% 13
treatment_medica/no_without/was/05CDE041$there was no treatment¿% 11
treatment_medica/prolong_difficult/05CDE044$the treatment is¿% 11
care_avoid_protect/hurt_damage/05CDE030$avoid to hurt himself¿% 12
care_avoid_protect/must/05CDE053$must be careful¿% 9
discriin_isolated/patient/was/05CDE056$the patient was discriminated¿% 8
cure_curable/disease/05CDE057$the disease is curable¿% 8
_explain/patient/must be/05CDE058$the patient must be explained¿% 8
```

FIG. VIII.14 - AN EXAMPLE OF PHRASE DICTIONARY

VIII.8 - Learning the semantics by being told

Whenever a natural or artificial system acquires some knowledge it may be told how to modify it. JARGON learns words, and asks the user to help it to refine D; it learns word strings, and dialogues with the user to discover the restrict semantics of these phrases.

At the end of the PN training phase, JARGON knows the following:

VIII.15a) the dictionary D of the jargon J(L) used in the NLDB;

VIII.15b) the synonym relations between the terms of D;

VIII.15c) the most frequent verb/complement clusters in the

NLDB, and optionally, if it was taught a syntax of L, it also knows

VIII.15d) the most frequent verb/category clusterings in the NLDB.

Besides this, JARGON created data bases about this knowledge. In the same way, natural MPNNs control biological actuators (e.g, muscles, glands, etc.) in order to manipulate the external world W, JARGON uses the computational facilities provided by SMART KARDS(c) (see Chapter IX) to organize data bases about the knowledge it acquired. At the end of the PN training phase, JARGON already created the following data bases:

VIII.16a) DICTIONARY: a set of cards is assigned to each term of D. All information discovered by JARGON in the NLDB about germs and haloes, frequency, etc. of these terms are recorded in these cards, together with the knowledge obtained from the user about synonym relations, the syntax of L, etc., and

VIII.16b) PHRASES: every phrase encountered in the NLDB is written in a special card (phrase card) assigned to all verbs of D and to any other special term chosen by the user. Any NLDB phrase containing one of these terms is written in the corresponding phrase card. This data base contains all training examples used to build PN, indexed according to the PN modules they helped to craft.

The system uses all this knowledge to dialogue with the user in order to learn about the restricted semantics of the phrases represented in the PN modules. JARGON presents to the user each verb/complement or verb/category cluster it learned in the NLDB together with the corresponding training examples and asks the user to classify the cluster as:

VIII.17a) very well formed: if the words in the cluster unequivocally define a specific meaning in J(L). In this case, the user must provide a phrase JARGON will use to refer itself to this cluster. Once articles, prepositions, etc., may be eliminated from D in the WN traning phase, the well formed string JARGON learned may be linguistically incomplete. For example, the cluster

HURT_AFFECT/WHICH/SKIN/NERVES

in the phrase dictionary shown in Fig. VIII.14 clearly maps into the phrase WHICH AFFECTS THE SKIN AND NERVES, but the article THE and the conjunction AND were eliminated from the dictionary shown in FIG. VIII.10;

VIII.17b) well formed: if there is one training phrase which best specifies the most frequent meaning assigned to the cluster in J(L). In this case, the user has to point this phrase. This is the case of the cluster

HURT_DAMAGE/PATIENT/MAY

associated to the phrase THE PATIENT MAY HURT HIMSELF, which was the most frequent example in the phrase training set asssociated with the verb HURT_DAMAGE;

VIII.17c) ambiguous: if more than one meaning can be currently assigned to the cluster being shown. In this case, the user has to provide JARGON with a phrase which may provide the best approximation of these meanings, even if it is composed of conflicting information. An example of ambiguity could appear in the case of a cluster of the type

PATIENT/HEIGHT/

In this case

THE HEIGHT OF THE PATIENT IS:
SHORT (10%), MEDIUM(70%), TALL(20%)

would be a phrase provided by the user after the inspection of the phrase training set. Ambiguity and restricted jargon are contradictory concepts. As the restriction of the semantics of J(L) increases, the possibility of ambiguous utterances in J(L) must decrease. No ambiguity was observed in the LDB analyzed in figs. VIII.10 to 14.

VIII.17d) badly formed: if the cluster is syntactically or semantically incorrect. In this case, JARGON will remove the corresponding module from PN.

Whenever the decision of the user is VIII.17b or c, JARGON makes a note of this decision, so that if inquired in the future about the meaning of its utterances, it can show the user the phrase training set used to define the semantics of these clusters. Also, the frequency of these decisions is used as measures of the degree of the restriction of the jargon J(L) used in the NLDB.

VIII.9 - Recodifying NLDB

Once JARGON has learned the restricted semantics of the jargon J(L) in the NLDB, it can use this knowledge to rewrite the NLDB using the phrases the user assigned to the PN modules. This has the advantage of creating a copy of the NLDB where the information is in a standard format, which may favor many data base functions. The recodified NLDB (RNLBD) together with the Dictionary and Phrase data bases described in VIII.16 provides a new description of the NLDB, which serves many different purposes.

Although JARGON was initially designed as a tool to summarize the NLDB, its users discover many roles for it. For instance, for one of its users, the Phrase data base

ORIGINAL TEXT

hanseniasis is a diseach which was called leprosy in the past
it is a disease which hurts the nerves and sking and
provokes anesthesia
because of this the patient has to be careful
with the type of his activity
it is known today that the disease is poorly contaminative
and curable
the treatment is prolonged by efficient

RECODED TEXT

hanseniasis is a disease
which hurts the skin and nerves
provokes anesthesia
it was called leprosy
it is known today
it is poorly contaminative
the treatment is efficient
the treatment is prolonged
must be careful

FIG. VIII.15 - THE NLDB RECODIFICATION

```
  PN         0         1         2
MODULE   12345678901234567890123456789012345

TEXT:  001  *** **  * ****** *
TEXT:  002  ****           ** **
TEXT:  003  ** ****** *       **
TEXT:  004  ** *  * ***     **** *
TEXT:  005  ** *** **        **
TEXT:  006  ** **  * ** **    *
TEXT:  007  **           * ** *
TEXT:  008  * * **   * * **
TEXT:  009  ** *   *  * *** *
TEXT:  0010 *    * **      *  *
TEXT:  0011 *** ** ******* * **
TEXT:  0012 ** ****  ** * * **
TEXT:  0013 * * * *    * **
TEXT:  0014 ****** *  ** *    **
TEXT:  0015 *         * * *
TEXT:  0016 **** ** ***  * * *
TEXT:  0017 *****    ***     * *
TEXT:  0018 ** * * **** *   *
TEXT:  0019 *****   * *  ** ***
TEXT:  0020 ***   *  **** * * **

         12345678901234567890123456789012345
             0         1         2
```

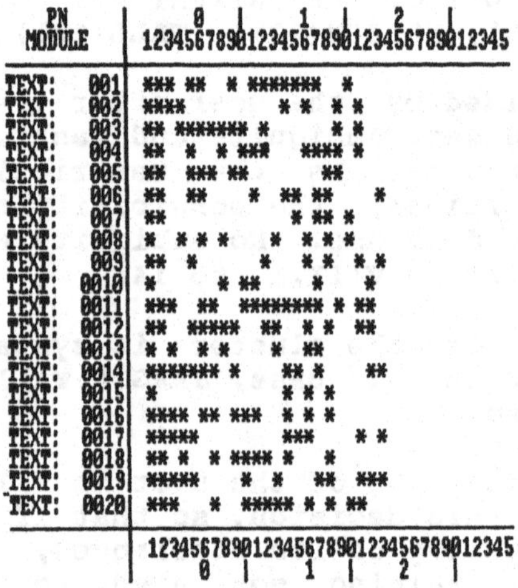

Total codified phrases in the NLDB: 190
Mean number of codified phrases per text: 9.5
Total phrases in the NLDB: 309
Mean number of phrases per text: 15.45
Recovering index: .614

FIG. VIII.16 - THE NLDB RECODIFICATION STATISTICS

became the key product produced by JARGON concerning his
NLDB, because it allowed him to filter some numeric data
related to some technical procedures described in natural
language. These data were considered very important for
future planning of the activities of his company. To
discover and analyze the same data without help of JARGON
was a very dull activity involving many people without
obtaining the same precision in the results.

JARGON rewrites each text in the NLDB using the knowledge encoded in PN. One example of this recoding in the case of the LDB is shown in Fig. VIII.15. At the same time this job is done, a basic statistic about the recoding is processed, so that at the end of this phase, JARGON provides the user with (Fig. VIII.16):

VIII.18a) a raster histrogram showing the distribution of the PN codes in the NLDB;

VIII.18b) the mean number of phrases in the NLDB;

VIII.18c) the mean number of recoded phrases, and

VIII.18d) an index about the efficacy of the recoding, which here is called recovering index.

This index serves two purposes:

VIII.18e) to be a measure of the closure of the jargon J(L) used in the NLDB, and

VIII.18f) to be a measure of the quality of the knowledge provided by the user about the restricted semantics of J(L).

JARGON allows the user to improve the quality of the taught semantics through an iterative process, by means of which the user may start again the training of each of its nets after analyzing the quality of the output of the system. The user can analyze this output and discover mistakes he did in the teaching of JARGON, or discover a better way to encode the restricted semantics of J(L). The quality of the RNLDB can be improved by this iterative approach. This improvement is quantified by the evolution of the recovering index. JARGON uses the RNLDB to try to process summaries about the contents of the NLDB.

VIII.10 - The text net TN

The structural entropy of PN is used by JARGON as a guideline to discover the possible summaries in the NLBD. The strongest PN modules are chosen as thematic modules. The number of the thematic PN modules determine the number of modules to be created in the text net TN. In other words, JARGON creates as many TN modules as the possible themes of the NLDB. These themes correspond to the most frequent PN modules. The PN modules exibiting intermediate structural entropy are used to define the possible rhemes in the NLDB, and the weakest (complementary) PN modules provides complementary information to specify both the chosen themes and rhemes. The semantic of the fuzzy quantifiers strongest, intermediate and weakest is defined by the user, who must specify two different thresholds to classify the PN modules. The number of intermediate layers in the TN modules varies according to the complexity of the clustering of the PN modules in the RNLDB. Before inspecting the RNLDB, JARGON creates as many TN modules as the thematic PN modules with the following initial structure (Fig. VIII.17):

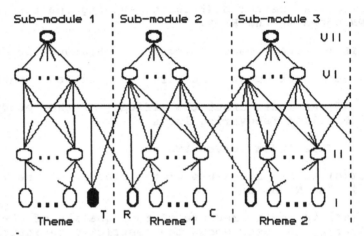

FIG. VIII.17 - THE STRUCTURE OF THE TN MODULE

VIII.19a) the first input TN node (T in Fig. VIII.17) is linked to the theme PN module and defines the theme of the TN module;

VIII.19b) the required numbers of rheme input TN nodes (type R in Fig. VIII.17) are created to be linked to the existing rheme PN modules. As a matter of fact, the rheme nodes define submodules in the TN module, in the same way microcolumns are identified in the cortical columns (Chapter VI, section VI.2);

VIII.19c) the required numbers of input neurons (type C in Fig. VIII.17) are created to be linked to the theme and rheme PN modules, and

VIII.19d) five intermediate layers are created, besides the input and output layers.

      In the sequence, JARGON learns (see e.g. Figs. VIII.20):

VIII.19e) the most frequent associations between the complementary phrase and the theme or rheme phrases, representing them in neurons in the first intermediate layers of the corresponding submodules of the TN module, and

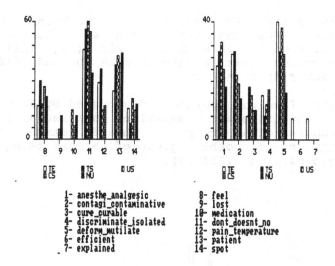

1- anesthe_analgesic
2- contagi_contaminative
3- cure_curable
4- discriminate_isolated
5- deform_mutilate
6- efficient
7- explained

8- feel
9- lost
10- medication
11- dont_doesnt_no
12- pain_temperature
13- patient
14- spot

FIG. VIII.18 - VARIABLE TERMS IN THE LEPROSY DICTIONARIES

VIII.19f) the most frequent associations between the theme and rhemes in the NLDB, representing them in neurons in the upper intermediate layers of the corresponding TN module.

At the end of the TN training phase, all unused associative layers are removed from the TN modules. No automatic pruning is used in TN, but the user may define a module strength threshold to kill the weakest TN modules. The value of this threshold defines the minimum confidence in the final summaries produced by JARGON.

VIII.11 - Handling the leprosy data base

JARGON was used to calculate the summaries of the LDB studied by Theoto et al., 1987, 1990. These summaries were compared with the "mean" text obtained with the graph methodology described in section VIII.3. The results of this analysis is presented and discussed in this section.

The dictionaries obtained for the 5 different populations of decoders (Untrained Students (US), Classroom Students (CS), Trained Students (TS), Teachers (TE) and Nurses (NU)) were mostly composed of the same words. As a matter of fact, a common dictionary with 35 words obtained from all words with a frequency higher than 6 in each population dictionary (PD) showed 21 terms being equally frequent in these PDs, and only 14 terms exhibiting a more variable distribution in these dictionaries (Fig. VIII.18). Some of these words occurred only in some PDs and did not appear in any other dictionaries.

The most frequent phrases discovered by JARGON in the texts of the different populations supplied a phrase dictionary composed of a group of phrases common to all texts (Fig. 21), and of another group of phrases being used only by some groups (Fig. VIII.19).

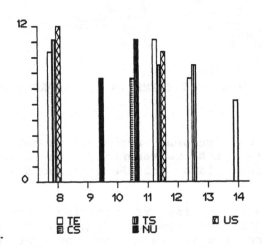

1- does not feel pain and temperature   4- there was no treatment in the past
2- contagious disease   5- hanseniasis has cure
3- looses the sensitivity   6- the treatment is efficient

FIG. 19 - VARIABLE PHRASES IN THE LBD

These results point to some consensus about the contents of the comprehension of the same text by the different populations, which is flavored by the distinct knowledges about Leprosy each population has. This different knowledge backgrounds assign different confidences to some of the phrases of the original text, which results in selecting different information as relevant for the definition of the chosen rhemes. All groups chose the same theme: Hansen's disease, and mostly agreed with some rhemes: duration of the treatment; sumptoms, etc., while disagreeing in respect to some other rhemes: discrimination; contagioun; cure, etc.

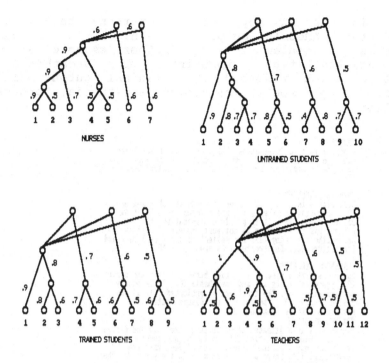

FIG. VIII.20 - TN MODULES GENERATED BY THE DIFFERENT POPULATIONS

The TN modules generated different topologies about the LDB (Fig. VIII.20) which reflect both the consensus in the studied populations about the contents of the text, and the variation imposed upon this common understanding by their distinct knowledge backgrounds. The summaries produced from these TN modules are shown in Fig. VIII.21.

The summaries correspond almost exactly to the output provided by JARGON. Only minor editings to avoid word repetition were done by one of us. An example of such editing is the phrase Hanseniasis is a (contagious) disease in the summary of the nursing texts, which replaced the JARGON's output: Hanseniasis is a disease, Hanseniasis is contaminative. Another example are the phrases of the type: provoking spots, anesthesia and deformation which replaced the original output of the type: provoking spots, provoking anesthesia, provoking deformation.

NURSES
Hanseniasis is a (contagious (2)) disease (1) which hurts the skin and nerves (3), provoking spots (4) and anesthesia (5). The treatment is prolonged (6). It was called Leprosy (7).

UNTRAINED STUDENTS
Hanseniasis is a disease (1) which hurts the skin and nerves (2), provoking anesthesia (3) and deformation (4). It was called Leprosy (5). It is poorly contaminative (6). The patient has to take care to not hurt himself (7). It may be hurt without knowing (8). The patient does not fell pain and temperature (9). The treatment is prolonged (10).

CLASSROOM STUDENTS
Hanseniasis is a disease (1) which hurts the skin and nerves (2), provoking spots (3) and anesthesia (4). It is poorly contaminative (5). The treatment is prolonged (6). It was called Leprosy (7). The patient was discriminated (8) because of the deformation (9).

TRAINED STUDENTS
Hanseniasis is a disease (1) which hurts the skin and nerves (2), provoking deformation (3), spots (4) and anesthesia (5) It was Leprosy (6). The patient does not feel pain and temperature (7). There was no treatment in the past (8). The patients were discriminated.

THEACHERS
Hanseniasis is a disease (1), called Leprosy (2). There was no treatment in the past (3). It hurts the skin and nerves (4), provoking spots (5) and anesthesia (6). I causes deformation (7). We know today (8) that the disease is poorly contaminative (9). Because of the anesthesia (10) the patient does not feel pain and temperature (11). The treatment is prolonged (12).

FIG. 21 - THE SUMMARIES PROVIDED BY JARGON

The input node numbers in Fig VIII.20 correspond to the phrase numbers between brackets in Fig VIII.21.

The comparison of the summaries provided by JARGON with the mean texts obtained by Theoto, 1990, using the graph technology can be done by comparing Fig. VIII.6 in with Fig. VIII.21. This analysis shows that JARGON is able to produce summaries that are very close to the real contents of the data base. These summaries are shorter than the mean text produced by the graph technology, because the recovering index for the recoding of the LDB was around .7 in the different populations. However, there was no atempt here to improve the quality of the teaching about the restricted semantics by means of the iterative process discussed in section VIII.9, in order to give an idea of the capacity of JARGON to work with unfriendly users.

VIII.12 - JARGON's multifunctions

JARGON was originally designed as a tool to extract knowledge from data base encoded in natural language, such as reports about specialized activities, contents of interviews about defined subjects, etc. It has been successfully used in this context, to analyze:

VIII.20a) Expert Data Bases: 4 medical data bases; 1 data base of reports about well offshore completion operations (Miura et al., 1991), and 1 data base on blast furnace control;

VIII.20b) Specialized Data Bases: a set of laws about health politics; a set of abstracts of articles in nursing informatics, and

VIII.20c) Interview Data Bases: on Leprosy health problems (2), nursing techniques (1) and media news (2).

These different types of NLDB exhibit a wide range of variation of the closure of the semantic of their jargons J(L). The summaries provided by JARGON were (Rocha et al, 1992):

VIII. 21a) well structured if the restriction of the semantics tended to be high as in the case of the expert and specialized data bases, and they became

VIII.21b) fuzzy as the semantic restriction decreased as in the case of some of the interviews. But even in this situation JARGON was able to provide the user with useful descriptions of the contents of the data base, as in the case of LDB.

The usefulness of the summaries was assessed by means of different strategies. In the case of the expert data bases, it was evaluated by:

VIII.22a) the capacity of JARGON to provide different summaries of texts associated with different classification, and

VIII.22b) the degree of agreement between the contents of these summaries and the knowledge required by the experts to support the corresponding diagnosis (Theoto et al, 1989) or the corresponding knowledge in engineering technical reference manuals (Miura et al, 1991).

VIII.22c) the success of using these summaries as knowledge data base for expert systems (see Chapter IX, section 8).

In the case of interviews, the quality of the summaries was assessed by 1) checking its contents with the knowledge the experts have about the complaints of the patients or by 2) their capacity of providing meaningful information about the subject of the interview (Theoto, 1990). Finally, in the case of specialized data bases, JARGON provides the dictionaries to be used to index the data base, and the quality of these dictionaries is evaluated by the degree of satisfaction provided by the retrieved material in response to the queries imposed upon the system (Theoto and Rocha, 1992).

The quality of the summaries provided by JARGON is:

VIII.23a) directly related with the degree of restriciton of the employed jargon, and

VIII.23b) dependent of the quality of the selective pruning in PN.

Almost all word strings JARGON discovers in a high restricted language carry a specific meaning easily identified by the user. On the contrary, low restricted jargons supply poorly characterized word strings, demanding a close look in the phrase training sets either to choose one of them to define its meaning or to orient the user to build the semantics from his own knowledge of the subject. Therefore, the decoding of the texts provided by JARGON becomes more subjective as the restriction of the used semantics decreases. Because of this, JARGON is now also being used as a tool for the study of how different users may decode the same texts in different ways.

The knowledge acquired by Jargon in expert data bases is now being used to build expert systems for medical diagnosis (Machado et al, 1991, 1992); to orient completion operations in offshore platforms (Miura, et al, 1991), and to blast furnace fuzzy control (Fernandes and Gomide, 1991), because it allows both the NLDB and the information provided by the user to be recoded into the same jargon the expert system uses to reason.

The most interesting finding arising from the use of JARGON by different users, is the new functions these users are discovering for it. JARGON is being pointed as a important tool to analyze the institutional memory of companies. According to Miura et al, 1991:

"The institutional memory is one of the major properties of a company. It has been kept by editing and reviewing the technical documents, such as procedural manuals, safety guideline manuals, etc. Editing these technical documents, one should have primary knowledge about actual procedures and fault modes, which can be obtained throughout daily report analyses, which are stored in natural language data bases. "

JARGON proved to be a powerful tool in extracting this institutional knowledge from these reports. Besides, it has also been used as a friendly filter to localize numeric and other types of information encoded in natural language. Because of this, JARGON is now part of SMART KARDS, an intelligent environment using the notions of Object Oriented Programming and Multipurpose Neural Nets.

ACKNOWLEDGEMENT

The development of JARGON was closely guided by the experimental results obtained by M. Theoto on language understanding. Her questions on practical domains were very influential in defining most of the basic operations of Jargon, and were very important issues accounting for the capabilities the system has to cope with some complex issues in knowledge acquisition.

I am in debt with the hundreds of students who volunteered to the experiments we performed during the last 10 years about language understanding. The students of the undergraduate course of Neurophysiology, Institute of Biology, UNICAMP, year 1991, used JARGON as tool for experiencing with language understanding.

I have also to thanks my former students Edson Françozo for his patience in teaching me some basic concepts in Linguistics, and the late G. Greco for doing some very important research about the brain activity during language decoding. Ivan R. Guilherme has worked part of the programs composing JARGON.

Kazuo Miura, student of the post-graduate course of Engineering Applied to Petroleum used JARGON to analize a data base on Offshore Plataform Operations provided by Petrobrás. The results of this analyzis is part of his master thesis.

The financial support of CNPq and FAPESP are deeply appreciated.

CHAPTER IX

SMART KARDS(c):
OBJECT ORIENTED MPNN ENVIRONMENT

IX.1 - MPNN systems and object oriented programming

The basic features of MPNN systems (MPNNS) are:

IX.1a) multinet structure: each MPNNS is composed of a family of MPNN nets;

IX.1b) modular structure: each net of MPNNS is composed of a variable number of subnets which share some common initial structure. Each module is used to represent part of the knowledge the MPNNS learned and it is used to calculate part of the solution of the problem;

IX.1c) hierarchy: the distinct nets of the MPNNS are hierarchically organized as a strategy to reduce any possible combinatorial explosion in the attempt to model complex environments;

IX.1d) neuronal message exchanging: the MPNN neurons exchange both numeric and symbolic information by means of their synapses. This synaptic message exchange is supported by a formal language L(G) (see Chapter III) founded on a set of transmitters (t) or pre-synaptic labels; a set of receptors (r) or post-synaptic labels, and a set of controllers (c) or active labels. The concatenation properties of these labels

$$t \; \hat{} \; r \; » \; c$$

define the syntax of L(G). The controllers are triggers used to activate defined MPNN neural functions, which define the semantics of L(G);

IX.1e) inheritance: the structure of the MPNN modules are programmed by means of the specification of the L(G) used by its neurons. The characteristics of this L(G) are specified by its genetics G (see Chapter III). The modules of a given MPNN net shares some common properties, or in other words, they inherit a common L(G);

IX.1f) message distribution: messages are distributed in the MPNNS by means of three different systems:

IX.1f1) mail system: neurons address messages to other specific neurons by means of their axonic branching. Axons are used as phone cables (address matrices) to deliver information at specified addresses;

IX.1f2) broadcast system: hormones (a special type of transmitter) are released in the blood stream (blackboard) to be captured by any neuron or effector system interested in a given type of information. This mechanism may be implemented in Artificial Neural Nets by sending this type of message to a blackboard or bus, from where it may be read by any neuron; and

IX.1f3) partial broadcasting: modulators (another special type of transmitter) are released by special axonic systems or by local vessel nets to exert their actions over the neurons in defined areas of the brain. This mechanism may be implemented in artificial MPNNs by restricting the reading of these messages in the blackboard or bus.

Object Oriented Languages (OOL) are proposed as programming tools for hierarchical parallel processing (e.g. Cox, 1987), and they introduce a new paradigm, called Object Oriented Paradigm (OOP), for programming artificial systems. The basic assumptions of this paradigm are (Cox, 1987):

IX.2a) objects are autonomous computational structures: an object is a computational entity dedicated to perform some specified processing with defined variables. Each object has its specific methods to treat these variables;

IX.2b) classes are families of objects sharing some common variables or methods: classes are used to hierarchically organize objects sharing common properties;

IX.2c) inheritance: objects belonging to the same class inherit the properties of this class;

IX.2d) objects exchange messages: objects exchange information by means of

IX.2d1) mail systems: one object directs the results of its processing to some specific object it knows. In other words, one object mails the information to the objects it has the addresses of, and

IX.2d2) broadcasting systems: objects write messages in blackboards or release them into communication buses, from where the messages are read by other objects interested in that kind of information;

IX.2e) modular programming: classes, objects and messages pathways compose a program in an OOL. First objects must be idealized to solve specific tasks, and are organized into classes according to the variables and methods they share. Second, the message network servicing these objects and classes is designed to organize the solution of the problem as a sequential and parallel distribution of tasks, and

IX.2f) local maintenance: once objects are autonomous computational structures, the maintainance of the system is achieved mainly by locally modifying specific tasks, without major changes in the structure of the program. However, because objects inherit properties, the size of the neighborhood affected by the maintenance activities is directly dependent of the strength of the inheritance defined in the system.

The close correlation between the purpose of OOP and the characteristics of the MPNN selects OOL as the adequate tool for programming MPNNS. The present chapter briefly describes SMART KARDS(c) an Object Oriented Environment to program intelligent systems supported by MPNN, and comments on some applications developed with this system. The present book is an example of such an application. It was edited and processed in the environment provided by SMART KARDS(c), taking advantage of some of its intelligent features, such as the use of JARGON to learn and process its remissive index.

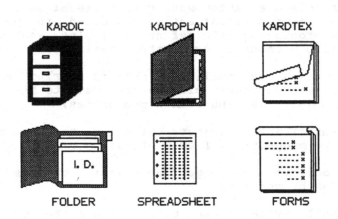

FIG. IX.1 - SMART KARDS(c)

IX.2 - Introducing SMART KARDS(c)

SMART KARDS(c) is a modular computational environ-ment (Fig. XI.1) integrating data base techniques (the subsystem KARDIC), spreadsheet facilities (KARDPLAN) and text management (KARDTEXT) for both standard and intelligent Object Oriented Programming. In this latter case, it is used to implement MPNNS and it takes advantage of the learning capability of JARGON to extract knowledge from text data bases produced by itself. SMART KARDS(c) is running in a MS-DOS environment.

The basic objects handled by SMART KARDS(c) are:

IX.3a) CARDS: it is the basic object used to build any SK (short notation for SMART KARDS(c)) data base. The cards referring to the same user form the FOLDER of this user (Figs. IX.1 and 2). The folders of different users are placed in the SLOTS of an electronic CABINET. KARDIC is the KS subsystem handling cabinets;

IX.3b) SHEET: it is the basic object used to build any SK spreadsheet. The different sheets associated with the different topics of a given subject form a SUBJECT, and related subjects are combined into a BINDER (Fig. IX.1 and 7). KARDPLAN is the KS subsystem to handle files, and

IX.3c) FORMS: it is the basic object used to build a SK hipertext. The different forms about the same rheme compose a REPORT, and the different reports about the same theme are combined into a TEXT. KARDTEX is the KS subsystem for producing and processing texts.

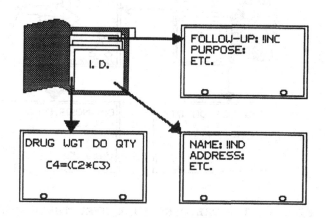

FIG.2 - THE FOLDER

IX.2a - Kardic

The basic computational structure handled by KARDIC is the FOLDER. The folder is the class of CARDS which belongs to the same user. For example, the folder of a patient (Fig. IX.2) contains his identification, follow-up, drug, etc. cards. Each of these cards contains specific information about the user (e.g. ID and follow-up cards) or performs defined processings about these kinds of information (e.g. drug card).

From the formal point of view, each card is a set of variables (e.g.: name, address, drug, etc.) and a set of methods (indexing, arithmetic calculations, etc.) which can be applied to process these variables. For example, the method !IND associated to the variable NAME in the ID card (Fig. IX.2), is used by KARDIC to organize the FOLDER in the SLOTS of the CABINET; arithmetic methods (CAL) are assigned to the cards in Fig. IX.4 to program the calculation of sellings, taxes and payment parceling.

The key feature of the CARDS handled by KARDIC is that they are structured objects. This means that internal relations between variables are defined by means of a graph or net (Fig. IX.3). This net is used to address the message flux inside the card, among its variables. The purpose of this message organization inside the card is discussed below.

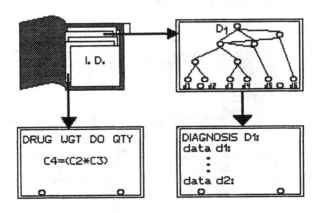

FIG. IX.3 - THE CARD AS A STRUCTURED OBJECT

Each card is an autonomous computational entity, so that any modification of the program is done as local alterations in the contents of each card. This is called the encapsulation of the processing (Fig. X.4). For example, if the percentage of the tax changes from one to another state, the adequate modification of its value is made in card 4, and the entire processing will be adjusted to it through the object message exchange.

A special card which may be included in the folder is the TEXT card, providing a screen of up 1KB of free text. This type of card allows the creation of language data bases in the cabinets. Another special type of card is the table card (card 3 in Fig. IX.4). In this type of card, the contents of a variable may be divided into fields which may be read and processed independently.

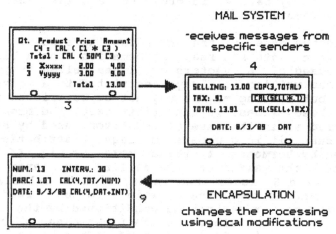

FIG. IX.4 - MAIL SYSTEM

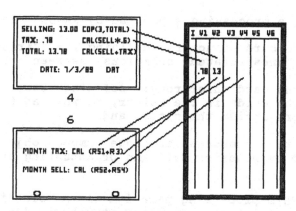

USING THE BLACKBOARD OR BUS

FIG. IX.5 - THE BROADCASTING SYSTEM

Cards of the same or different Folders are allowed to exchange messages either by mail or broadcast systems:

IX.4a) in the case of the mail system, the address of the sender is registered in the receiver card, so that when this object is instantiated it asks the required information from the corresponding object. The method COP assigned to card 4 in Fig. IX.4 asks the total of the selling to card 3. Some methods may directly ask information to specified cards. For instance, the method CAL assigned to the item PARC of card 9 in Fig. IX.4 asks the value of the TOTAL = SELLING + TAX calculated by card 4, in order to calculate the parcelation of the payment;

IX4b) in the case of the broadcasting system, cards write information into and read information from a blackboard or bus composed of 6 vectors or lines called V_1 to V_6 (Fig. IX.5). The writing is always restricted to a given line or vector, but reading may or may not be restricted to one of a few of these lines. In the first case, the broadcasting is restricted as in the case of modulators used by MPNN, in the second case it is a general message distribution like that provided by hormones. In the example of Fig. IX.5, the card 4 writes the values of the selling and taxes in V_1 and V_2, and card 8 reads this information to calculate monthly selling and taxes. The writing in and reading of the blackboard may or may not be conditioned by the position of the folder in the slot. In the first case, a hierarchy of the message passing is implemented. This is another example of restricted broadcasting.

The processing supported by KARDIC is dependent of:

IX.5a) the card structure: defining its variables, methods and the encapsulated relations between these elements;

IX.5b) the card hierarchy: defining the basic relations between the cards in the folder, as well as the relations of the folders inside the slots; and

IX.5c) the message exchange among objects: which is processed by means of mail and broadcasting systems.

This processing paradigm is a classic OOP, but it also supports the processing structure described for the MPNNSs since:

IX.6a) each KS card may be programmed as a MPNN module: if each of its variables is associated with a neuron; the methods used to handle these variables are actions defined by the semantics of L(G) (see Chapter III), and the connectivity among these neurons is described by the graph associated with the card. This graph describes the internal mail system supporting the neural message exchange inside the MPNN module;

IX.6b) each KS folder can contain all MPNN modules of a given net: since the message exchange between these modules is guaranteed by both the addressed mail and the broadcasting systems discussed above, and

IX.6c) the folders may be hierachically organized in the slots.

The general scheme of any processing performed by KARDIC is illustrated in Fig. IX.6. The user chooses a Cabinet, opens one of its Slots, takes a Folder and handles a Card to process. KARDIC supports different kinds of processing of a cabinet: filling and modifying the contents of a folder; searching folders according to their contents; printing folders; reasoning, etc. Some of these processings will be discussed later.

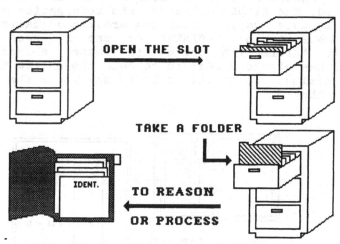

FIG. IX.6 - HANDLING A CABINET

KARDIC can operate either as

IX.7a) a traditional data base system: if its cards are not taken as structured entities; or as

IX.7b) an intelligent environment: if some of its cards are taken as MPNN modules, while others are assumed to be effector or input objects to be handled by a neural processing. This kind of approach will be discussed in section IX.7, 8, 10 and 11.

IX.2b - Kardplan

The primary object handled by KARDPLAN is the spreadSHEET (Fig. IX.7). 20 KB memory is allocated to each sheet to handle lines and rows. Methods are assigned to rows, individual lines or blocks of lines. Cards can write data on sheets. Sheets can write data on the Blackboard, too. Also, the contents of the Blackboard at a given moment may be frozen in a sheet. In this way, the dinamics of SMART KARDS(c) can be stored in sheets. Different sheets can be organized into a SUBJECT. For example, the CARDs of the CABINET SELLINGS may write total dayly sellings, taxes, etc. into sheets, so that these sheets can totalize monthly results. The different sheets Sellings, Taxes, etc., of a given month composes the SUBJECT = MONTH. Different Subjects can be included into a BINDER. In the example being discussed, the BINDER collects the information of an entire year. In this way, BINDERS can provide summaries about the data bases in the CABINETS or can be used to process specific information stored in these cabinets. Another special use of the Binder is to hold the entire history of a KS processing, since its sheets may store the dynamics of the Blackboard.

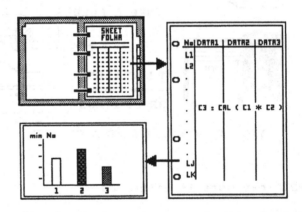

FIG. IX.7 - THE BINDER

A special kind of sheet is the GRAPHICsheet, which allows KARDPLAN to show, print and store X/Y and X/Y/Z graphics and histrograms about the contents of data stored in any sheet of a binder, or about the contents stored in the cabinets (Fig. IX.7). The KS Sheet has a distinctive feature concerned with the traditional spreadsheet systems. The KS Sheet can have a graph assigned to it, in the same way Cards can be structured by means of MPNNs.

The general scheme of any processing performed by KARDPLAN is illustrated in Fig. IX.8. The user takes a Binder, chooses one of its Subjects, and select one of the Sheets speaking about this subject. KARDPLAN supports different kinds of processing of a binder: filling and modifying the contents of a sheet; printing sheets, subjects or binders, etc.

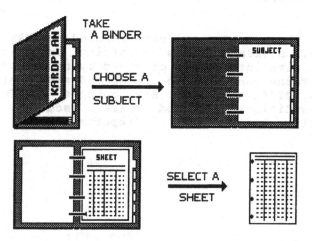

FIG. IX.8 - HANDLING THE BINDER

At first glance, no conceptual difference can be noticed between the structure of Cards and Sheets, despite the amount of memory allocated to each of these structures: 1 KB for cards and 20 KB for sheets. However:

IX.8a) Folders were idealized as a set of very repetitive small autonomous modules or cards, in the same way the cortex is composed by a high number of columns, and

IX.8b) Subjects were designed to be a small collection of large integrated plans or sheets, to be used

IX.8b1) to summarize the contents of Cabinets: in this way, the Binder may be viewed as an organized traditional spreadsheet system; or

IX.8b2) to organize a complex processing of the contents of Cabinets: in this approach, the Binder may be viewed as a collection of general reasonings to with some specific actions involving SMART KARDS(c).

In this latter approach, the graphs associated with the KS sheets may be viewed as a general mail system to spread information from or to specific KS cards and for or to the external user. A special application of the KS sheets is to encode questionnaires implementing specific intelligent processings of the contents of the cabinets. The special binder containing these questionnaires is called QUEST. It will be the subject of section IX.10 and 11.

IX.2c) Kardtext

The primary objects handled by KARDTEX are FORMS, that is, standard texts containing specific sites to be filled with information provided by cards or sheets (Fig. IX.9). Forms contrary to cards and sheets are unstructured objects. Different forms may compose a REPORT. Assignment and conditioning methods are used to organize the information inside forms and reports.

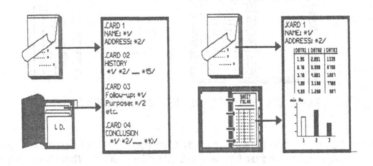

FIG. IX.9 - FORMS

Assignment methods specify the source of information to be filled in. For example, the method .CARD N in Fig. IX.9 specifies the card N to provide the information to fill the places marked by *n/ with the contents of its variable n. Conditioning methods are used to determine the forms to compose a report. The choice of these forms can be conditioned by information stored in defined cards. In this way, the contents of a Report may vary according to a particular processing performed by SMART KARDS(c). The use of these conditioned reports will be discussed in section X.10 and 11. The report is an important tool used by SMART KARDS(c) to speak about itself.

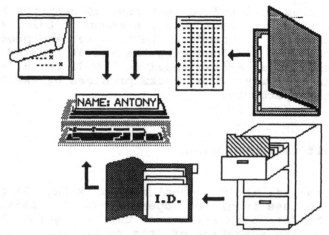

FIG. IX.10 - FILLING REPORTS

Printing methods are used to format the output of the reports. They may control size and type of letters; quality of the printing; size of the printed pages and select different types of reports, etc.

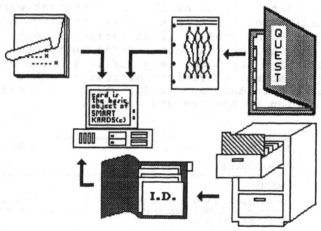

FIG. IX.11 - READING A KS TEXT

Reports are produced by telling SMART KARDS(c) to open a cabinet and/or binder, to select a card or sheet to provide information required to produce the report (Fig. IX.10). Different reports about the same subject may be combined into a TEXT. In this way, forms and reports play the same roles assigned to sections and chapters to organize the information inside a text. Because of this, KARDTEX may be used also as a standard text editor and processor. The present book was produced with the aid of KARDTEX. QUEST can also be used as an active index to the KS texts (Fig. IX.11). In this way, the books produced by KARDTEX can be handled as hypertexts.

IX.2d) Self-referred system

One of the main features of SMART KARDS(c) is its capacity to use itself to speak about its functions. The SMART KARDS(c) technical and user's guide books were produced taking advantage of the facilities provided by the system itself. For example, technical references, methods description and syntax, etc. were stored in the folders of the cabinet HELP (e.g. Fig. IX.12).

The   consultation of this   cabinet   is context sensitive, providing local information to the user according to the activities he is involved in. Whenever help is desired, the user press ALT-H and is provided access to a specific folder and slot of the HELP cabinet which contains theinformation related to his actual activity. Any KS screen can be saved to be used as illustration in texts about SMART KARDS(c) and its applications. These texts can be produced with KARTEX. For instance, Fig. IX.12 is the actual output of the HELP cabinet providing information about the module KARD to handle folders and cards, which was saved to be used as illustration in the present chapter.

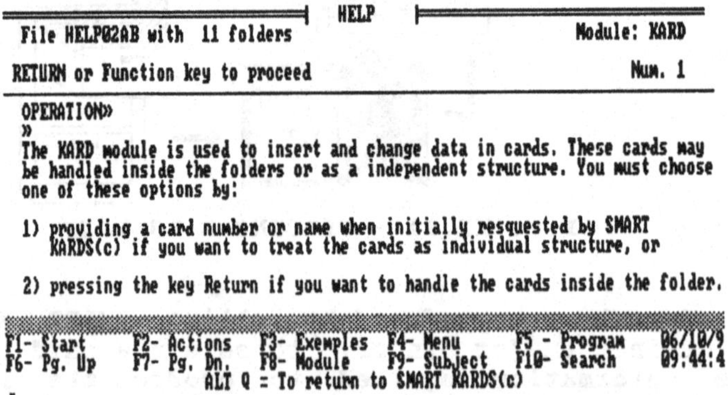

```
========================|  HELP  |========================
 File HELP02AB with  11 folders              Module: KARD

 RETURN or Function key to proceed           Num. 1
─────────────────────────────────────────────────────────
 OPERATION»
   »
 The KARD module is used to insert and change data in cards. These cards may
 be handled inside the folders or as a independent structure. You must choose
 one of these options by:

 1) providing a card number or name when initially resquested by SMART
    KARDS(c) if you want to treat the cards as individual structure, or

 2) pressing the key Return if you want to handle the cards inside the folder.

 F1- Start    F2- Actions   F3- Exemples  F4- Menu     F5 - Program    06/10/9
 F6- Pg. Up   F7- Pg. Dn.   F8- Module    F9- Subject  F10- Search     09:44:4
                    ALT Q = To return to SMART KARDS(c)
```

FIG. IX.12 - HELPING THE USER

Whenever required, the same HELP cards can provide information to print reports about specific tasks performed by SMART KARDS(c). The printing output can be stored in specific files, to compose a KS text. QUEST can be used to orient the reading of this information (Fig. IX.11). In this way, specific reference books can be organized according to defined applications of SMART KARDS(c), and specific readings of these books can be programmed to help the user.

Another key feature of SMART KARDS(c) is its capacity of being programmed as both a standard integrated data base-spreadsheet-text processing system and as MPNNS. In this way, the standard system may be used as both:

IX.9a) a friendly interface between the user and the intelligent system supported by the MPNNS: the data base structure provides both a sophisticated source of information for learning and reasoning, and a strong support for storing the results of this very same learning and reasoning. Besides this, the KS text processing capability provides a powerful tool for communicating and explaining the learning and reasoning done by SMART KARDS(c); and

IX.9b) the actuator system being controlled by the MPNNS: providing this MPNNS with complementary powerful standard computational tools. This type of integration is the subject of the next sections.

IX.3 - The expert environment

The overall activities of the expert's environment may be organized with SMART KARDS(c) because they involve:

IX.10a) data base activities: necessary to store information about the cases handled by the expert. For example, in the case of Health Sciences, this implies storing data about the patients: identification items, address, the history of the illness, laboratory test results, etc.;

IX.10b) decision making: using information from the data base, the expert reasons about diagnosis, treatment and prognosis, and

IX.10c) report making: reports are produced to ask for and to inform about specific laboratory tests; to inform people about the conditions of defined cases in the data base or about the decision made.

Let Health Sciences be an example of this kind of application. Identification and other cards may be defined in the Folder of a Cabinet, to be distributed into its Slots according to the name of the patient (primary index) (Figs. IX.13 and 14). Secondary addressing may be used to define related data banks associated with the patient's follow-up, diagnosis, etc (Fig. IX.13). These information is useful for the management of the expert environment.

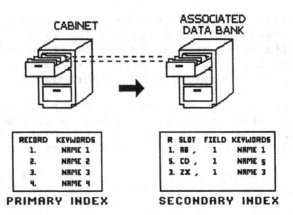

FIG. IX.13 - CABINET AND DATA BANKS

The medical consultation may be organized by the following special cards (Figs. IX.14) (Rocha, A.F., 1990):

IX.11a) History - inheriting methods for hypothesis triggering,

IX.11b) Hypothesis - inheriting decision methods for choosing the best hypotheses either for investigating or for final decisions, and

IX.11c) Diseases - inheriting approximate reasoning methods for calculating their acceptance according to the data gathered to support them.

After the identification of the patient, the physician collects initial data about the history of the illness or the patient's complaints (Fig. IX.14). During this passive phase, the expert waits for data which might trigger some diagnostic hypothesis. The elicited hypotheses are ranked in the passive phase, before the expert starts to investigate additional data to support each one of the possible diseases (Eddy and Clanton, 1982).

The patient's complaints are filled into a text card, to which triggering methods are assigned. These methods scan the text searching for triggering words or word-combinations like: fever, loss of weight, etc. These data are the components of the initial clusters of the knowledge graph provided by the expert (see Chapter V). If there is enough confidence in these data the associated hypothesis is taken into consideration for further investigation. The hypotheses under consideration are written in the hypothesis card according to the rank of the confidence in them.

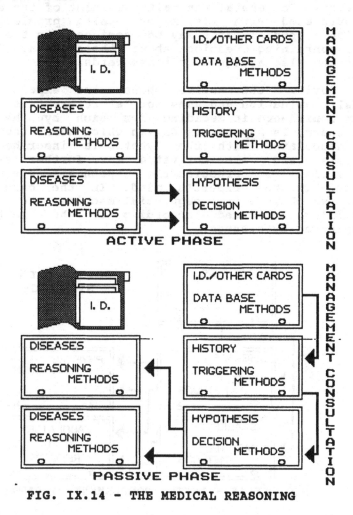

FIG. IX.14 - THE MEDICAL REASONING

The passive phase ends when either (Eddy and Clanton, 1982):

IX.12a) the expert triggers 3 or more hypotheses for future investigation, or

IX.12b) the patient stops supllying information.

Sometimes, the expert submit the patient to a

general questioning at the end of the passive phase. This is done to disclose any other associated disease, which is not the major concern of the patient. This general questioning is very important in some medical specialization, e.g, geriatrics. In general the main complaint of the elderly is not correlated with its major health problem. In this situation, special cards may be assigned to the patient's folder, containing questions about these data. Triggering methods are also assigned to these cards.

During the active phase, the expert performs physical examination and asks for laboratory tests in order to gain confidence in deciding for each hypothesis. This active search is hypothesis driven which means that specific cards associated with the hypothesis inscribed in the hypothesis card, are activated for further exploration (Figs. IX.14). In practice the hypotheses are paralleled activated in the expert's mind. In the case of SMART KARDS(c) which runs in a sequential machine, these cards are sequentially activated according to the rank of the corresponding hypotheses.

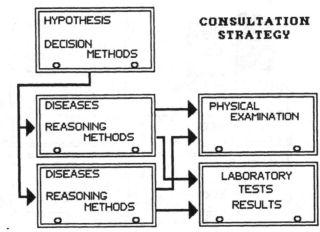

FIG. IX.15 - ASKING COMPLEMENTARY DATA

The variables assigned to the disease cards prompt the system to inquiry about specific data on physical examination and laboratory investigation (Fig. IX.15). This inquiry is performed according to the procedural knowledge provided by the expert (see Chapter V). These data modify the actual confidence on the hypotheses under consideration. It implies reordering the Diseases in the Hypothesis cards according to their calculated new acceptance (Fig. IX.14). The final decision making is supported by the fuzzy reasoning discussed in Chaper V, sections 6, 7 and 8.

In many intances, some of the information about the actual patient must be asked by means of appropriate forms, as in the case of the request for laboratory tests. SMART KARDS(c) provides the facilities for the automatic request of this data, by producing special reports (Fig. IX.16) filled with data from the patient's I.D., hypothesis, etc. cards. The forms used in these reports may be both standard forms to ask for standard tests and conditional forms selected according to the hypotheses under consideration.

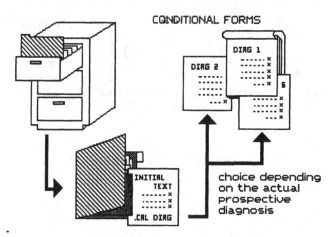

FIG. IX.16 - FILLING REPORTS ABOUT THE PATIENT

The information about the requested laboratory tests are provided to the expert with a time delay. Because of this, the consultation is a discontinued process. SMART KARDS(c) registers the actual status of the consultation process, writing the progress of the decision making about each diagnosis in the hypothesis cards as: accepted, rejected or pending.

The knowledge graph provided by the expert about each diagnosis is associated with the corresponding disease card (Fig. IX.17) and they are used to organize the SK reasoning. Decision about request of new information as well as the calculation of the actual confidence enjoyed by the diagnosis are supported by these graphs. The fuzzy deduction at each of its non-terminal nodes are assumed to be attained by means of the Extended Modus Ponens described in Chapters II, V and X. The actual $\top$-norm used in the aggregation as well as the actual implication relations may be chosen from the library of SMART KARDS(c) methods. In the same way, the matching functions of the input neurons or nodes may be chosen from the same library.

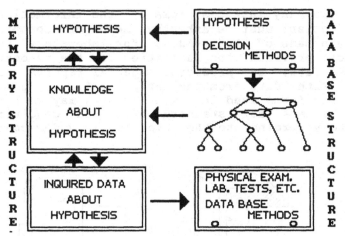

FIG. IX.17 - KNOWLEDGE GRAPH SUPPORTS SMART KARDS(c) REASONING

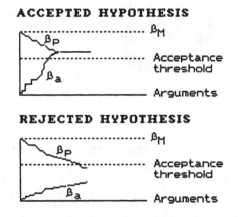

FIG. IX.18 - THE SK THRESHOLD REASONING

After all the hypotheses are explored, one or more diagnoses are chosen taking into consideration the threshold reasoning discussed in Chapter V (Fig. XI.18). In this way:

IX.13a) every hipothesis exhibiting confidence greater than the acceptance threshold may be taken into consideration in the case of multiple fault diagnosis; or

IX.13b) the highest confidence determines the winning

hipothesis in the case of single fault systems; and

IX.13c) every hypothesis enjoying a confidence below the acceptance threshold is rejected.

This implements the threshold reasoning discussed in Chapter V, section V.10.

All information acquired about the patient during the consultation is stored in the data base supported by the Patient Cabinet. In this way, it is available for future refinement of SMART KARDS(c) knowledge by means of inductive or deductive learning. Any datum obtained during the consultation which is not explained by any hypothesis considered, is written in a special card, named Ignorance card (IC). IC is an important source for future modification of the SMART KARDS(c) knowledge.

JARGON can be used to disclose the knowledge stored in the Patient Cabinet, since most of the data is written in natural language. This enables SMART KARDS(c) to check the knowledge provided by the expert from time to time and to change their initial knowledge to accommodate new findings or new patterns of diseases. Also, keeping track of the amount of ignorance in the Patient Cabinet and of its performance in solving the cases, SMART KARDS(c) can decide to use the information stored in IC to modify their actual knowledge by means of deductive learning. In this case, the graphs of the accepted diagnosis are used to generate new knowledge graphs by addition of the information in IC. These mutations can be guided by any knowledge about the type of data in IC and/or about the time these informations are obtained. In this way procedural knowledge can be acquired.

Any explanation or report may be provided about the decision process or the actual status of the patient by selecting and filling forms according to the contents of the Patient Cabinet (Fig. IX.16). In this way, SMART KARDS(c) provides the user with a very strong tool to organize these activities.

IX.4 - Leprosy: an example of application

Leprosy is a infectious disease caused by the bacterium Mycobacterium leprae which affects the skin, mucous membranes and nerves. The disease is, therefore, characterized by anesthesia in localized portions of the body, and by dermatologic signs such as discoloration or erythema, maculae, plaques, papules and enlarged peripheral nerves. Leprosy, also called Hansen's Disease (HD), can be classified in 4 main types according to the so called Madrid classification: Indeterminate (HDI), Tuberculoid (HDT), Lepromatous (HDV) and Borderline (HDB).

This classification is to some degree fuzzy, because
individuals in the Borderline group may migrate toward two
different poles, resembling the forms HDT or HDV,
respectively. This migration depends on the immunological
state of the patient, the progress of the disease and the
success of the treatment. The diagnosis of the type of the
disease is important for prognosis and self-care education,
as well as to decide for the chemotherapy.

```
============================= FOLDERS =============================
File HANS01AF with 41 slot                              M E D C O
Card:
                                                        Num. 1
PATIENT'S I.D.
  1-Name: ANTONY JOSEPH
  2-Address: 132 Wilshire Av.
DIAGNOSIS and TREATMENT
  3-Type: HDB
  4-Scheme: 01
TESTS
  5-Mitsuda: NEGATIVE
  6-Histamine (Pilocarpine): INCOMPLETE
  7-Baciloscopy: POSITIVE

F1- Starts   F2- Item   F3- Save   F4- Menu    F5 - Print   11/10/9
F6- Review   F7- Erase  F8- Slot   F9- Folder  F10- Search  08:43:0
        ALT I = Return to program    ESC = DOS   ALT H = HELP
============================= FOLDERS =============================
File HANS01AF with 41 slot                              M E D C O
Card:
                                                        Num. 1
PHYSICAL EXAMINATION

ERITEMATHOUS PLAQUES IN THE FACE AND EARS, NEAT LIMITS.»
ERITEMATHOUS MACULAES, NEAT LIMITS, IN THE ARMS AND LEGS.»
THIKEN NERVE: RIGHT CUBITAL NERVE, DISCRETE.»
TERMIC ANESTHESIA IN THE PLAQUES AND MACULAES.»
INFILTRATION IN THE TRUNK.»

F1- Starts   F2- Item   F3- Save   F4- Menu    F5 - Print   11/10/9
F6- Review   F7- Erase  F8- Slot   F9- Folder  F10- Search  08:43:0
        ALT I = Return to program    ESC = DOS   ALT H = HELP
============================= FOLDERS =============================
File HANS01AF with 41 slots                             M E D C O
Card:
                                                        Num. 1
BIOPSY

SKIN FRAGMENT SHOWING EPIDERMIC ATROPHY. ALSO LINPHOHISTIOCITIC INFLAMATORY
INFILTRATION AROUND THE NERVES AND GLANDS IN THE DERMIS. PRESENCE OF
VACUOLIZED HISTIOCYTES. SMALL GRANULOMA AT THE DERMIS-HYPODERMA TRANSITION.
SMALL NUMBER OF GRANULATE BAARs.
HANSEN'S DISEASE - TYPE HDB.

F1- Starts   F2- Item   F3- Save   F4- Menu    F5 - Print   11/10/9
F6- Review   F7- Erase  F8- Slot   F9- Folder  F10- Search  08:43:0
        ALT I = Return to program    ESC = DOS   ALT H = HELP
```

FIG. IX.19 - THE PATIENT'S FOLDER

The specialized language used by the dermatologist
handling Leprosy is strongly based on descriptors for signs,
laboratory tests, symptoms, etc., and fuzzy or crisp
qualifiers and quantifiers. The intersection of the sets of
symbols and qualifiers is non-empty, because some symbols

can be transformed into qualifiers. For example,
infiltration can be a sign or can be transformed into the
qualifier infiltrated. The verbs play a secondary role in
the dermatology jargon because the specialized language in
this case is descriptive.

IX.5 - The patients

The following data from 140 patients were stored
into a SK Cabinet (Fig. IX.19):

IX.14a) physical examinations performed during the first
consultation of the patients in a Dermatologic Clinic;

IX.14b) laboratory test results: the results of 3 different
tests were registered: Histamine test (HT), Mitsuda test
(MT) and baciloscopy (BA);

IX.14c) the description of the histological examination of
biopsies of some of the skin lesions, and

IX.14d) the prospective diagnosis.

The data from IX.14a and b were fed into text SK
cards, whereas the information provided by IX.14b and the
diagnosis were stored in the corresponding field of the I.D.
card (Fig. IX.19). The type of the disease was used as a
secondary index to access the cabinet. If this secondary
index is used, only the folders of the patients having the
chosen diagnosis are accessible.

JARGON used this information to learn the possible
datum patterns associated with the different types of
Hansen's disease. It inspected the different cards of the
slots in the cabinet and extracted the most common patterns
of data associated with each diagnosis, which was used as
the index to access the cabinet. The patterns of physical
examination, laboratory tests and histopathology JARGON
learned in the case of HBD are shown in Figs. IX.20, IX.21
and IX.22, respectively.

The knowledge about the disease obtained by
inspecting the data base was encoded in the following data
bases:

324

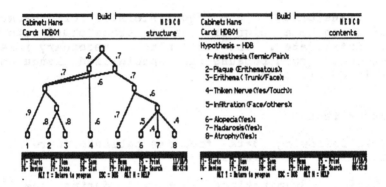

FIG. IX.20 - THE HDB PHYSICAL EXAMINATION PATTERN

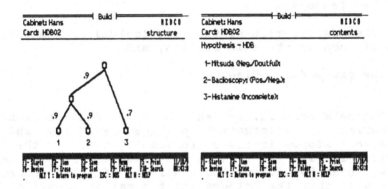

FIG. IX.21 - THE HDB LABORATORY TESTS PATTERN

IX.14b) DICTIONARY: the words and phrases JARGON learned as the jargon about Leprosy were stored in a cabinet called Dictionary; and

IX.14a) PATIENT: the learned graph and contents of the standard patterns about the different types of Leprosy were stored in special cards of the folder of the cabinet patient (Figs. IX.20 to 22) containing the standard descriptions of the physical examination, laboratory test results and histopathology of each Leprosy type. These cards read data from the corresponding ID, Physical Examination and Biopsy of the patient's folder to calculate a degree of macthing between the actual patient's data and the prototypical patterns JARGON learned in the same data base. The details of this processing will be discussed in section IX.8. The expected information to be obtained in this reading is shown between brackets in Figs. IX.20 to 22).

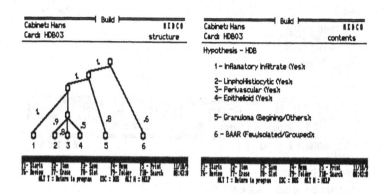

FIG. IX.22 - THE HDB HISTOPATHOLOGY PATTERN

The cabinet Dictionary is consulted when SMART KARDS(c) need to translate the actual patient's data into the standard phrases and words of the jargon it learned. The results of this translation can be stored in some other cards of the folder of the cabinet PATIENT (Fig. IX..23). The translated data are used by SMART KARDS(c) to reason about Leprosy e.g. to calculate the degree of matching discussed above or to cr eate secondary indices about defined words or jargon phrases, etc.

IX.6 - The experts

4 experts in Leprosy were interviewed about the description (declarative knowledge) of each Leprosy's form and about the way they get the data (procedural knowledge) to establish the corresponding diagnosis. Other experts in histopathology provided the knowledge necessary to classify the biopsies into the same Leprosy types. The methodology described in Chapter V, sections V.2 and 3 was used to conduct these interviews. Each interview began focusing on one diagnosis and progressed throughout the others whenever all the data about the actual hypothesis were obtained. Because the expert was a busy person, many interviews were distributed along different sessions on different days. The

jargon list and the knowledge graphs provided by the experts
supplied a knowledge about Leprosy which was compared with
the knowledge JARGON learned from the data base PATIENT.

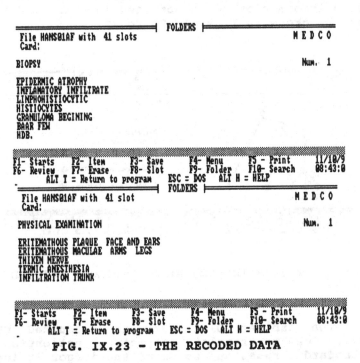

FIG. IX.23 - THE RECODED DATA

The different knowledge graphs obtained from these
experts for each diagnosis were aggregated into a population
graph following the technique discussed in Chapter V,
section V.3. Fig. IX.24 shows the population graphs obtained
in the case of HDB for both the clinical (left) and
histopathology (right) diagnoses. The distribution of the
clinical data over the terminal nodes of these graphs
revealed the expected disagreement between the experts,
because no data exhibited a predominant frequency in some of
these nodes, which were labeled V in the graphs. In the case
of Fig. IX.24, the node V may be filled wth any of the
following signs: maculae, or alopecia, or madarosis.

The comparison between the knowledge graphs obtained
from the experts and the standard patterns learned by
JARGON from the patient's cabinet revealed that:

IX.15a) there is a very close correspondence between both
types of knowledge, as it may be verified by comparing the
standard patterns shown in Figs. IX.20 to 22 with the expert
knowledge displayed in Fig. IX.4, in the case of HDB;

IX.15b) the standard descriptions were richer than the knowledge graph, showing that the expert privileges some data as key information, to keep the reasoning simple (property V.32a of the expert reasoning in Chapter V, section V.12), but also supporting the knowledge disagreement between the experts. For example, the contents variable node V in Fig. X.24 closely correlates with the nodes 5, 6 and 7 in Fig. IX.20; and

IX.15c) the values of relevance assigned to the arcs of the knowledge graphs do not correlate well with the arc frequency learned by JARGON from the data base in the case of the standard graphs. This agrees with the fact that inductive learning is a kind of qualitative statistics, which can be influenced by factors other than the real conditional frequencies associated with the process. Emotion, Causal Knowledge, etc., are examples of these extra factors modifying the relevance learned by the expert.

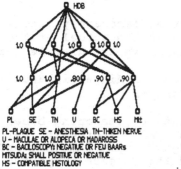
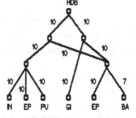

PL-PLAQUE SE – ANESTHESIA TN–THIKEN NERVE
U – MACULAE OR ALOPECA OR MADAROSIS
BC – BACILOSCOPY: NEGATIVE OR FEU BAARs
MITSUDA: SMALL POSITIVE OR NEGATIVE
HS – COMPATIBLE HISTOLOGY

IN – INFLAMATORY INFILTRATE EP – EPITHELIOD
PV – PERIVASCULAR GI – GRANULOMA: BEGINING
RE – DO NOT REACH THE EPIDERMIS
BA – PRESENCE OF FEU BAARs

FIG. IX.24 - THE EXPERT KNOWLEDGE ON HDB

IX.7) Reasoning with the expert knowledge

The expert knowledge encoded in the population graphs about the different types of Leprosy was stored in special cards of the folder of the cabinet patient similar to that used to keep the corresponding standard descriptions (Fig. IX.20 to 22) learned by JARGON. These cards correspond to the disease cards discussed before (see Figs. IX.14), and they were able to read data from the corresponding ID, Physical Examination and Biopsy cards in the patient's cabinet, too. They were used to implement the expert reasoning in this cabinet.

The most important information to trigger hypothesis driven reasoning discussed in section IIX.3 is that provided by the physical examination. Because of this, triggering methods were assigned to this card instead of the habitual card History. The triggers found in the Physical Examination Card were used to activate the corresponding Disease Card, which can read information in the other (data) cards of the folder. These cards used the informations about the Leprosy Jargon stored in the cabinet DICTIONARY to understand the contents of the data cards. If the required information was found, it was used to reason about the confidence in the existence of the disease being considered, using the information provided by the knowledge graph associated with the disease card. Otherwise, the disease card stopped the processing and wrote the achieved confidence with the actual data in the Hypothesis card, and marked the corresponding hypothesis with the label pending.

After all hypotheses were investigated, SMART KARDS(c) looked at the hypothesis cards to make one of the following decisions:

IX.16a) to print forms requesting laboratory tests and/or biopsies: this decision is based on the actual status of the consultation and/or the reasoning in the disease cards; or

IX.16b) to calculate the prospective diagnosis and to write it in the patient's I.D. card; and/or

IX.16c) to print a report about its reasoning based on the contents of the data, disease and hypothesis cards. The layout of this report was designed with KARDTEX, and might contain extra information about the disease supported by any theoretical knowledge of the disease. Because of the high specific contents of these reports, it was considered useless to present, here, any examples of it concerning Leprosy.

The matching functions σ assigned to the input neurons of the knowledge graph were discrete matching functions of the type:

$$\sigma : T(X) \longrightarrow [0,1]$$

where T(X) is a set of linguistic terms (shown between brackets in Figs. IX.20 to 22) used to classify the linguistic variable represented in the node under consideration. These functions assign a discrete value in the closed interval [0,1] to each term of the list T(X). This is because the data used by the dermatologist is qualitative. For instance, the different values of the variable Plaque are Erythematous, Hypochromic, Violaceous, etc., Absent. Each of these values triggers different degrees of confidence depending on the hypothesis being considered. For example, Erythematous is the most prototypical characteristic of the plaque in the case of HDB, while Hypochromic is the primary choice in the case of HDT. These discrete functions were implemented by means of the adequate tables.

The reasoning in the knowledge graph was supported by the Extended Modus Ponens, for which:

IX.17a) the aggregation and projection functions assigned to each non-terminal neuron of the knowledge graph were combined into a single function of the type proposed by Rocha et al., 1990d, and

IX.17b) the σ_c was used as a measure of confidence on the knowledge encoded in the consequent.

SMART KARDS(c) used the expert knowledge to classify all the patients in the cabinet PATIENT having a prospective diagnosis well established by the physicians of the hospital. The main results were:

IX.18a) the percentage of right decisions reached 95%;

IX.18b) most of these decisions were achieved using all kinds of information (physical examination, lab. tests and histopathology); only in a few cases SMART KARDS(c) were able to reach a decision using only data from physical examination; and

IX.18c) laboratory tests were very influential in the case of HDT and HDI; pathology was very important for deciding in the case of HDV, and both kinds of information were frequently used to decide in the case of HDB.

IX.8) Reasoning with standard patterns

The data of another set of 20 patients were introduced in the cabinet PATIENT, whose diagnoses were not well established by the experts. The patients were supposed to have the disease, but the type was not clearly decided.

JARGON was used to try to learn any standard pattern for this set of patients, but if failed since the patient's symptomatology was poor and variable. The standard patterns previously discovered by JARGON in the data base were, then, used in an attempt to classify these patients. The standard patterns instead of the expert knowledge were used to reasoning about these cases because the experts failed in classifying these patients. The cards containing the description of these patterns (Fig. IX.20 to 22) were used as the reasoning cards in the place of the expert knowledge cards. The type of reasoning was the same implemented for the expert reasoning discussed in the previous section.

Reasoning with the standard patterns, SMART KARDS(c) could decide for a defined type of the disease in 50% of the cases. In some of these instances, the decision was supported by some very specific information such as the difference between a erythematous or hypochromic plaque. In some other instances, histopathology was determinant in the choice of the Leprosy's type assigned to the patient. In the remaining 50% of the cases, SMART KARDS(c), like the experts, were unable to make a clear decision. The interesting point was to compare the SMART KARDS(c) decision using the data of the first consultation with the diagnosis finally assigned to these patients in later consultations. 90% of these expert later classifications agreed with the initial decisions of SMART KARDS(c).

IX.9) The outpatient service

The main activities of the Outpatient Service of Leprosy are Diagnosis, Treatment and Management. The main goals in this environment are (Theoto and Rocha, 1990):

IX.19a) to achieve efficiency in these tasks to reduce risks to patient and community;

IX.19b) to get better use of special drugs and laboratory tests, and

IX.19c) to enhance productivity, which means to augment the number of patients being accepted by the Service and to minimize the duration of the treatment.

The problems posed to these goals are derived from the fact that Leprosy:

IX.20a) requires a very long treatment (at least 2 years), not always well accepted by the patient;

IX.20b) is an infectious disease;

IX.20c) is not taught in all medical schools, and

IX.20d) carries a very particular status in the society as a stigmatizing illness, adding extra psychologic and social pressures upon the patient.

All of this imposes:

IX.21a) a tight control of the patient's follow-up and of the continuity of the treatment;

IX.21b) health control of people having a close contact with the patient;

and results in

IX.21c) many people being initially misdiagnosed and referred to the Service, and

IX.21d) unnecessary extra attention to be paid to the patient, because psychologic and social pressures may not be quickly recognized.

Standard routines of procedures are proposed by the World Health Organization for diagnosing and treating Leprosy patients. Because of this:

IX.22) Management in the Service is mainly concerned with organizing these standard and other conditional procedures into optmized Standard and Conditional Routines for Diagnosis, New Patients, Old Patients, etc., taking into account the particularities of the local environment.

The tools required for a efficient automatization of the management of this environment are:

IX.23a) a data base system to store the patients' records, in order to provide, among other things, an efficient follow-up of the patient and the health control of their relatives, as well as to calculate the adherence of the patient to the proposed treatment. This is a key point for the goal of minimizing t he duration of the treatment, because any interruption of the drug intake restarts necessarily the entire treatment, and

IX.23b) a decision supporting system to help to organize procedures into the best management routines.

But the system has to be very friendly to be handled easily by untrained people and to facilitate the new staff personnel to adhere quickly to the routines under use. Because of this, it is also desirable that the system be able to:

IX.23c) handle the specialized language (jargon) of the environment and to accept its activities to be commanded with the use of natural language,

IX.23d) operate standard as well conditional routines as sets of the above phrase commands.

SMART KARDS(c) were used to develop a system to cope with the majority of the tasks required by the management of the environment of the Outpatient Service of Leprosy.

IX.10 - Programming actions

The SMART KARDS(c) processing is based upon messages exchanged between objects as well as objects and the outside world or the user. The flux of these messages is controlled by special functions such as: Handling, Searching, Printing, etc. and by questioning about specific data. SMART KARDS(c) can store all message exchanges concerning a given processing in a specific data base called ACTIONS, whose semantics it inquires from the user. In other words, it requires the user to assign a name to each processing it stores in the data base. Keeping track of the different processings in the cabinet ACTIONS, the system is able to discover frequent processings with can define standard procedures or actions.

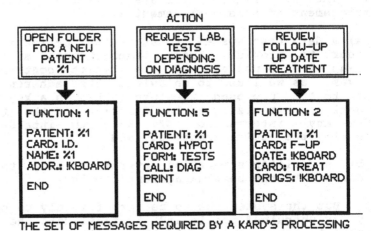

THE SET OF MESSAGES REQUIRED BY A KARD'S PROCESSING

FIG. IX.24 - ACTIONS

Any standard ACTION is a processing for which some of the messages are constant and others are variable. (Fig. IX.24). For example, any consultation of the file PATIENTS requires the opening of a slot of this cabinet, the choice of a folder and the use of one of its cards to work on it. The variable messages in this procedure are the names of the patient and of the card in his folder to be handled. In this

condition, an standard Action called Take a Card may be
programmed to control the use of SMART KARDS(c) by defining
the SK functions to be used; providing the standard messages
(e.g. the name of the cards in Fig. IX.25) to guide one
specific SK processing, and prompting the system to receive
information about the variable messages (marked with % or
!KBOARD in fig. IX.25) from the user.

The actions implemented for the management of the
Outpatient Service correspond to the basic procedures
defined in this environment, such as: open a folder for a
new patient; request lab tests; get data from clinical
consultation, etc. (Fig. IX.25). Because of this, actions
are tightly linked to the jargon used by the experts
(physicians, nurses, etc.) running the environment. As a
consequence, jargon phrases not only trigger actions but can
also provide the argument (e.g. the patient's name, date of
follow-up, etc.) for the variable messages. In this way,
jargon phrases can be assigned to the actions SMART KARDS(c)
discovered in order to make their interaction with the user
very friendly (Fig. 26). Also, additional explanation may be
programmed to be displayed by the object ACTION.

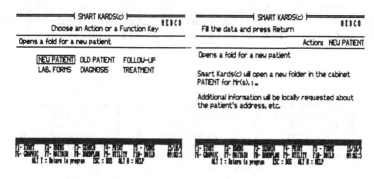

FIG. IX.25 - USING AN ACTION

IX.11 - Implementing routines

Actions may be combined into Forms, as a set of
jargon phrases. Forms are used to define Standard Procedures
(Fig. IX.26) such as: Admission of a New Patient, Follow-up
Lab Tests, etc. But actions can also be triggered by
Decision Graphs, in the case of conditional procedures (Fig.
IX.26).

Smart Kards can handle both crisp and fuzzy decision rules. Crisp calculations are used for simple decisions such as sending a prospective patient to clinical consultation or other services in the case of misdiagnosis. Fuzzy rules are used to mimic the expert reasoning in choosing the set of lab tests necessary to establish the final diagnosis. Treatment is based on standard procedures sensitive to the final diagnosis, thus it is dependent of both fuzzy and crisp decisions. These different types of reasoning are implemented by choosing the adequate method to be assigned to the SK objects, or by calling an adequate SK Action.

The type of reasoning required to manage the Outpatient Service (Fig. IX.27) is handled by a special SK Binder called QUEST (Figs. IX.28 and 29). Each sheet of this binder describes one of the conditional ROUTINES necessary to manage the environment. Related routines compose a PLAN of work. In this way, a conditional routine may be a subgraph of the entire decision graph. This will correspond to the PLAN. This type of implementation is a consequence from the necessity to segment the reasoning according to the discontinuity of the tasks in this type of environment. For example, a period of time elapses between the first and second consultations when the laboratory tests are requested and their results are available. In this way, the first conditional routine controlling the first consultation ends with the request of the laboratory tests, and the second routine begins when the results are available.

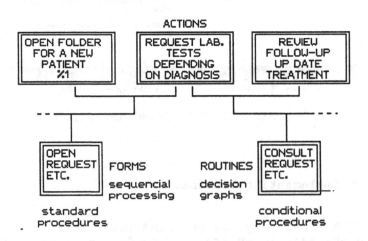

FIG. IX.26 - ROUTINES

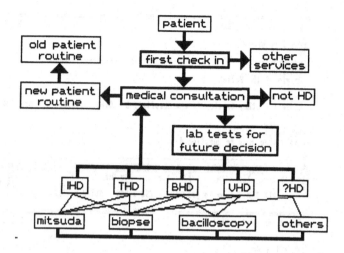

FIG. IX.27 - DECISION GRAPH

The implementation of the questionnaire or decision net supporting the First Consultation Routine is illustrated in Figs. IX.28 and IX.29. The structure of the questionnaire is exemplified in Fig. 28 and a sample of its use is shown in Fig. 29.

Each node of the decision net is an object for which the following must be defined:

IX.24a) the contents of the question to be proposed to the user or to the system itself: the answer to any question is provided either by the external user or by any other SK object, independently of the complexity of this object. In the example in Figs. IX.28 and IX.29, the answer to the first question is provided by the user, while the datum required by item 4 is the result of the complex action handling the medical consultation (Fig. IX.30). This action involves another type of decision which is supported by the expert reasoning;

IX.24b) the contents of some explanation of the question being proposed or its purpose: these messages will be displayed to help or to inform the user;

IX.24c) the methods used to obtain the answer to the proposed question: if this answer is to be provided by any object other than the external user (e.g. Call Consultation in Fig. IX.28);

```
========================| QUEST |=========================
  Questionnaire: GERE01                               M E D C O

 1000 ; record size»
 /»
 ;----------------------------------------------------------»
 ;QUESTION 1»
  Origin of the Patient: »
 /»
 ; coments»
  Chose 1 - Other Service in the Hospital, 2 - Other Service, 3 - Others»
 /»
 ; methods»
 !DEC (3=4)
 ; if 3 then jump to question 4
 /
 ;----------------------------------------------------------»
 F1- Start    F2- Item    F3- Save    F4- Menu    F5 -        01/06/90
 F6- Print    F7-         F8-         F9-         F10-        10:10:50
          ALT I = Return to program    ESC = DOS   ALT H = HELP

 ========================| QUEST |=========================
  Questionnaire: GERE01                               M E D C O
 /»
 ;----------------------------------------------------------»
 ;QUESTION 4»
  Diagnosis: »
 /»
 ; coment»
  Call Consultation
 /»
 ; method»
 !ACT CONSULTATION»
 /»
 ;----------------------------------------------------------»
 ;QUESTION 5
  Lab. Tests: »
  /»
  ;coment»
   Call Tests
```

FIG. IX.28 - IMPLEMENTING A QUESTIONNAIRE

IX.24d) the methods to be used to process the information received: in many instances the information obtained as the answer for a question must be processed before it is useful to any decision about the net navigation, and

IX.25b) the conditions of navigation of the net: the methods to be used to decide the pathway to be followed must be assigned to the decision nodes of the questionnaire (e.g. question 1 in Fig. IX.28).

The above specifications are supported by the MPNNS. In this way, the decision net is a particular type of MPNN, which supports the chosen questionnaire. This type of net is used to inquire the environment about information to make decisions.

The different Plans to handle a given environment composes the Binder of that environment, e.g. the plan for the patient's consultations. Similar plans are developed to manage the patient's relatives consultations, required to control the spread of the disease in the community. In this way, the binder QUEST may be viewed as the Strategy used to handle a given environment. This strategy is supported by a set of related MPNNs.

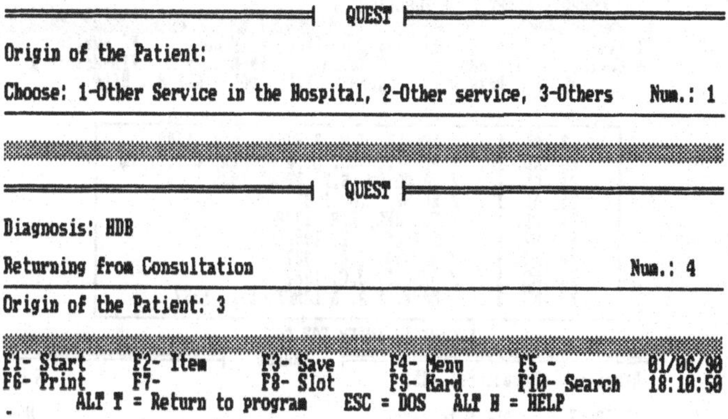

```
========================| QUEST |========================

Origin of the Patient:

Choose: 1-Other Service in the Hospital, 2-Other service, 3-Others    Num.: 1

▓▓▓▓▓▓▓▓▓▓▓▓▓▓▓▓▓▓▓▓▓▓▓▓▓▓▓▓▓▓▓▓▓▓▓▓▓▓▓▓▓▓▓▓▓▓▓▓▓▓▓▓▓▓▓▓▓▓▓▓▓

========================| QUEST |========================

Diagnosis: HDB

Returning from Consultation                                  Num.: 4

Origin of the Patient: 3

▓▓▓▓▓▓▓▓▓▓▓▓▓▓▓▓▓▓▓▓▓▓▓▓▓▓▓▓▓▓▓▓▓▓▓▓▓▓▓▓▓▓▓▓▓▓▓▓▓▓▓▓▓▓▓▓▓▓▓▓▓
F1- Start    F2- Item    F3- Save    F4- Menu    F5 -            01/06/90
F6- Print    F7-         F8- Slot    F9- Kard    F10- Search     18:10:50
       ALT I = Return to program   ESC = DOS   ALT H = HELP
```

FIG. IX.29 - USING THE QUESTIONAIRE

Another strategy similar to the one already discussed is now being implemented to help the management of a different environment: The Offshore Completion Well Activities.

IX.12 - Learning indices

SMART KARDS(c) uses JARGON to learn the indices which are useful to retrieve information from data bases stored in cabinets. A very useful application of this SK capability is to create the codewords or phrases to be used by the external users to access information in cabinets storing technical information in text cards. Let a data base about articles on nursing informatics illustrate this application (Theoto and Rocha, 1992).

The cabinet REFERENCES hold folders containing the following cards:

IX.26a) I.D.: the name of the author; the name of the article, the journal, etc. of a given paper is registered in this card;

IX.26b) Abstract: the contents of the abstract of the paper is recorded in this card, and

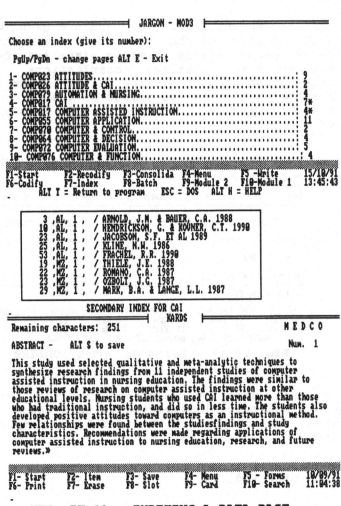

FIG. IX.31 - INDEXING A DATA BASE

IX.26c) Comments: any comments the user wants to make on the paper are stored in this card.

JARGON inspects the Abstract card in order to discover the most significative words in this abstract. The user refines this initial dictionary and JARGON learns the most frequent word associations, among which the user selects those most significant words and phrases to be used as the index of the data base (Fig. IX.31).

The user may select defined subjects in this index to search for defined information in the cabinet. In the example of Fig. IX.32 the chosen itens are CAI and Computer Assisted Instruction, which are two possible ways to speak about the same subject.

The selected itens are used by SMART KARDS(c) to generate a secondary index to reach the folders containing the articles which speak abou the chosen subject.

The abstract shown in Fig. IX.31 belongs to one of the folders SMART KARDS(c) selected as speaking about CAI.

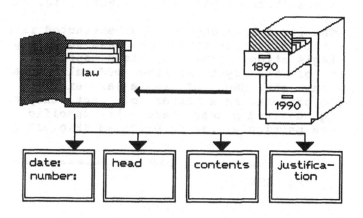

FIG. IX.32 - A LAW DATA BASE

This automatic SK indexing is very useful because keyword creation:

IX.27a) is a time consuming task when performed by the data base manager, since he has to read and classify all the information being stored in this data base; or

IX.27b) relies on a subjective classification of each author if this information is requested by the editor and provided by the journal.

The keyword learning capability of SMART KARDS(c) not only reduces the effort to get this type of information, but also allows any user to create his own set of key words:

IX.28a) by inspecting the initial word dictionary provided by JARGON to choose the word he is interested in, and

IX.28b) by selecting the meaningful word associations in the phrase dictionary according to his own beliefs.

Since JARGON stores the phrases it used to learn the word associations, it can show these phrases to the user in order to facilitate the above activities.

IX.13 - Learning forms

Depending on the structure of the technical data base, SMART KARDS(c) can discover patterns of information to be used to orientate the user in the search of information. For example, SMART KARDS(c) is being used to handle a data base (Cabinet LAWS) of one century of laws about public health in the State of São Paulo, Brazil (Fig. IX.32).

The first knowledge JARGON extracted from this data base is the different types of laws composing the public health legislation, that is, laws speaking about: the structure of the systems; the medical procedures; the vaccination, etc. Each of these laws exhibit a different linguistic pattern. As a matter of fact, they are as forms containing defined places where the specific information related toa particular law is provided (Fig. IX.33).

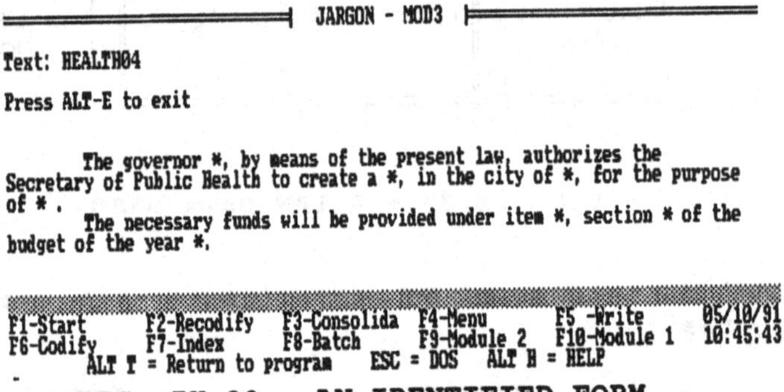

==================================| JARGON - MOD3 |==================================

Text: HEALTH04

Press ALT-E to exit

          The governor *, by means of the present law, authorizes the
     Secretary of Public Health to create a *, in the city of *, for the purpose
     of * .
          The necessary funds will be provided under item *, section * of the
     budget of the year *.

F1-Start   F2-Recodify   F3-Consolida F4-Menu   F5 -Write   05/10/91
F6-Codify   F7-Index   F8-Batch   F9-Module 2   F10-Module 1 10:45:43
          ALT T = Return to program   ESC = DOS   ALT H = HELP

FIG. IX.33 - AN IDENTIFIED FORM

The learned patterns in the data base may serve two different purposes:

IX.29a) to provide explanations to the user about the contents of the data base, so that he may decide how to get the information he is interested in, and

IX.29b) to provide forms to be used as SK objects to improve the SMART KARDS(c) processing capacity.

If a syntax founded over the notion of Case Gramatics proposed by Fillmore, 1968 is used as the t concatenation syntax (see Chapter VIII, section VIII.7) to generate the phrase dictionary, then the above forms can be very useful structures to be used by SMART KARDS(c) to find the answers to specific queries proposed by the user.

The case grammar proposes a syntax founded on the concepts of WHO, WHAT, WHERE, WHEN, WHY, HOW, etc., to define the case compatibilities between words in a phrase and phrases in a text. In this line of reasoning, the variable places in the form of Fig. IX.33 point to the answer of the following queries: WHO authorizes WHAT, WHERE and WHY.

All the information JARGON learns about the cabinet LAWS it may store in another cabinet called EXPLANATION and the strategy to handle the cabinet LAWS taking into consideration the information in EXPLANATION may be written in the binder QUERY (Fig. IX.34.

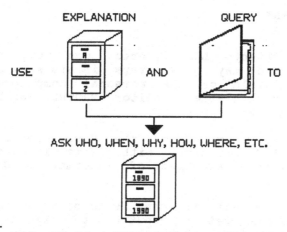

FIG. IX.34 - HANDLING THE DATA BASE

IX.14 - An intelligent MPNN environment

SMART KARDS(c) is being developed as an integrated intelligent environment (Fig. IX.34), where:

IX.30a) the thinking activities are supported by MPNNS;

IX.30b) the actuator systems are computational structures to handle data-bases, spreadsheets and text editing and processing;

IX.30c) the Oriented Object Paradigm is used to build and to program the system, and

IX.30d) the computational structures in IX.30b are used both
to organize the MPNNs into a hierarchic neural system MPNNS,
and as the main actuators controlled by the MPNNS to
manipulate a specific environment.

SMART KARDS(c) did not develop all its capabilities,
yet. But even in their infancy they may be considered very
smart in comparison with many of the intelligent systems
described in the AI literature. Like any other natural
nervous system, SMART KARDS(c) is supposed to acquire other
capabilities as it grows up in a rich environment, where
learning is encouraged. Unlike the natural nervous system,
SMART KARDS(c) may greatly increase its intelligence in the
future by being programmed by the authors to handle other
reasoning and computational methods.

ACKNOWLEDGEMENT

Juan Fuente Los Huertos was very friendly in
teaching during may many stays in his service, how to
explore the world of the personal computers. He also
provided us with his text editor PROTEX, which became the
nucleus of KARDTEX.

I deeply appreciated the discussion I had about
Object Oriented Languages with Pietro Torasso and Luca
Console, from the Department of Informatics, University of
Turin.

I am in debt with many users of SMART KARDS(c) with
helped us with suggestions about how to make the system user
friendly. Among these users:

1) The Leprosy Outpatient Service from the Clinic Hospital,
Faculty of Medicine, USP: provided us with a very
interesting environment to test some of the SMART KARDS
capabilities, and

2) Rogerio Zimmermann and Silvio Augusto Margarido: our
partners in the Law Data Base Project

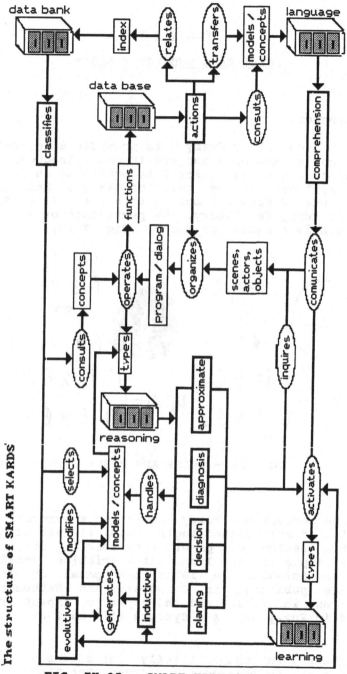

FIG. IX.35 - SMART KARDS(c)

CHAPTER X

FUZZY SETS AND FUZZY LOGIC

X.1 - Introduction

The concept of Fuzzy Sets provides a generalization of the classic idea of a set and it was introduced by Zadeh in a seminal paper on Fuzzy Sets published in Information and Control, 1965. In the case of Classic Set Theory, an element x does or does not belong to a set A (Fig. X.1a). In the case of Fuzzy Set Theory, the pertinence of an element x to the set A is a matter of degree (Fig. X.1b).

$$f_A(x) = \begin{cases} 1 \text{ if } x \in A \\ 0 \text{ if } x \notin A \end{cases} \qquad u_A(x) \rightarrow \begin{cases} 1 \text{ if } x \in A \\ 0 \text{ if } x \notin A \end{cases}$$

FIG. X.1 - CRISP AND FUZZY SETS

The concept of Fuzzy Logic is a generalization of the notion of Multivalued logic, and it was developed by Zadeh in a series of papers, the most important articles being the 3 papers on linguistic variables published in Information Sciences in 1975; the paper on linguistic quantifiers published in Computers and Mathematics with Applications in 1983, and the paper on expert reasoning published in Fuzzy Sets and Systems also in 1983.

The present chaper briefly introduces and discusses the basic concepts and tools in Fuzzy Set Theory and Fuzzy Logic, which are of interest for the comprehension of the entire book. It does not intend to give an extensive coverage of the subject, but only a commented index for the reader interested in enhancing his knowledge in the fuzzy field. Because of this, the contents of the following sections are in the most extent abstracts of the most

popular papers concerned with the basic concepts in Fuzzy
Sets and Fuzzy Logic.

X.2 - Fuzzy sets

Given any subset A of the universe of discourse U,
and any element x of U, there exists a function f (Fig.
X.1a)

$$f_A : U ---> A \qquad (X.1a)$$

so that $f_A(x)=1$ if x belongs to A, otherwise $f_A(x)=0$.
This function is called the characteristic function A, since
it defines the elements of this subset. If A is a finite
subset of U, it can also be defined by the list of its
elements:

$$A = \{ x_1, ..., x_n \} \qquad (X.1b)$$

Definition X.1b is equivalent to:

$$A = \{ 1/x_1 + ... + 1/x_n \} \qquad (X.1c)$$

or

$$A = \sum_{i=1}^{n} x_1 \qquad (X.1d)$$

or

$$A = \sum_{U} x_i \qquad (X.1e)$$

where $1/x_i$ expresses the fact that if x_i belongs to A,
then $f_A(x)=1$, and + stands for the union operator.

Let A be a fuzzy set of the universe of discourse U.
The pertinence of any element x of U to A is a matter of
degree. This means that any x shares some membership $\mu_A(x)$
with A. If this membership is measured in the closed
interval [0,1] then (Zadeh, 1965, 1975):

$$\mu_A : U ---> [0,1] \qquad (X.2a)$$

so that if $\mu_A(x) --->1$ then x tends to belong to A (Fig.
X.1b), otherwise if $\mu_A(x) --->0$ then x tends to not belong
to A. If A is a finite fuzzy subset of U then

$$A = \{ \mu_1/x_1 + ... + \mu_n x_n \} \qquad (X.2b)$$

or

$$A = \sum_{U} \mu_i/x_i \qquad (X.2c)$$

where μ_1/x_i expresses the fact that the grade of
membership of x_i in A is μ_1, and + denotes the union
operator.

It follows from definition X.2 that both fuzzy sets and fuzzy logic are tools for dealing with both closed and partially closed sets or worlds. In the first case, μ_i assumes values 1 or 0 and defines a crisp mathematics. In the second case, μ_i takes values in the entire interval [0,1] according to the closure (unclosure) of the environment. Any cell is a partially closed system because its membrane is selectively permeable to different materials. This is because fuzzy set is the adequate tool for modelling the physiology of the neuron. For instance, the sodium is an extracellular ion, what means that it is not absent inside the cell, but predominates in the extracellular space. In this way, the potassium is mainly an intracellular ion.

The following are some useful definitions in fuzzy sets:

X.3a) the height h(A) of a fuzzy set A is the maximum V membership of the elements x of U in A:

$$h(A) = \underset{U}{V} \; \mu_i/x_i$$

X.3b) a fuzzy set A is said to be normal if h(A) = 1, otherwise it is said to be subnormal.

X.3c) A is a subset of B or is contained in B if and only if
$$\mu_A(u) \leq \mu_B(x)$$
for all x ϵ U.

X.3d) The cardinality card(A) or the Σ-count(A) of A is

$$card(A) = \Sigma\text{-count}(A) = \underset{U}{\Sigma} \; \mu_i$$

In other words, the cardinality of a fuzzy set A is equal to the summation of the membership of the elements x of U in A. In the case of the crisp sets, μ_i is either 1 or 0 for all the elements of U. Then the cardinality of the crisp set B becomes

$$card(B) = \Sigma\text{-count}(B) = n$$

where n is the number of elements x of U belonging to B.

X.3e) the α-cut set A_α of a fuzzy set A is the crisp set

$$A_\alpha = \{ \; x \; | \; \mu_A(x) \geq \alpha \; \}$$

In other words, $f_{A\alpha}(x) = 1$ if $\mu_A(x) \geq \alpha$, otherwise $f_{A\alpha}(x) = 0$.

X.3f) the β-level set A_β of a fuzzy set A is the fuzzy set

$$A_\beta \;=\; \{\; u_{\beta i}/x_i \;\mid\; \mu_{\beta i} \;=\; \mu_A(x_i) \;\geq\; \beta \;\}$$

In other words, $\mu_\beta(x) = \mu_A(x)$ if $\mu_A(x) \geq \beta$, otherwise $\mu_\beta(x) = 0$.

X.3g) the support S(A) of a fuzzy set A is its α-cut set A_α for which $\alpha = 0$.

X.3h) a fuzzy singleton is the fuzzy set whose support is a single point of U. A fuzzy set is a set of singletons.

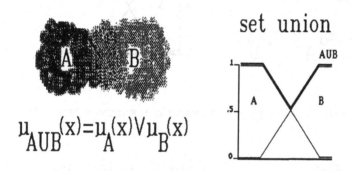

$$\mu_{AUB}(x) = \mu_A(x) \vee \mu_B(x)$$

FIG. X.2 - UNION OF FUZZY SETS

The following are important properties of fuzzy sets as proposed by Zadeh, 1965:

X.4a) the union AUB of two fuzzy sets A and B (Fig. X.2) is

$$\mu_{AUB}(x_i) = \mu_A(x_i) \vee \mu_B(x_i)$$

$$AUB = \sum_U \mu_{AUB}(x_i)/x_i$$

In other words, the union AUB of two fuzzy sets is given by the elements of U having the maximum ($\vee$) membership in A or B.

X.4b) the intersection $A \cap B$ of two fuzzy sets A and B (Fig. X.3) is

$$\mu_{A \cap B}(x_i) = \mu_A(x_i) \cap \mu_B(x_i)$$

$$A \cap B = \sum_{U} \mu_{A \cap B}(x_i)/x_i$$

In other words, the intersection A∩B of two fuzzy sets is given by the elements of U having the minimum (∩) membership in A or B.

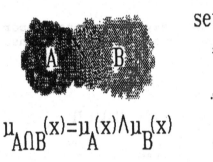

set intersection

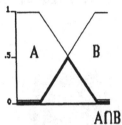

$$\mu_{A \cap B}(x) = \mu_A(x) \wedge \mu_B(x)$$

A∩B

FIG. X.3 - INTERSECTION OF FUZZY SETS

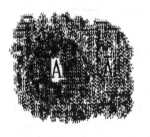

complementary set

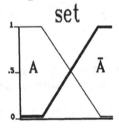

$$\mu_{\bar{A}}(x) = 1 - \mu_A(x)$$

FIG. X.4 - COMPLEMENTARY SET

X.4c) the complementary set ˜A of a fuzzy set A (Fig. X.4) is

$$\mu_{\tilde{A}}(x_i) = 1 - \mu_A(x_i)$$

$$\tilde{A} = \sum_{U} \mu_{\tilde{A}}(x_i)/x_i$$

The reader can easily see that the above definitions are equivalent to the definitions of union, intersection and complementary sets in classic set theory when $\mu_A(x)$ or $\mu_B(x)$ are either equal to 1 or zero for any element x of U, as is the case of the crisp sets.

Let U_n be the Cartesian product of n universes $U_1, \ldots , U_n$ (Zadeh, 1975):

$$U_n = U_1 \times \ldots \times U_n \qquad (X.5a)$$

The fuzzy n-ary relation R in U is the fuzzy set

$$R = \sum_U \mu_R(u_1, \ldots , u_n)/(u_1, \ldots , u_n) \qquad (X.5b)$$

where μ_R is the membership of R. Definition X.5b is equivalent to

$$R : U_n \dashrightarrow [0,1] \qquad (X.5c)$$

Common examples of binary fuzzy relations R are: much greater than, resembles, is close to, etc.

If R is a relation from U to V, or equivalent, a relation in U x V

$$R : U \times V \dashrightarrow [0,1] \qquad (X.6a)$$

and S is a relation from V to W

$$S : V \times W \dashrightarrow [0,1] \qquad (X.6b)$$

then the composition R $\cdot$ S is a fuzzy relation from U to W

$$R \cdot S : U \times V \dashrightarrow [0,1] \qquad (X.6c)$$

For example

$$R \cdot S = \sum_{U \times W} V (\mu_R(u,v) \cap \mu_S(v,w))/(u,w) \qquad (X.6d)$$

is known as the max-min compositional rule.

X.3 - $\top$ -norms and $\top$ -conorms

The basic operators introduced by Zadeh to manipulate fuzzy sets were the maximum V and minimum $\cap$ operators, which closely correlated, respectively, with the

operators OR and AND in classic logic. Initially, this
assured fuzzy sets as an extension of the classic set
theory, because it preserved all basic properties of the
crisp sets in the fuzzy sets. Very soon, however, many
authors (e.g. Zimmermann and Zysno, 1980; Greco et al.,
1984; Greco and Rocha, 1987) began to discover that fuzzy
reasoning included many other semantics for conjunction and
disjunction. Dubois and Prade, 1982, introduced the concepts
of $\top$ -norms and S-norms in the Fuzzy Set Theory, to cope
with these other semantics of the ORing and ANDing
operations. From these authors we quote the following:

A triangular $\top$ -norm is a two place real-valued
function whose domain is the unit square $[0,1] \times [0,1]$, and
which satisfies the following conditions:

X.7a) $\top (0,0)=0;\ \top (a,1) = \top (1,a) = a$ (boundary conditions)

X.7b) $\top (a,b) \leq \top (c,d)$ if $a \leq c;\ b \leq d$ (monotonicity)

X.7c) $\top (a,b) = \top (b,a)$ (symmetry)

X.7d) $\top (a,T(b,c) = \top (T(a,b),c)$ (associativity)

Let A and B be two fuzzy sets over the universe U.
The membership function $\mu_{A\hat{}B}$ of the intersection $A\hat{}B$ of
these sets can be point-wisely defined as

$$\mu_{A\hat{}B} = I(\mu_A,\mu_B)\qquad\text{(X.7e)}$$

It is easily seen that X.7a to d are reasonable requirements
for the mapping I. Thus, the intersection operation of fuzzy
sets may be generalized as a $\top$ -norm operation. Simple, but
important $\top$ -norms are the minimum operator (min(a,b)), the
product operator (a.b), the so-called T_M operator (max
(0,a+b-1)), the so-called T_W operator, etc. The following
inequality holds

$$T_W(a,b) \leq \max(0,a+b-1) \leq a.b \leq \min(a,b)\qquad\text{(X.7f)}$$

Moreover, for any $\top$ -norm it holds:

$$T_W(a,b) \leq \top (a,b) \leq \min(a,b)\qquad\text{(X.7g)}$$

The Archimedian $\top$ -norm satisfies

$$\top (a,a) \leq a\qquad\text{(X.7h)}$$

and it enjoys the following property

$$\top (a,b) = f^{-1}(f(a)+f(b))\qquad\text{(X.7i)}$$

where

$$f : [0,1] \longrightarrow [0,\infty]\qquad\text{(X.7j)}$$

and f_{-1} is the pseudoinverse of f. The product is an Archimedian $\top$-norm, and it can be calculated if f is the logarithmic function and f_{-1} is the power function. The function f is called the additive generator of T.

A mapping S satisfying

X.8a) $S(1,1) = 1; S(0,a) = S(a,0) = a$ (boundary conditions)

X.8b) $S(a,b) \leq S(c,d)$ whenever $a \leq c; b \leq d$ (monotonicity)

X.8c) $S(a,b) = S(b,a)$ (symmetry)

X.8d) $S(a,S(b,c) = S(S(a,b),c)$ (associativity)

is called a S-norm or a $\top$-conorm. Any S-norm can be generated from a $\top$-norm through the transformation

$$S(a,b) = 1 - \top (1-a,1-b) \qquad (X.8e)$$

The tranformation of the $\top$-norms discussed above yields:

X.8f) $\min(a,b)$ ----> $\max(a,b)$

         $a.b$ ----> $a+b - ab$ (probabilistic sum)

    $\max(0,a+b-1)$ ----> $\min(1,a+b)$ (bounded sum)

The membership function μ_{AUB} of the union AUB of two fuzzy sets A and B can be point-wisely defined as

$$\mu_{AUB} = M(\mu_A,\mu_B) \qquad (X.8g)$$

It is easily seen that X.8a to d are reasonable requirements for the mapping M. Thus, the union of fuzzy sets may be generalized as a S-norm operation.

Assuming the union of fuzzy sets to be defined by the S-norm $\cdot$ and the intersection of the same sets to be defined by the $\top$-norm $\cap$, then the relational composition R $\cdot$ S defined in X.6d becomes:

$$R \cdot S = \sum_{UxW} \cdot (\mu_R(u,v) \cap \mu_S(v,w))/(u,w) \qquad (X.9)$$

Eq. X.9 is called the $\cdot$-$\cap$ compositional rule, and it generalizes the max-min compositional rule in X.6d. Also, each pair $\cdot$-$\cap$ identifies a type of calculus in Fuzzy Set Theory, since it defines different properties for the set union and intersection.

Trillas, 1980, proposed the extension of the usual fuzzy set complementation

$$\mu^{-}{}_A(x_i) = 1 - \mu_A(x_i) \qquad (X.10a)$$

to be the mapping

$$C: [0,1] \longrightarrow [0,1] \qquad (X.10b)$$

so that:

X.10c) $C(0) = 1$;

X.10d) C is involutional, that is $C(C(a)) = a$;

X.10e) C is strictly decreasing, and

X.10f) C is continuous.

The same author proved that for any negation C, there is a mapping t so that

$$C(a) = t_{-1}(1-t(b)) \qquad (X.10g)$$

where

$$t : [0,1] \longrightarrow [0,\infty] \qquad (X.10h)$$

Also,

X.10i) there is always a unique number $s \in [0,1]$ so that $C(s) = s$. In this case $s = t_{-1}(t(1)/2)$.

The concepts of $\top$ -norms, $\top$ -conorms and Complement operation discussed above turn Fuzzy Set Theory in more than a generalization of the Classic Set Theory, since this theory holds as a special case of the fuzzy sets defined when • is the supremum, ∩ is the minimum operator and the complement of A is obtained as $1 - \mu_A(u)$ for all elements of the universe of discourse U.

X.4 - Fuzzy variables and possibility theory

A variable is characterized by a triple

$$\{X, U, R(X:u)\} \qquad (11.a)$$

in which X is the name of the variable; U is the universe of Discourse; u is the generic name for the elements of U, and R(X:u) is a subset of U which represents the restriction on the values of u imposed by X (Zadeh, 1978). For convenience, let R(X) be the abbreviated form of R(X:u) and x to denote the elements of X. This variable is associated with an assignment equation:

$$x = u:R(X) \text{ or } x = u \text{ if } u \in R(X) \qquad (X.11b)$$

which represents the assignment of a value u to x subject to the restriction R(X). For example, the variable Human Age is a variable defined in the universe of discourse taken in the set of integers 0, 1, 2, ..., if R(X) is the subset 0, 1, 2, ..., 100.

A fuzzy variable X is defined if R(X) is a fuzzy restriction. R(X) defines a fuzzy set of U. Let this fuzzy set be called F and $\mu_F(u)$ be the membership of u in F. In this condition:

$$x = u: \mu_F(u) \qquad (X.11c)$$

$$R(X) = F \qquad (X.11d)$$

The membership $\mu_F(u)$ is interpreted as the degree to which the constraint represented by F is satisfied when u is assigned to the fuzzy variable X. For example, the Human Age is defined as a fuzzy variable in the universe of discourse taken in the set of integers 0, 1, 2, ..., if R(X) is the fuzzy subset SMALLER THAN 200, so that the following could be true:

$$x = 30: \mu_F=.9; \quad x = 100: \mu_F=.5; \quad x = 200: \mu_F=0 \qquad (X.11e)$$

This could mean that 30 years is really a common human age; 100 is achieved by some people and it is impossible to the human being to reach the age of 200 years.

A fuzzy restriction R(X) may be interpreted as a possibility distribution, with its membership function playing the role of a possibility distribution function. In this context, a fuzzy variable X is associated with this possibility distribution in much the same manner as a random variable is associated with a probability distribution.

Let X be a variable taking values in U, and let F act as a fuzzy restriction R(X) associated with X. Then the proposition "X is F" which translates into (Zadeh, 1978)

$$R(X) = F \qquad (X.12a)$$

associates a possibility distribution Γ_X with X

$$\Gamma_X = R(X) \qquad (X.12b)$$

Correspondingly, the possibility distribution function associated with X is denoted π_X and is defined to be numerically equal to the membership function of F:

$$\pi_X = \mu_F \qquad (X.12c)$$

Thus, the possibility $\pi_X(u)$ that X=u is postulated to be equal to $\mu_F(u)$. In the example X.11e, this means that there is a high possibility of 30 years to be the age of people; that the possibility of 100 years to be the age of a human being is .5 and it is impossible to an individual to be 200 years old. In short, the compatibility of a value of u with Human Age becomes converted into the possibility of that value to be the age of a human being.

According to Zadeh, 1978, the mathematical apparatus of the theory of fuzzy sets provides a basis for the manipulation of possibility distribution by the rules of this calculus. But since the semantic of the set operations is generalized by the concepts of $\top$ -norms and S-norms, this same author pointed the necessity of a great deal of empirical work to provide us with a better understanding of the ways in which possibility distributions are manipulated by humans. This is because most of the propositions in the previous chapters are heavily supported by experimental data about the expert reasoning.

An important point to be stressed is the difference of the concepts of probability and possibility. A high degree of possibility does not mean a high probability, although a low possibility implies low values of probability. To illustrate this relation, consider the statement discussed in Zadeh, 1978:

"Hans ate X eggs for breakfast"   (12d)

with X taking values in $U = \{1, 2, 3, 4, \ldots\}$. A possibility distribution Γ_X with X is associated with the ease with which Hans can eat eggs. The probability distribution with X can interpret $P_X(u)$ as the probability of Hans eating u eggs for breakfast. Let it be assumed that Hans generally eats 3 eggs each morning. Thus, although the possibility $\pi_X(1) = 1$, $P_X(1)$ is low while $\pi_X(3) = 1$ and $P_X(3)$ tends to 1. But let us consider now that John hate eggs. The possibility Γ_X associated with

"John ate X eggs for breakfast"   (12e)

has to imply $\pi_X(u)$ tending to zero does not matter the value of u. In this condition, it is wise to expect $P_X(u)$ tending to zero for any value of u, too.

Another point of difference to stress is that in the case of the probability theory:

$$\sum_U P_X(u) = 1 \qquad (X.13a)$$

This means that the probability measure is additive. On the contrary, in the case of possibility theory:

$$\bigvee_U \pi_X(u) \leq 1 \qquad (X.13b)$$

that is the maximum possibility is equal or smaller than 1, and

$$\sum_U \pi_X(u) \geq 0 \qquad (X.13c)$$

can assume any value. Thus, the possibility measure is not necessarily additive.

X.5 - Linguistic variables

One of the most striking features of Fuzzy Logic is its capability to deal with propositions in either natural or formal languages. For example, the proposition

"IF Fever is $\leq$ 38º Celsius THEN Fever is Low" (X.14a)

is an acceptable proposition in Fuzzy Logic.

In the antecedent part of the proposition X.14a, FEVER is a fuzzy variable X defined in the universe of discourse T of the temperatures by the restriction 37ºC Celsius $\leq$ t. This restriction defines a fuzzy subset F of T. In this condition:

$$x = t : \mu_F(t) \qquad (X.14b)$$

$$R(X) = 36ºC \leq t \leq 38º \qquad (X.14c)$$

$\mu_H(t)$ being as in Fig. X.5A.

In the consequent part of the proposition X14a, the variable Fever has a different structure. It does not take values in any universe of numbers, on the contrary it takes values that are words or sentences in a given language, such as Fever equal to: Absent, Low, Moderate, High, etc. This kind of variable is called Linguistic Variable.

A linguistic variable is characterized by a quintuple

$$(X, T(X), U, G, M) \qquad (X.15a)$$

in which X is the name of the variable; U is the universe of discourse; T(X) is a set of terms of a natural or artificial language used to speak about X; G is the syntatic rule used to generate the terms of T(X); and M is the semantic rule defining the meanings of T(X). This semantics associates each term x of T(X) with the base variable u according to the compatibility $\mu_{Rx}(u)$ of u with the fuzzy set T(x) (Zadeh, 1975). Each fuzzy set T(x) is defined by the corresponding restriction $R_x(X)$ associated with each term x of T(X). Fig. X.5B shows a definition for Fever Absent, Low, Moderate and High. In this context, the actual value of x subject to the actual t is

if $\alpha_i \leq t \leq \alpha_{i+1}$ then $x_i = \mu_{R(x_i)}(t)/T(x_i)$ (X.15b)

or

$$x = \sum_{T(X)} \mu_{R(x_i)}/T(x_i) \qquad \text{(X.15c)}$$

In the case of Fig. X.5, the value of X subject to t = 38.5º C is:

x = { 0/absent, .4/low, 1/moderate, 0/high }   (X.15d)

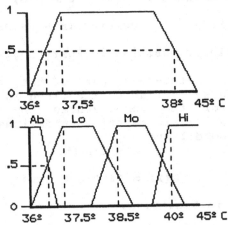

FIG. X.5 - FUZZY AND LINGUISTIC VARIABLES

A linguistic modifier is defined as an operator used to change the meaning of linguistic variables. Examples of these modifiers are: very, quite, more-or-less, not-very, etc. (Zadeh, 1975). They are used to create new terms in T(X), such as: very high; more-or-less high, quite low, not-very high, etc. The role played by the linguistic modifiers is to change the compatibility $\mu_{R(x)}(u)$ of u with the fuzzy set T(x) (Zadeh, 1975). For example, Zadeh proposed the following actions for the linguistic modifiers:

X.16a) $\mu_{R(vx)}(u) = (\mu_{R(x)}(u))^2$

that is, in the case of very x (vx) the compatibility of $\mu_{R(vx)}(u)$ of u with the fuzzy set very T(x) being equal to the power function of the compatibility $\mu_{R(x)}(u)$ of u with the fuzzy set T(x) (Fig. X.6a), and

X.17b) $\mu_{R(\pm x)}(u) = (\mu_{R(x)}(u))^{1/2}$

that is, in the case of more or less x (±x0) the
compatibility of $\mu_{R(\pm x)}(u)$ of u with the fuzzy set very
T(x) being equal to the square root of the compatibility
$\mu_{R(x)}(u)$ of u with the fuzzy set T(x) (Fig. X.6b).

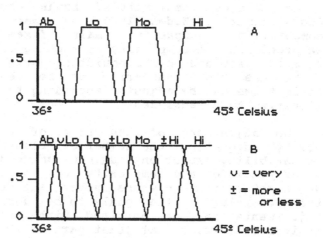

FIG. X.6 - THE ROLE OF THE LINGUISTIC MODIFIERS

In this context, it may be proposed that given a set
L of linguistic terms and a set M of linguistic modifiers,
and a sintax ϕ, the set T(X) in X.15a can be obtained as

$$\phi : M \times L \times T(X) ---> [0,1] \qquad (X.18)$$

In other words, T(X) is the set of productions in the
language defined by the sintax ϕ. For example, this sintax
may generate very high, more-or-less high, very low, etc. as
productions of T(x), while forbidding expressions like
more-or-less medium, less medium, etc. to be included in
T(X).

One of the fundamental tenets of modern science is
that a phenomenon cannot be claimed to be well understood
until it can be characterized in precise quantitative terms
(Zadeh, 1975). This paradigm has been very successful in the
field of the so called hard-sciences, like physics,
chemistry, engineering, etc., and very inadequate when
applied to the so called soft-sciences as psychology,
biology, sociology, etc. Natural language continues to be
the tool used for modeling in the soft-sciences in contrast
with the use of mathematical models in the hard-sciences.
Zadeh, 1975, argues that it seems that one of these fields -
hard sciences - took precision as its specialization,
whereas the other field - soft-sciences - elected complexity

as its main subj ect. If the computational capacity of the processing machines remains fixed, high precision and high complexity arise as incompatible concepts. This principle of incompatibility would explain why conventional techniques of system analysis and computer simulation - so successful in hard-sciences - are intrinsically incapable of coming to grips with the great complexity of human thought processes and decision making (Zadeh, 1975). As a matter of fact, if the computational capacity remains fixed, a trade-off between precision and complexity is required. This trade-off mechanism is provided by linguistic variables and fuzzy logic, because the granularity of the semantics of the linguistic terms can be adjusted according to the complexity of the problem to be analyzed.

The adjustment of the size of the granule is obtained by adequating the number of elements in $T(X)$ and the compatibility functions $\mu_{R(x)}(u)$ with the required degree of precision and complexity of the calculus. $T(X)$ provides a set of filters whose number and properties are adjusted according to the goals to be achieved. Most of theseadjustments can be obtained with the use of the linguistic modifiers, but at least part of them are achieved by redefining the size of the universe of discourse. The bigger the universe of discourse, the smaller the precision and greater the complexity of the subject under investigation. Thus, the semantics of $T(X)$ is a matter of learning and the choice of the compatibility functions is part of the analysis or simulation. Both requirements are supported by MPNNs.

X.6 - Linguistic quantifiers

The term linguistic quantifiers denotes a collection of quantifiers in natural languages whose representative elements are: several, most, much, many, at least n, few, etc. (Zadeh, 1983). The semantic of the linguistic quantifiers is very dependent of the concept of cardinality or Σ-counting of the fuzzy sets to which they are applied. Linguistic modifiers can be used to change this semantics in the very same way they adjust the meaning of the linguistic variables.

There are two basic types of linguistic or fuzzy quantifiers (Fig. X.7):

X.19a) absolute fuzzy quantifier Q: its semantic is referred to the absolute Σ-counting of the corresponding supporting fuzzy set A:

There are Q A's <--> Σ-count(A) is Q

Common examples of this type of quantifier are: several, few, many, approximately n, larger than n, etc.; and

X.19b) relative fuzzy quantifier Q: its semantics is referred to the relative Σ-counting of the supporting fuzzy sets A and B:

$$Q \text{ A's are B's} <\text{---}> \Sigma\text{-count}(B/A) \text{ is } Q$$

Common examples of this type are: most, many, often, much of, etc.

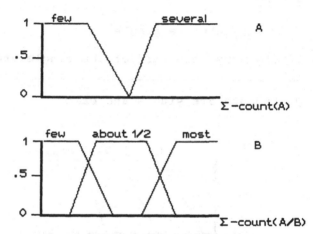

FIG. X.7 - LINGUISTIC QUANTIFIERS

It follows from the above examples that some quantifiers, e.g. many, can be classified as both types depending on the context of use.

The meaning of any fuzzy quantifier Q is determined in two steps (Kacprzyck, 1986a,b; Zadeh, 1983; Yager, 1990b):

X.20a) first the Σ-count(A) or Σ-count(A/B) is calculated, and then

X.20b) the meaning μ_Q of Q is obtained as

$$\mu_Q : \Sigma\text{-count} \text{---}> [0,1]$$

In other words, μ_Q measures the compatibility of Σ-count with the prototypical knowledge about Q (Fig. X.7).

Fuzzy quantifiers are combined according to the type

of syllogism they are involved in (Zadeh, 1983, 1988), an
instance of which is the intersection/product syllogism:

Q_1 A's are B's

Q_2 (A's and B's) are C's

$Q_1* Q_2$ A's are (B's and C's)   (X.21a)

where * is the fuzzy arithmetic product. Given the following
example, the results of $Q_1* Q_2$ are shown in Fig. X.8
(Zadeh, 1983).

most students are single

little more than half of the single students are male

Q students are single and male   (X.21b)

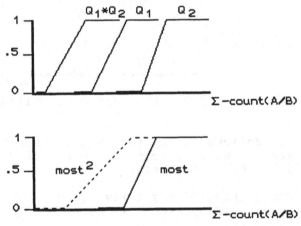

FIG. X.8 - COMBINING LINGUISTIC QUANTIFIERS

X.7 - Fuzzy logic

According to Zadeh, 1988:

"Fuzzy logic may be viewed as an extension of classic logic
and multivalued logic. Its uses and objectives are, however,
quite different. Thus, the fact that fuzzy logic deals with
approximate rather than precise modes of reasoning implies
that, in general, the chains of reasoning in fuzzy logic are

short in length, and rigor does not play as important a role as it does in classical logical systems."

It could be added that fuzzy logic is related to partially closed worlds, in contrast with other logics requiring the closeness of the universe of discourse. Since intelligence implies non-deterministic modelling (Wah et al., 1989), it follows that fuzzy logic is better equipped to describe the human reasoning than these other types of logic. Another strong property of fuzzy logic is its capability of dealing with propositions in natural language. The meaning of an imprecise proposition is represented as an elastic constraint on a (linguistic) variable, and the answer to a query is deduced through a propagation of elastic constraints.

A proposition p in a natural or synthetic language may be viewed (Zadeh, 1988) as a collection of elastic constraints $C = \{ C_1, \ldots, C_k \}$ which restrict the values of a collection of variables $X = \{ X_1, \ldots, X_n \}$. These propositions may be:

X.22a) simple: $p \equiv X_i$ is C_i;

X.22b) conjunctive: $p \equiv X_1$ is C_1 and $\ldots$ and X_n is C_k;

X.22c) disjunctive: $p \equiv X_1$ is C_1 or $\ldots$ or X_n is C_k, and/or

X.22d) conditional: $P \equiv$ IF X_i is C_i then Z is C_p

where the meaning of $\equiv$ is defined as.

Equation X.22 implies that a possibility distribution Γ_X is associated with X according to the constraints C. Thus, the possibility $\pi_p(u)$ being true for a given value of u in the universe of discourse U is

$$\pi_p(u) = \mu_C(u) \qquad (X.23a)$$

so that in the case of the:

X.23b) conjunctive proposition:

$$\pi_p(u) = \mu_{C1}(u) \cap \ldots \cap \mu_{Ck}(u)$$

where $\cap$ is a $\top$ -norm;

X.23c) disjunctive proposition:

$$\pi_p(u) = \mu_{C1}(u) * \ldots * \mu_{Ck}(u)$$

where $*$ is a $\top$ -conorm, and

X.23d) conditional proposition:

$$\pi_P(u) = \mu_{C_P}(u) = f(\mu_{C_i}(u))$$

where

$$f : C_i \times C_p \dashrightarrow [0,1]$$

is a relation between the fuzzy sets C_i and C_p.

There are some important rules of inference in Fuzzy Logic:

X.24a) entailment principle:

X is A

A (B

_____

X is B

In other words, the entailment principle asserts that from the proposition X is A it is possible to infer a less specific proposition X is B. This principle may be regarded as a generalization to fuzzy sets of the inheritance principle widely used in knowledge representation systems, since X inherits from A the properties of B;

X.24b) extension principle: let X be a variable taking values in the universe of discourse U and being constrained by X is A, and let the mapping be f from U to V so that X is mapped into $f(V)$, then

X is A

_____

$f(X)$ is $f(A)$

where the membership of $f(A)$ is defined by

$$\mu_{f(A)}(v) = \sup_u \mu_A(u)$$

subject to

$$v = f(u),\ u \in U \text{ and } v \in V$$

The extension principle plays an important role in fuzzy logic by providing a mechanism for computing induced contraints (Zadeh, 1988);

X.24c) extended modus ponens:

IF X is A then Y is B

X is A'

_____

Y is B'

where

$$B' = A' \circ (A \dashrightarrow B)$$

which means that B' is obtained as the composition of the relations $R_{A'}(X)$ defining A' and the relation f defining the implication between the fuzzy sets A and B. Thus, from eq. X.9:

$$\mu_{B'}(v) = \underset{u}{\circ} (\mu_F(u,v) \cap \mu_{A'}(u))$$

given that X is a variable in the universe of discourse U and B is a variable in V, $\circ$ is a $\top$ -conorm (in general the supremum) and $\cap$ is a $\top$ -norm.

The solution of EMP is obtained in 4 steps (Zadeh, 1983a):

V.25a) Matching: the compatibility σ between A and A' is the measure of the equality [A≡A'] between the fuzzy sets A and A' (Pedrycz; 1990a,b), so that the matching between (X is A') to (X is A) (Godo et al, 1991) is calculated as:

$$(X \text{ is } A') \equiv (X \text{ is } A) \text{ is } \sigma$$

Since $A \equiv A'$ implies

$$A (A' \quad \text{and} \quad A') A$$

$$\mu_A(x) \leq \mu_{A'}(x) \qquad \text{and} \qquad \mu_{A'}(x) \leq \mu_A(x)$$

the assessment of the value of σ means to evaluate how equal are these two fuzzy sets taking into account their elements. As pointed out by Pedrycz, 1990a, the choice is rather free and can vary from

X.25a1) pessimistic: A and A' are equal if they are equal in all elements of X

$$\sigma = \underset{X}{\min} [A \equiv A'](x), \text{ or}$$

X.25a2) optimistic: A and A' are equal if they are equal in at least one element of X

$$\sigma = \underset{X}{\max} [A \equiv A'](x)$$

Maybe the best approach could be

X.25a3) realistic: A and A' are equal if they are equal in Q of their elements

$$\sigma = \underset{X}{Q} [A \equiv A'](x)$$

where Q is a linguistic quantifier of the type MOST, AT LEAST N, etc.

X.25b) Aggregation: all compatibilities σ_i assigned to the atomic propositions in the antecedent part of the implication are aggregated into a unique value representing the compatibility of the antecedent σ_a

$$\sigma_a = \overset{n}{\underset{i=1}{\Theta}} (\sigma_i)$$

X.25c) Projection: the compatibility σ_c of the consequent is obtained as function of the aggregated value σ_a (Delgado et al, 1990b; Diamond et al, 1989; Godo et al, 1991; Katai et al, 1990a,b) $\sigma_b = f(\sigma_a)$

σ_b measures the compatibility of (Y is B') with (Y is B).

X.25d) Inverse-Matching: given σ_b and B, it is necessary to obtain B' as a subset of B so that:

$$[B' \equiv B](y) \leq \sigma_b$$

Again, 3 different strategies may be used:

X.25d1) pessimistic: B' is composed by the elements of B providing a compatibility equal to σ_b:

$$B' = \{ y \mid \mu_{B'}(y) = \mu_B(u) = \sigma_b \}$$

X.25d2) optimistic: B' is composed by the elements of B providing a compatibility equal or smaller than σ_b

$$B' = \{ y \mid \mu_{B'}(y) \leq \mu_B(u) \leq \sigma_b \}$$

X.25d3) realistic: B' is composed by Q elements of B providing a compatibility equal or greater than σ_b

$$B' = Q\{ y \mid \mu_{B'}(y) \geq \mu_B(u) \geq \sigma_b \}$$

The otimistic approach is that proposed by Zadeh, 1983, and it is widely used. The realistic approach has also been proposed in the literature (Pedrycz, 1990a,b; Katai et al, 1990a,b; Gomide et al, 1991)

X.25e) Deffuzification: sometimes it is necessary to find the most representative singleton b' of B' to replace this fuzzy set as the solution of the fuzzy implication. It is the case of many applications of fuzzy logic in control.

Many approaches have been proposed in the literature to implement deffuzification (e.g. Castro and Trillas, 1990;

Delgado et al, 1990a; Diamond et al, 1989; Gomide et al, 1991; Katai et al., 1990a,b; Mandani, 1974; Mantaras et al, 1990; Mizumoto and Zimmermann, 1982; Mizumoto, 1989; Soula and Sanchez, 1982; Yager, 1984; Zadeh, 1983a). The most popular method is the center of area, but means and powered means have also been proposed as tools for defuzzification, and it seems that for practical purposes they are equivalent to the center of the area. (Gomide et al, 1991).

To implement these other methods the following procedure is proposed. Given $[(Y$ is $C)$ is $\sigma_c]$ and an equality index α, the inverse-matching is to find those $y \in Y$ which result into (Pedrycz, 1990a,b):

$$[C \equiv C'](y) \geq \alpha \qquad (V.5g)$$

The solution is a family I of constraint-intervals (Katai et al, 1990ab):

$$I_i = \{y \in Y \mid [C \equiv C'](y) \geq \alpha, \ \mu_{C'}(y) = v_i, \ v_i \in [\alpha, 1]\}$$
(V.5h)

The mean points of these intervals are calculated, and the singleton is obtained as the average of these mean points. This last averaging may be powered by v_i.

The inference mechanisms (inference engine) used in Fuzzy Logic markedly differ, according to the field of application. The inference engine is very simple in the case of Fuzzy Logic Control (FLC) in comparison with that used in the case of expert systems (ES) (Lee, 1990). In the first case, rule chaining is an exception, whereas in the second case, rule grouping and rule group chaining are basic strategies used to organize the knowledge base (see Chapter V).

A typical data base in FLC is:

IF X_1 is C_1 and .... and X_m is D_1 .... THEN Z is F_1
.
.   (X.26a)
.
IF X_1 is C_n and .... and X_m is D_n .... THEN Z is F_n

That is, the antecedents and consequents of the different rules are composed by the same variables taking different linguistic values in $T(X)$ or $T(Z)$. For example:

IF Velocity is High and Acceleration is High THEN Brake is High
.
.   (X.26b)
.
IF Velocity is Low and Acceleration is Low THEN Brake is Low

A typical inference chain in FLC is:

X.26c)   the actual   values of the variables X_i are provided
in such a way that their matching with the knowledge encoded
in the antecedent linguistic variables are calculated;

X.26d)   the   aggregation   of   the antecedent   is   performed;

X.26e)   the   different F_i' are   obtained   by   calculating
$\mu_{Fi'}$ according to the procedure in X.24c and by applying
the optimistic approach of X.25d2;

X.26f)   the different rules of the data base are combined by
the connective ELSE, so that the final F' is obtained as the
union or intersection of the different F'is, and finally

X.26g)   a singleton $f \in F'$ is calculated - in general with
the application of the center of gravity approach - as the
solution of the controlling variable Z. E.g. a possible
output in the case of the example X.26b could be Brake = .8.

Typical knowledge bases in ES are knowledge nets KN
composed of a family of related knowledge graphs KG (see
Fig. V.1, Chapter V, section V.2). In other words, they are
structured bases of fuzzy conditional rules. Deduction in
this case is the solution of a nonlinear program (Zadeh,
1983, 1988) to maximize the output at the root nodes of the
knowledge graphs subject to the restrictions input at their
terminal nodes (see Chapter II, sections II.6 and 7).

ACKNOWLEDGMENT

        I am in debt with Alejandro B. Engel who introduce
me to Fuzzy Sets.

REFERENCES

AIKINS, J.S. (1983) Prototypical knowledge for expert systems. Artificial Intelligence 20/2   p.163-210

ALLEN, G.I. and TSUKAHARA, N. (1974) Cerebrocerebellar communication systems. Physiological Rev. 54/4 p.957-1006

ANDERSON, J.A. and ROSENFELD, E. (Eds.) (1989) Neurocomputing: foundations of research Third Printing 729p The MIT Press, Cambridge, Mass.

ANDERSON, J.R. (1989) A theory of the origins of human knowledge. Artificial Intelligence 40/1-3   p.313-51

ARCH, S. and BERRY, R.W. (1989) Molecular and cellular regulation of neuropeptide expression: the bag cell model system. Brain Research Rev. 14/2   p.181-201

ARMSTRONG, C.M. (1981) Sodium channels and gating currents. Physiological Rev. 61/3 p.644-83

BALLARD, D.H. (1986) Cortical connections and parallel processing: structure and function. The Behavioral and Brain Sciences 9/1   p.67-120

BARTOLIN, R.; BONNIOL, V. and SANCHEZ, E. (1988) Inflammatory protein variations: medical knowledge representation and approximate reasoning. In: BOUCHON, B.; SAITTA, L. and YAGER, R.R. (Eds.) Uncertainty and Intelligent Systems v.313 - Lecture Notes in Computer Science p.306-13 Springer-Verlag, Berlin

BASSANI, J.W. (1979) A measurement of the cell excitability. Master Thesis (in portuguese) UNICAMP Faculty of Engineering of Campinas Campinas, Brazil

BLACK, I.B.; ADLER, J.E.; OREYFUS, C.F.; JONAKAIT, G.M.; KATZ, D.M.; LaGAMMA, E.F. and MARKEY, K.M. (1984) Neurotransmitter plasticity at the molecular level. Science 225/4668   p.1266-70

BLOMFIELD, S. (1974) Arithmetical operations performed by nerve cells. Brain Research 69/1   p.115-24

BOBROW, D.G. (1980) Editor's Preface. Artificial Intelligence 13/1-2   p.1-4

BOOKER, L.B.; GOLDBERG, D.E. and HOLLAND, J.H. (1989) Classifier systems and genetic algorithms. Artificial Intelligence 40/1-3   p.235-82

BROWN, W.S.; MARSH, J.T. and SMITH, J.C. (1976) Evoked potential waveform differences produced by the perception of different meanings of an ambiguous phrase. Eletroencephalograph. Clin. Neurophysiol. 41 p.113-23

BUNGE, M. (1977) Emergence and the mind. Neuroscience 2/4 p.501-9

BUNO, W.Jr.; FUENTES, J. and SEGUNDO, J.P. (1978) Crayfish stretch-receptor organs, effects of length-steps with and without perturbations. Biological Cybernetics 31 p.99-110

BURNSTOCK, G. (1976) Do some nerve cells release more than one transmitter?. Neuroscience 1/4 p.239-48

BYRNE, J.H. (1987) Cellular analysis of associative learning. Physiological Rev. 67/2 p.329-439

CASTRO, J.L. and TRILLAS, E. (1990) Logic and fuzzy relations. In: VERDEGAY, J.L. and DELGADO, M. (Eds.) Approximate Reasoning Tools for Artificial Intelligence p.3-20 Verlag TUV, Germany

CHANDRASEKARAN, B.; GOEL, A. and ALLEMANG, D. (1988) Connectionism and information. AI Magazine p.25-34

CHEN, S.; KE, J. and CHANG, J. (1990) Knowledge representation using Fuzzy Petri Nets. IEEE Transactions on Knowledge and Data Engineering 2/3 p.311-319

COON, D.D. and PERERA, A.G.U. (1989) Integrate-and-fire coding and Hodgkin-Huxley circuits employin silicon diodes. Neural Networks 2 p.143-52

COTMAN, C.W.; NIETO-SAMPEDRO, M. and HARRIS, E.W. (1981) Synapse replacement in the nervous system of adult vertebrates. Physiological Rev. 61/3 p.684-784

COWAN, W.M.; FAWCETT, J.W.; O'LEARY, D.O.M. and STANFIELD, B.B. (1984) Regressive events in neurogenesis. Science 225/4668 p.1258-65

COX, B.J. (1987) Object Oriented Programming: an evolutionary approach Addison-Wesley, Reading, Mass.

DAVIS, M. (1980) The mathematics of non-monotonic reasoning. Artificial Intelligence 13/1-2 p.73-80

DELGADO, M.; VERDEGAY, J.L. and VILA, M.A. (1989) A general model for fuzzy linear programming. Fuzzy Sets and Systems 29/1 p.21-9

DELGADO, M.; MORAL, S. and VILA, M.A. (1990a) A new view of generalized modus ponens. Proc. International Conference on Fuzzy Logic & Neural Networks 2   p.963-8   Iizuka, Japan

DELGADO, M.; TRILLAS, E.; VERDEGAY, J.L. and VILA,
M.A. (1990b)   The generalized "modus ponens" with linguistic
labels. Proc. International Conference on Fuzzy Logic &
Neural Networks 2   p.725-8   Iizuka, Japan

DIAMOND, J.; McLEOD, R.D. and PEDRYCZ, W. (1989)   A fuzzy
cognitive system: foundations and VLSI implementation.
Proc. 3rd IFSA Congress   p.396-9   Seattle, Washington

DONCHIN, E., RITTER, W. and McCALLUN, C. (1978)   Cognitive
psychophysiology: The endogenous components of the ERP. In:
CALAWAY E; TUETING, P. and KOWLOW, S. H. (Eds.) Event
Related Potentials in Man   p.349-411   Academic Press, New
York

DUBOIS, D. and PRADE, H. (1982)   A class of fuzzy measures
based on triangular norms. A general framework for the
combination of uncertain information. Int. J. General
Systems 8/1   p.43-61

ECCLES, J.C. (1981)   The modular operation of the cerebral
neocortex considered as the material basis of mental
events. Neuroscience 6/10   p.1839-56

EDDY, D. and CLANTON, C.H. (1982)   The art of diagnosis:
solving the clinicopathological exercise. New England J.
Medicine 21/306   p.1263-8

EDELMAN, G.M. (1987)   Neural Darwinism: the theory of
neuronal group selection   371p. Basic Books, New York

FERNANDES, C.A.de C. and GOMIDE, F.A.C. (1991)   A real time
expert supervisory process control system. The WORLD
CONGRESS on Expert Systems   Orlando, Florida   to appear

FIGUEIREDO, M.; GOMIDE, F.; ROCHA, A.F. and YAGER, R.R.
(1991)   Comparison of Yager's level set method for fuzzy
logic control with Mamdani's and Larsen's methods. submitted

FILLMORE, C.J. (1968)   The case for case. Universals in
linguistic theory. In: BACH and HARMS (Eds.) Holt, Rinehart
and Winston Inc.,   New York

FLORKIN, M. (1974)   Concepts of molecular biosemiotics and
of molecular evolution. Reprinted from Comprehensive
Biochemistry 29/Part A   p.1-124   Elsevier, Amsterdam

GAINES, B.R. and KOHOUT, L.J. (1975)   The logic of automata.
Int. J. General Systems 2/4   p.191-208

GALLANT, S.I. (1988)   Connectionist expert systems.
Communications of ACM 31/2   p.152-69

GLASSER, E. and RUCHKIN, D.S. (1976) Principles of neurobiological signal analysis. **Academic Press, New York**

GODO, L.; JACAS, J. and VALVERDE, L. (1991) Fuzzy values in fuzzy logic. Int. J. Intelligent Systems 6 p.199-212

GOLDMAN-RAKIC, P.S. (1988) Topography of cognition: parallel distributed networks in primate association cortex. Ann. Rev. Neuroscience 11 p.137-56

GOMIDE, F., GUDWIN, R., ROCHA, A.F., SILVA, M.A., ALMEIDA, H.J. and RIBEIRO, I.C. (1991) Fuzzy control engineering: a computer aided tool. submitted

GOMIDE, F.A.C. and ROCHA, A.F. (1991) Neurofuzzy controllers. submitted

GOODMAN, C.S.; BASTIANI, M.J.; DOE, C.Q.; LAC, S.du; HELFAND, S.L.; KUWADA, J.Y. and THOMAS, J.B. (1984) Cell recognition during neuronal development. Science 225/4668 p.1271-94

GORDON, J. and SHORTLIFFE, E.H. (1984) The Dempster-Shafer theory of evidence. In: BUCHANAN, B.G. and SHORTLIFFE, E.H. Rule-based expert systems: the MYCIN experiments of the Stanford Heuristic Programming Project **Addison-Wesley, Reading, Mass.**

GRAFSTEIN, B. and FORMAN, D.S. (1980) Intracellular transpor in neurons. Physiological Rev. 60/4 p.1167-283

GRECO, G., ROCHA, A.F. and ROCHA, M.T. (1984) Fuzzy logical structure of a text decoding. Proc. 6th International Congress of Cybernetics and Systems 1 p.193-8 Paris

GRECO, G. and ROCHA, A.F. (1987) The fuzzy logic of a text understanding. Fuzzy Sets and Systems 23/3 p.347-60

GRECO, G. and ROCHA, A.F. (1988) Brain activity and fuzzy belief. In: ZETENYI, T. (Ed.) Fuzzy Sets in Psychology p.297-319 Elsevier, Amsterdam

HALL, L.O.; SZABO, S. and KANDEL, A. (1986) On the derivation of memberships for fuzzy sets in expert systems. Information Sciences 40/1 p.39-52

HALL, L.O. and ROMANIUK, S.G. (1990) A hybrid connectionist, symbolic learning system. Proc. AAAI-90 8th National Conference on Artificial Intelligence p.783-8 Boston, Mass.

HANDELMAN, D.A. and STENGEL, R.F. (1987) An architecture for real-time rule-based control. presented at The American Control Conference Minneapolis

HINTON, G.E. (1989) Connectionist learning procedures.
Artificial Intelligence 40/1-3   p.185-234

HODGKIN, A.A. and HUXLEY, A.F. (1952) A quantitative
description of membrane currente and its application to
conduction and excitation in nerve. J. Physiology 116
p.500-44

HOLLAND, J. (1975) Adaptation in Natural and Artificial
Systems. University of Michigan Press, Ann Arbor, Michigan

HOLTZMAN, E. (1977) The origin and fate of secretory
packages, especially synaptic vesicles. Neuroscience 2/3
p.327-55

HOPFIELD, J.J. (1982) Neural networks and physical systems
with emergent collective computational abilities. Proc.
National Academy of Sciences 79   p.2554-8

HYVARINEN, J. (1982) Posterior parietal lobe of the primate
brain. Physiological Rev. 62/3   p.1060-129

KACPRZYK, J. (1985) Zadeh's commonsense knowledge and its
use in multicriteria, multistage and multiperson decision
making. In: GUPTA, M.M.; KANDEL, A.; BANDLER, W. and KISZKA,
J.B. (Eds.) Approximate Reasoning in Expert System
p.105-22   Elsevier, Amsterdam

KACPRZYK, J. (1986a) Group decision making with a fuzzy
linguistic majority. Fuzzy Sets and Systems 18/2 p.105-18

KACPRZYK, J. (1986b) Towards "human-consistent" multistage
decision making and control nodels using fuzzy sets and
fuzzy logic. Fuzzy Sets and Systems 18/3   p.299-314

KACPRZYK, J. (1988) Fuzzy logic with linguistic
quantifiers: A tool for better modeling of human evidence
aggregation processes?. In: ZETENYI, T.(Ed.) Fuzzy Sets in
Psychology p.233-63   Elsevier, Amsterdam

KACPRZYK, J.; FEDRIZZI, M. and NURMI, H. (1990) Group
decision making with fuzzy majorities represented by
linguistic quantifiers. In: VERDEGAY, J.L. and DELGADO, M.
(Eds.) Approximate Reasoning Tools for Artificial
Intelligence p.126-45   Verlag TUV, Koln, Germany

KANDEL, E.R. and SCHWARTZ, J.H. (1982) Molecular biology of
learning: modulation of transmitter release. Science 218/
4571   p.433-43

KASSIRER, J.P. and GORRY, G.A. (1978) Clinical problem
solving: a behavioral analysis. Annals of Internal
Medicine 89   p.245-55

KATAI, O.; IDA, M.; SAWARAGI, T. and IWAI, S. (1990a)

Treatment of fuzzy concepts by order relations and constraint-oriented fuzzy inference. Proc. NAFIPS'90 p.300-3   Toronto, Canada

KATAI, O.; IDA, M.; SAWARAGI, T. and IWAI, S. (1990b) Fuzzy inference rules and their acquisition from constraint-oriented perspectives. Proc. International Conference on Fuzzy Logic & Neural Networks 1 p.211-16 Iizuka, Japan

KENNEY, R.M. (1981) Between never and always. New England J. Medicine 305/18   p.1097-8

KICKERT, W.J.M. and MANDANI, E.H. (1978) Analysis of a fuzzy logic controller. Fuzzy Sets and Systems 1 p.29-44

KLIR, G. (1989)   Probability-possibility conversion. Proc. 3rd IFSA Congress   p.408-11 Seattle, Washington

KOHN, A. F.; ROCHA, A. F.; SEGUNDO, J. P. (1981) Presynaptic irregularity and pacemaker inhibition Biological Cybernetics 41/1   p.5-18

KOHONEN, T. (1982) Self-organized formation of topologically correct feature maps. Biological Cybernetics 43/1   p.59-69

KOLONDER, J.L. (1983) Towards an understanding of the role of experience in the evolution from novice to expert. Int. J. Man-Machine Studies 19/4   p.497-518

KONG, A.; BARNETT, G.O.; MOSTELLER, F. and YOUTZ, C. (1986) How medical professionals evaluate expressions of probability. New England J. Medicine 315/12   p.740-4

KUNO, M. (1971) Quantum aspects of central and ganglionic synaptic transmission in vertebrates. Physiological Rev. 51/4   p.647-678

KUTAS, M. and HILLYARD, S.A. (1980) Reading senseless sentences: brain potentials reflect semantic incongruity. Science 207/4427   p.203-4

LADURON, P.M. (1987) Axonal transport of neuroreceptors: possible involvement in long-term memory. Neuroscience 22/3   p.767-79

LaGAMMA, E.F.; ADLER, J.E. and BLACK, I.B. (1984) Impulse activity differentially regulates (Leu)enkephalin and catecholamine characters in the adrenal medulla. Science 224/4653   p.1102-4

LAM, D.M.; SU, Y-Y T. and WATT, C.B. (1986) The self-regulating synapse: a functional role for the co-existence of neuroactive substances. Brain Research Rev. 11/3   p.249-57

LANE, S.H.; HANDELMAN, D.A. and GELFAND, J.J. (1990) Can robots learn like people do?. Proc. SPIE Conference on Applications of Artificial Neural Networks Orlando, Florida

LARKIN, J.P.; McDERMOTT, J.; SIMON, D.P. and SIMON, H.A. (1980) Expert and novice performance in solving physics problems. Science 208/4450 p.1335-42

LAUGER, P. (1987) Dynamics of Ion transport systems in membranes. Physiological Rev. 67/4 p.1296-331

LEÄO, B.F. and ROCHA, A.F. (1990) Proposed methodology for knowledge acquisition: A study on congenital heart disease diagnosis. Methods Information Medicine 29/1 p.30-40

LEE, C.C. (1990) Fuzzy Logic in Control Systems: Fuzzy Logic Controller - Part I. IEEE Transactions on Systems, Man and Cybernetics 20/2 p.404-18

LEE, C.C. (1990) Fuzzy Logic in Control Systems: Fuzzy Logic Controller - Part II. IEEE Transactions on Systems, Man and Cybernetics 20/2 p.419-35

LESMO, L. and TORASSO, P. (1987) Prototypical knowledge for interpreting fuzzy concepts and quantifiers. Fuzzy Sets and Systems 23/3 p.361-70

LEVESQUE, H. and MYLOPOULOS, J. (1979) A procedural semantics for semantic networks. In: FINDLER, N.V. (Ed.) Associative networks: representation and use of knowledge by computers Academic Press, New York

LIEBERMAN, P. (1967) Intonation, perception and language 210p. The MIT Press, Cambridge

LOFGREN, L. (1977) Complexity of descriptions of systems: a foundational study. Int. J. Gen. Systems 3/4 p.197-214

LURIA, A.R. (1974) Cerebro y lenguage: la afasia traumatica: sindromes, exploraciones y tratamiento 553p. Fontanella, Barcelona

LYNCH, G. and BAUDRY, M. (1984) The biochemistry of memory: a new and specific hypothesis. Science 224/4653 p.1057-63

MACHADO, R.J. and ROCHA, A.F. (1989) Handling knowledge in high order neural networks: the combinatorial neural model. Technical Report CCR-076 22p. IBM Rio Scientific Center, Rio de Janeiro, Brazil

MACHADO, R.J. and ROCHA, A.F. (1990a) The combinatorial neural network: a connectionist model for knowledge based systems. In: BOUCHON-MEUNIER, B.; YAGER, R.R. and ZADEH, L.A. (Eds.) Uncertainty in knowledge bases v.521 - Lecture

Notes in Computer Science **p.578-87 Springer-Verlag, Paris**

MACHADO, R.J.; ROCHA, A.; RAMOS, M.P. and GUILHERME, I.R. (1990) Inference and inquiry in fuzzy connectionist expert systems. Proc. Cognitiva'90 p.97-101 Madrid, Spain

MACHADO, R.J.; ROCHA, A.F. and LEÄO, B.F.(1990b) Calculating the mean knowledge representation from multiple experts. In: KACPRZYK, J. and FEDRIZZI, M. (Eds.) Multiperson Decision Making Using Fuzzy Sets and Possibility Theory **p.113-27 Kluwer Acad. Publ., Netherlands**

MACHADO, R.J.; DUARTE, V.H.A.; DENIS, F.A.R.M. and ROCHA, A.F. (1991a) NEXT - The neural expert tool. Technical Report CCR-120 57p. IBM Rio Scientific Center, Rio de Janeiro, Brazil

MACHADO, R.J., FERLIN, C., ROCHA, A.F. and SIGULEM, D. (1991b) Combining Semantic and Neural Networks in Expert Systems. The WORLD CONGRESS on Expert Systems Orlando, Florida to appear

MACHADO, R.J.; ROCHA, A.F. and GUILHERME, I.R. (1991c) FRANK: a hibrid fuzzy connectionist and bayeasian expert systems. Proc. IFSA'91 Brussels Artificial Intelligence p.125-8 Brussels - Belgium

MACHADO, R.J. and ROCHA, A.F. (1992) A hybrid architecture for fuzzy connectionist expert systems. Intelligent Hybrid Systems. In: KANDEL, A. and LANGHOLZ, G. CRC Press Inc., USA in press

MAEDA, H. and MURAKAMI, S. (1988) A fuzzy decision-making method and its application to a company choice problem. Information Science 45/2 p.331-46

MAEDA, H. and THEOTO, M. (1990) Theoretical and experimental results on confidence. Proc. International Conference on Fuzzy Logic & Neural Networks 1 p.147-50 Iizuka, Japan

MANDANI, E.H. (1974) Applications of fuzzy algorithms for control of a simple dynamic plant. Proc. IEEE 12/1 p.1585-8

MANTARAS, R.L.; GODO, L. and SANGUESA, R. (1990) Connective operator elicitation for linguistic term sets. Proc. International Conference on Fuzzy Logic & Neural Network 2 p.729-33 Iizuka, Japan

McCALLUN, W.W.; CURRY, S.J.; POCOCK, P.V. and PAPAKOSTOPOULOS, D. (1983) Brain event related potentials as indicators of early selective processes in auditory target localization. Psychophysiology 20 p.1-17

McCARTHY, J. (1980) Circumscription - A form of non-monotonic reasoning. Artificial Intelligence 13/1-2 p.27-39

McCLELLAND, J.L. and KAWAMOTO, A.H. (1986) Mechanisms of sentence processing: assigning roles to constituents of sentences. In: McCLELLAND, J.L.; RUMELHART, D.E. and PDP RESEARCH GROUP Parallel Distributed Processing - Explorations in the Microestructure of Cognition 2: Psychological and Biological Models p.272-325 The MIT Press, Cambridge, Mass.

McCONNELL, S.K. (1988) Development and decision-making in the mammalian cerebral cortex. Brain Research Rev. 13/1 p.1-23

McCULLOCH, W.S. and PITTS, W. (1943) A logical calculus of the ideas immanent in nervous activity. Bull. Mathematical Biophysics 5/4 p.115-33

McDERMOTT, D. and DOYLE, J. (1980) Non-monotonic Logic I. Artificial Intelligence 13/1-2 p.41-72

McILWAIN, H. (1977) Extended roles in the brain for second-messenger systems. Neuroscience 2/3 p.357-72

MILLER, G.A. and ISARD, S. (1963) Some perceptual consequences of linguistic rules. J. Verbal Learning and Verbal Behavior 2 p.217-28

MILLER, H.E.; PIERSKALLA, W.P. and RATH, G.J. (1976) Nurse scheduling using mathematical programming. Operations Research 24/5 p.857-70

MILLER, R.A. and MASARIE Jr., F.E. (1990) The demise of the "Greek Oracle" model for medical diagnostic systems - editorial. Methods Information Medicine 29/1 p.1-2

MILNE, R. (1987) Strategies for diagnosis. IEEE Trans. Systems, Man and Cybernetics 3 p.333-9

MINSKY, M. and PAPERT, S. (1969) Perceptrons: an introduction to computational geometry. 258p. The MIT Press Cambridge, Mass.

MIURA, K.; MOROOKA, C.K.; ROCHA, A.F. and GUILHERME, I.R. (1991) Knowledge acquisition from natural language data bases. Proc. LAIC-PEP'91 - Latin American Conference on Artificial Intelligence in Petroleum Exploration and Production p.133-41 Rio de Janeiro, Brazil

MIZUMOTO, M. and ZIMMERMANN, H.J. (1982) Comparison of fuzzy reasoning methods. Fuzzy Sets and Systems 8/3 p.253-83

MIZUMOTO, M. (1989) Improvement methods of fuzzy controls. Proc. 3rd IFSA Congress p.60-2 Seattle, Washington

MONTANA, D.J. and DAVIS, L. (1989) Training feedforward neural networks using genetic algorithms. Proc. International Joint Conference of Artificial Intelligence IJCAI-89 p.762-7

MOSKOWITZ, A.J.; KUIPERS, B.J. and KASSIRER, J.P. (1988) Dealing with uncertainty, risks and tradeoffs in clinical decisions - A cognitive science approach. Annals of Internal Medicine 108/3 p.435-49

MOUNTCASTLE, V.B.; POGGIO, G.F. and GERHARD, W. (1964) The relation of thalamic cell response to peripheral stimuli varied over an intensive continuum. J. Neurophysiology 27 p.807-34

MOUNTCASTLE, V.B. (1978) An organizing principle for cerebral function: the unit module and the distributed system. In: EDELMAN, G.M. and MOUNTCASTLE, V.B. The Mindful Brain - Cortical organization and the group-selective theory of higher brain function. p.7-50 The MIT Press, Cambridge, Mass.

MUHLENBEIN, H. (1990) Limitations of multi-layer perceptron networks - steps toward genetic neural networks. Parallel Computing 14 p.249-60

MURATA, T. (1989) Petri Nets: properties, analysis and applications. Proc. IEEE 77/4 p.541-80

MUSA, A.A. and SAXENA, V. (1984) Scheduling nurses using goal-programming techniques. IIE Transactions 16/3 p.216-21

MYLOPOULOS, J.; SHIBAHARA, T. and TSOTSOS, J.K. (1983) Building knowledge-based systems: the PSN experience. Computer 16/10 p.83-9

NATHANSON, J.A. (1977) Cyclic nucleotides and nervous system function. Physiological Rev. 57/2 p.157-256

NEGOITA, C.V. and RALESCU, D.A. (1975) Applications of fuzzy sets to systems analysis 191p. John Wiley & Sons, New York

NEVILLE, H. J.; KUTAS, M. and SCHMIDT, A. (1982) Event-related potential studies of cerebral specialization during reading: I. Studies of normal adults. Brain and Language 16/2 p.300-15

OLERON, P. (1980)   Social intelligence and communication:
introduction.   International J. Psycholinguistics 7-1/2/
17-18   p.7-10

OLSON, D. R. (1980) On language and literacy.
International J. Psycholinguistics 7-1/2/17-18   p.69-82

OZKARAHAN, I. and BAILEY, J.E. (1988)   Goal-programming
model subsystem of a flexible nurse scheduling support
system. IIE Transactions 20/3   p.306-16

PAUKER, S.G. and KASSIRER, J.P. (1980)   The threshold
approach to clinical decison making. New England J.
Medicine 302 p.1109-17

PEDRYCZ, W. (1990a) Relevancy of fuzzy models.
Information Sciences 52/3   p.285-302

PEDRYCZ, W. (1990b) Direct and inverse problem in comparison
of fuzzy data. Fuzzy Sets and Systems 34/2   p.223-35

PEDRYCZ, W. and ROCHA, A.F.   (1992) Fuzzy-set based models
of neurons. submitted

PLANT, R.E.   (1976)   The geometry of the Hodgkin-Huxley
model. Computer Programs Biomedicine 6/2   p.85-91

POGGIO, G.F. and MOUNTCASTLE, V.B. (1963)   The functional
properties of ventrobasal thalamic neurons studied in
unanesthetized monkeys. J. Neurophysiology 26   p.775-806

POPPER, K.R.   (1967)   El desarrollo del conocimiento
cientifico: conjeturas y refutaciones   463p.   Paidós,
Buenos Aires

POPPER, K.R. and ECCLES, J.C.   (1985) The self and its
brain   597p.   Springer, Berlin

PRADE, H.   (1982)   Fuzzy sets and their relations with
Lukasiewcz logic Possibility Sets.   Proc. 12th   IEEE
International Symposium on Multiple-Valued Logic p.223-7
Paris

RASMUSSEN, H. and GOODMAN, D.B.P.   (1977)   Relationships
between Calcium and cyclic nucleotides in cell activation.
Physiological Rev. 57/3   p.421-509

RASMUSSEN, H. and BARRET, P.Q.   (1984) Calcium messenger
system: an integrated view. Physiological Rev.   64/3
p.938-84

REITER, R. (1980) A logic for default reasoning.
Artificial Intelligence 13/1-2   p.81-132

RITTER, W., SIMSON, E. and WAUGHAN, H.G. (1983) Event related potential correlates of two stages of information processing in physical and semantic discrimination tasks. Psychophysology p.168-79

ROCHA, A.F. (1979) Brain's entropy partitions. Anais Academia brasileira Ciências 51/4 p.591-5

ROCHA, A.F. (1980) Temporal influences of the reticular formation on sensory processing. In: HOBSON, J.A. and BRAZIER, M.A.B. The reticular formation revisited: specifying function for a nonspecific system. p.105-15 Raven Press, New York

ROCHA, A.F.; FRANÇOZO, E. and BALDUINO, M.A. (1980) Neural languages. Fuzzy Sets and Systems 3/1 p.11-35

ROCHA, A.F. (1981a) Neural fuzzy point processes. Fuzzy Sets and Systems 5/2 p.127-40

ROCHA, A.F. (1981b) Neural encoding process. Post-Doctoral Thesis (in portuguese) UNICAMP Institute of Biology Campinas, Brazil

ROCHA, A.F. (1982a) Toward a theoretical and experimental approach of fuzzy learning. In: GUPTA, M.M. and SANCHEZ, E. (Eds.) Approximate Reasoning in Decision Analysis p.191-200 North-Holland, Netherlands

ROCHA, A.F. (1982b) Basic properties of neural circuits. Fuzzy Sets and Systems 7/2 p.109-21

ROCHA, A.F. and BASSANI, J.W.M. (1982) Information theory applied to the study of neural codes. Proc. 26th Annual Meeting of the Society for General System Research with the American Association Advancement of Science 2 p.528-33 Washington, D.C.

ROCHA, A. F. and BUNO, W.Jr. (1985) Sustained sensitivity modifications induced by brief length perturbations in the crayfish slowly adapting stretch receptor. J. Neurobiology 16/5 p.373-88

ROCHA, A.F. (1985) Expert sensory systems: initial considerations. In: GUPTA, M.M.; KANDEL, A.; BANDLER, W. and KISZKA, J.B. (Eds.) Approximate Reasoning in Expert Systems p.549-70 Elsevier, Amsterdam

ROCHA, A.F. and ROCHA, M.T. (1985) Specialized speech: a first prose for language expert systems. Information Sciences 37/1-2-3 p.193-210

ROCHA, A.F.; THEOTO, M. and TORASSO, P. (1988) Heuristic learning expert systems: general principles. In: GUPTA, M.M. and YAMAKAWA, T. (Eds.) Fuzzy Logic in knowledge-based

Systems, Decision and Control   p.289-306   Elsevier, Netherlands

ROCHA, A.F.; THEOTO, M.; RIZZO, I. and LAGINHA, M.P.R. (1989)   Handling uncertainty in medical reasoning. Proc. 3rd IFSA Congress   p.480-3   Seattle, Washington

ROCHA, A.F. (1990a)   Brain activity during language perception. In: SINGH, M.G. Systems & Control Encyclopedia Theory, Technology, Applications v.1 supplementary p.38-46 Pergamon Press, Oxford

ROCHA, A.F. (1990b)   K-neural nets and expert reasoning. Proc. International Conference on Fuzzy Logic & Neural Networks 1   p.143-6   Iizuka, Japan

ROCHA, A.F. (1990c)   The physiology of the neural nets. Tutorials International Conference on Fuzzy Logic & Neural Networks   p.135-71   Iizuka, Japan

ROCHA, A.F. (1990d)   Smart Kards(c): Object Oriented system for approximate reasoning. Proc. NAFIPS'90   p.71-4 Toronto, Canada

ROCHA, A.F.; LAGINHA, M.P.R.; SIGULEN, D. and ANÇÃO, M.S. (1990e)   Declarative and procedural knowledge: two complementary tools for expertise. In: VERDEGAY, J.L. and DELGADO, M. (Eds.) Approximate Reasoning Tools for Artificial Intelligence   p.229-53   Verlag TUV, Koln, Germany

ROCHA, A.F.; MACHADO, R.J. and THEOTO, M. (1990f) Complex neural networks. Proc. ISUMA'90 - 1st International Symposium on Uncertainty Modeling and Analysis p.495-9 Maryland, USA

ROCHA, A.F. (1991a)   The fuzzy neuron: biology and mathematics. Proc. IFSA'91 Brussels   Artificial Intelligence   p.176-9   Brussels, Belgium

ROCHA, A.F. (1991b) Fuzzy logics and neural nets: tools for expertise. Proc. International Fuzzy Engineering Symposium'91 1   p.482-93   Yokohama, Japan

ROCHA, A.F. and THEOTO, M. (1991) Searching fuzzy concepts in a natural language data base.In: FEDRIZZI, M.; KACPRZYCK, J. and ROUBENS, M. (Eds.) Interactive Fuzzy Optimization and Mathematical Programming Springer-Verlag, in press

ROCHA, A.F.; THEOTO, M. and THEOTO ROCHA, M. (1991) Investigating medical linguistic variables. submitted

ROCHA, A.F. and YAGER, R.R. (1992) Neural nets and fuzzy logic.In: KANDEL, A. and LANGHOLZ, G. Intelligent Hybrid Systems CRC Press Inc., USA   in press

ROCHA, A.F.; GUILHERME, I.R.; THEOTO, M.T.; MIYADAHIRA, A.M.K. and KOIZUMI, M.S. (1992) A neural net for extracting knowledge from natural language data bases. IEEE Transaction Neural Networks Special issue on fuzzy sets to appear

ROCHA, M.T. (1990) Decodification of a Leprosy's text by students, teachers, nurses and nurses-aid. PhD Thesis (in portuguese) University of São Paulo School of Public Health São Paulo, Brazil

ROMANIUK, S.G. and HALL, L.O. (1990) Towards a fuzzy connectionist expert system development tool. Proc. IJCNN Washington D.C.

ROSENBLATT, F. (1958) The perceptron: a probabilistic model for information storage and organization in the brain. Psychological Rev. 65 p.386-408

RUMELHART, D.E. and McCLELLAND, J.L. (1986) On learning the past tenses of english verbs. In: McCLELLAND, J.L.; RUMELHART, D.E. and the PDP RESEARCH GROUP Paralell Distributed Processing - Explorations in the Microestructure of Cognition 2: Psychological and Biological Models p.216-71 The MIT Press, Cambridge, Mass.

RUMELHART, D.E.; McCLELLAND, J.L. and the PDP RESEARCH GROUP (1986) Parallel Distributed Processing - Explorations in the Microestructure of Cognition 1: Foundations 547p. The MIT Press, Cambridge, Mass.

SAGER, N. (1987) Information formatting of medical literature. In: SAGER, N.; FRIEDMAN, C. and LYMAN, M.S. Medical Language Processing - Computer Management of Narrative Data p.197-220 Addison-Wesley, Reading, Mass.

SAGER, N. (1987) Computer processing of narrative information. In: SAGER, N.; FRIEDMAN, C. and LYMAN, M.S. Medical Language Processing - Computer Management of Narrative Data p.3-22 Addison-Wesley, Reading, Mass.

SAGER, N.; FRIEDMAN, C. and LYMAN, M.S. (1987) Medical Language Processing - Computer Management of Narrative Data 348p Addison-Wesley, Reading, Mass.

SAKATA, H.; SHIBUTANI, H. and KAWANO, K. (1980) Spatial properties of visual fixation neurons in posterior parietal association cortex of the monkey. J. Neurophysiology 43/6 p.1654-72

SANCHEZ, E. (1978) On possibility qualification in natural languages. Information Sciences 15/1 p.45-7

SANCHEZ, E. (1989) Importance in knowledge systems. Information Systems 14/6 p.455-64

SANCHEZ, E. and BARTOLIN, R. (1989) Fuzzy inference and medical diagnosis, a case study. First Annual Meeting of Biomedical Fuzzy System Association 18p. Kurashiki, Japan

SCHELLER, R.H. (1984) Neuropeptides: mediators of behavior in Aplysia. Science 225/4668 p.1300-8

SCHWARTZ, D.G. (1988) An alternative semantics for linguistic variables. In: BOUCHON, B.; SAITTA, L. and YAGER, R.R. (Eds.) Uncertainty and Intelligent Systems v.313- Lecture Notes in Computer Science p.87-92 Springer-Verlag, Berlin

SEGUNDO, J.P. and KOHN, A.F. (1981) A model for excitatory synaptic interactions between pacemakers. Its reality, its generality and the principles involved. Biological Cybernetics 40 p.113-26

SGALL, E.P.; HAJICOVA, E. and BENESOVA, P. (1973) Topic, focus and generative graar Scriptor Kronberg

SHAFER, G. (1976) A mathematical theory of evidence Princeton University Press

SHANNON, E.R. (1974) A mathematical theory of communication. In: SLEPIEN, D. (Ed.) Key Papers in the Development of the Information Theory p.5-29 IEEE Press

SHASTRI, L. (1988) A connectionist approach to knowledge representation and limited inference. Cognitive Science 12/3 p.331-92

SIDMAN, R.L. and RAKIC, P. (1973) Neuronal migration with special reference to developing human brain: a review. Brain Research Rev. 62/1 p.1-35

SIMPSON, R.H. (1963) Stability in meanings for quantitative terms: a comparison over 20 years. Quartely J. Speech 49 p.146-51

SMETS, P. (1981) Medical diagnosis: fuzzy sets and degree of belief. Fuzzy Sets and Systems 5/3 p.259-66

SMITHSON, M.J. (1987) Fuzzy sets analysis for behavioral and social sciences Springer Verlag, Berlin

SMOLIAR, S.W. (1989) Neural Darwinism: The theory of neuronal group selection (Book reviews). Artificial Intelligence 39 p.121-39

SOULA, G. and SANCHEZ, E. (1982) Soft deduction rules in medical diagnostic processes. In: GUPTA, M.M. and SANCHEZ, E. (Eds.) Approximate Reasoning in Decision Analysis p.77-88 North-Holland, Netherlands

STARKE, K.; GOTHERT, M. and KILBINGER, H. (1989) Modulation of neurotransmitter release by presynaptic autoreceptors. Physiological Rev. 69/3 p.864-989

SZENTAGOTHAI, J. (1975) The "Module Concept" in cerebral cortex architecture. Brain Research Rev. 95/2-3 p.475-96

SZENTAGOTHAI, J. (1978) The neuron network of the cerebral cortex: a functional interpretation. The Ferrier Lecture, 1977 Proc. Royal Society of London Series B 201/1144 p.219-48

TEICHBERG, V.I. (1991) The kainate receptor as a molecular switch and association detector. Neuroscience Facts 2/15 p.2

THEOTO, M. and KOIZUMI, M.S. (1987) The expert environment: a case study. Preprints of Second IFSA Congress 1 p.380-3 Tokyo, Japan

THEOTO, M.; SANTOS, M.R. and UCHIYAMA, N. (1987) The fuzzy decodings of educative texts. Fuzzy Sets and Systems 23/3 p.331-45

THEOTO, M.T. and ROCHA, A.F. (1989) Fuzzy belief and text decoding. Proc. 3rd IFSA Congress p.552-4 Seattle, Washington

THEOTO, M.; KOIZUMI, M.S.; MARGARIDO, L.T.M. and ROCHA, A.F. (1989) Comparing data base and the expert knowledge. RANI Technical Report 04 25p. Jundiai, Brazil

THEOTO, M.; ROCHA, A.F. and MACHADO, R.J. (1990) Approximate reasoning with partial data. Proc. ISUMA'90 - First International Symposium on Uncertainty Modeling and Analysis p.567-72 Maryland, USA

THEOTO, M. (1990) Text understanding on different populations: a technique for calculation of consensus. Proc. NAFIPS'90 p.75-8 Toronto, Canada

THEOTO, M. and KOIZUMI, M.S. (1990) Text decoding: an experimental and theoretical approach. Proc. International Conference on Fuzzy Logic & Neural Networks 1 p.139-42 Iizuka, Japan

THEOTO, M. and ROCHA, A.F. (1990) Smart objects for outpatient service management. Proc. NAFIPS'90 p.281-4 Toronto, Canada

THEOTO, M. and ROCHA, A.F. (1992) Data base intelligent indexing. RANI Technical Report 08 30p. Jundiai, Brazil

THOENEN, H. and BARDE, Y.A. (1980) Physiology of nerve

growth factor. Physiological Rev. 60/4 **p.1284-335**

TOOGOOD, J.H. (1980) **What do we mean by "usually?".** The Lancet 1/may 17 **p.1094**

TORASSO, P. and CONSOLE, L. (1989) Diagnostic problem solving: combining heuristic, approximate and causal reasoning. **Van Nostrand Reinhold & Kogan Page**

TRILLAS, E. and VALVERDE, L. (1987) **On inference in fuzzy logic.** Preprints of Second IFSA Congress 1 **p.294-7** Tokyo, Japan

URRY, D.W. (1971) **The Gramicidin A transmembrane channel.** Proc. National Academy of Sciences 63 **p.672-76**

VALVERDE, F. (1986) **Intrinsic neocortical organization:** some comparative aspects. Neuroscience 18/1 **p.1-23**

VERDEGAY, J.L. (1984) A dual approach to solve the fuzzy linear programming problem. Fuzzy Sets and Systems 14 **p.131-41**

VERDEGAY, J.L. (1989) **Fuzzy mathematical programming problem: resolution. In: SINGH, M.G.** Systems & Control Encyclopedia - Theory, Technology, Applications **p.1815-19** Pergamon Press, Oxford

WAH, B.W.; LOWRIE, M.B. and LI, G. (1989) **Computers for** symbolic processing. Proc. IEEE 77/4 **p.509-40**

WASHABAUGH, W. (1980) **The role of speech in the** construction of reality. Semiotica 31/3-4 **p.197-214**

WHITLEY, D.; STARKWEATHER, T. and BOGART, C. (1990) Genetic algorithms and neural networks: optimizing connections and connectivity. Parallel Computing 14 **p.347-61**

WIED, D. and JOLLES, J. (1982) **Neuropeptides derived from** pro-opiocortin: behavioral, physiological and neurochemical effects. Physiological Rev. 62/3 **p.976-1059**

WINOGRAD, T. (1980) **Extended inference modes in reasoning** by computer systems. Artificial Intelligence 13/1-2 p.5-26

WRIGHT, G. and AYTON, P. (1987) **Eliciting and modelling** expert knowledge. Decision Support Systems 3 **p.13-26**

YAGER, R.R. (1984) **Approximate reasoning as a basis for** rule-based expert systems. IEEE Transactions on Systems, Man and Cybernetics SMC-14/4 **p.636-43**

YAGER, R.R. (1988a) Prioritized, non-pointwise, nonmonotonic intersection and union for commonsense reasoning. In: BOUCHON, B.; SAITTA, L. and YAGER, R.R. (Eds.) Uncertainty and Intelligent Systems v.313 - Lecture Notes in Computer Science **p.359-65 Springer-Verlag, Berlin**

YAGER, R.R. **(1988b)** **On ordered weighted averaging aggregation operators in multi-criteria decision making.** IEEE Transactions on Systems, Man and Cybernetics 18 **p.183-90**

YAGER, R.R. (1990a) A set framework for default reasoning. In: **VERDEGAY, J.L.** and **DELGADO, M.** **(Eds.)** Approximate reasoning Tools for Artificial Intelligence **p.80-91 Verlag TÜV, Koln, Germany**

YAGER, R.R. (1990b) On a semantics for neural networks based on linguistic quantifiers. Technical Report MII-1103 **Machine Intelligence Institute 26p.** Iona College, New Rochelle, NY

YAGER, R.R. (1990c) **Decision making in mixed uncertainty environments: a look at importances.** Proc. **ISUMA'90 - First International Symposium on Uncertainty, Modeling and Analysis p.269-73** Maryland, USA

YAGER, R.R. (1990d) On the associations between variables in expert systems including default relations. Information Sciences 50/3 **p.241-74**

ZADEH, L.A. **(1965)** **Fuzzy Sets.** Information and Control 8/3 **p.338-53**

ZADEH, L.A. (1975) The concept of a linguistic variable and its application to approximate reasoning - I. Information Sciences 8/3 **p.199-249**

ZADEH, L.A. (1975) The concept of a linguistic variable and its application to approximate reasoning - II. Information Sciences 8/4 **p.301-57**

ZADEH, L.A. (1975) The concept of a linguistic variable and its application to approximate reasoning -III. Information Sciences 9/1 **p.43-80**

ZADEH, L.A. **(1978) Fuzzy sets as a basis for a theory of possibility.** Fuzzy Sets and Systems 1/1 **p.3-28**

ZADEH, L.A. (1979) A theory of approximate reasoning. Machine Intelligence 9 **p.149-94**

ZADEH, L.A. **(1983a)** **The role of fuzzy logic in the management of uncertainty in expert systems.** Fuzzy Sets and Systems 11/3 **p.199-227**

ZADEH, L.A. **(1983b) A computational approach to fuzzy quantifiers in natural languages.** Computer & Mathematics with Applications 9/1 **p.149-84**

ZADEH, L.A. (1985) Fuzzy logic in management of uncertainty in expert systems. In: GUPTA, M.M.; KANDEL, A.; BANDLER, W. and KISZKA, J.B. (Eds.) Approximate Reasoning in Expert Systems   p.3-31   Elsevier, Amsterdam

ZEMANKOVA, M. and KANDEL, A. (1985) Implementing imprecision in information systems. Information Sciences 37/1-2-3   p.107-41

ZIMMERMANN, H.J. (1979) Vesicle recycling and transmitter release. Neuroscience 4/12   p.1773-804

ZIMMERMANN, H.J. and ZYSNO, P. (1980) Latent connectives in human decision making. Fuzzy Sets and Systems 4/1   p.37-51

INDEX

AXON
 properties filtering
 13, 23, 38, 42-46, 53-54,
 166

 activity
 33, 36, 38, 43, 51, 53,
 240-241

CABINET
 folder
 314, 325, 328, 337

 specification
 310, 314

CARD
 cabinet
 307, 313, 325, 328

 disease
 209, 318, 323, 328-329

 hypotheses
 317-319, 329

 informormation
 259, 306, 308, 311-312, 323,
 329, 337

 method assignment
 308, 316, 318, 329

 specification
 306, 308, 311-312

 variable
 306, 308, 318

CHANNEL
 gate
 2, 4-6, 9

 ion
 1-4, 11, 31-32, 40

 activity
 2, 5, 36

CHEMICAL processing
 31, 58, 136, 219, 221

CONFIDENCE
 ordering
 192, 206

 decision making
 192, 208, 210, 241

 possibility
 206, 208

 space
 206, 207

CONORM
 35, 39, 42, 45, 48, 51, 57,
 65, 218-219, 241, 349, 351,
 352, 361, 363

CONTROL efferent
 26, 27, 29, 59, 70

CORTICAL
 layer_Layer
 213, 228

 parallel layer
 213, 228

CYCLE
 limit
 8-9, 12-13, 15-18

 point
 13, 15, 17

DATA
 base contents
 217, 245, 275, 298-299, 340

 card
 290, 305, 307, 309, 329, 337

 Knowledge
 240, 242, 245, 249, 256, 261
 273, 290, 299, 300, 305, 325
 327, 340,

 patient
 315, 319, 323, 325, 330, 333

Printing: Druckhaus Beltz, Hemsbach
Binding: Buchbinderei Schäffer, Grünstadt

Lecture Notes in Artificial Intelligence (LNAI)

Lecture Notes in Computer Science